The Digital Difference

The Digital Difference

MEDIA TECHNOLOGY AND THE THEORY
OF COMMUNICATION EFFECTS

W. Russell Neuman

John,
Here's the view of things
from back East
with warm regard
Russ

Harvard University Press

Cambridge, Massachusetts

London, England

2016

First printing

Library of Congress Cataloging-in-Publication Data

Names: Neuman, W. Russell, author.
Title: The digital difference : media technology and the theory of
communication effects / W. Russell Neuman.
Description: Cambridge, Massachusetts : Harvard University Press, 2016. |
Includes bibliographical references and index.
Identifiers: LCCN 2015039005 | ISBN 9780674504936 (alk. paper)
Subjects: LCSH: Information society. | Information networks. | Technology and
civilization. | Information technology—Social aspects. | Information
technology—Political aspects. | Mass media—Social aspects. | Mass
media—Political aspects.
Classification: LCC HM1206 .N477 2016 | DDC 303.48/33—dc23
LC record available at http://lccn.loc.gov/2015039005

In memory of Margaret W. Neuman

Contents

The Digital Difference

Prologue

The Information Revolution is causing a scientific revolution
in communication research.

—EVERETT ROGERS (1986)

Rather than waiting for media companies to deliver relevant content
at appropriate times, customers are increasingly reaching out to pull
content to them when they want.

—JOHN HAGEL AND JOHN SEELY BROWN (2005)

THIS IS A STUDY of a revolution in human communication: the digital difference. It examines how the computer-based media technologies are gradually but fundamentally transforming the relationship between the audience member and the media and among individuals in mediated social networks. In the media world it is a shift from "push" to "pull." Where once audience members could pick from only a few headlines or channels, they are now free to pose virtually any query imaginable in a search engine and in turn peruse a virtually unlimited global collection of articles, books, and videos. It is also a shift from one-way to two-way mass communication—that is, from broadcasting and publishing to social networking. Interpersonal and mass communication are increasingly intertwined. The distinction between face-to-face "interpersonal"

communication and mediated mass communication has been a fundamental divide in communication scholarship for half a century. Such distinctions are self-evident and essential to the older professors behind the
podium and are probably seen as more of a curious antiquity to the
younger students in front of the podium. The scholarly research paradigm
is lagging far behind the relentless pace of technical change.

Scholars have studied how the revolutions of speech and of written
language, printing, and broadcasting have each dramatically influenced the character of human existence in the domains of economics,
politics, culture, and social life. I would be remiss not to draw on these
insights in the pages ahead as I try to make sense of the revolution in
ubiquitous electronic communication swirling around us today. As a
result this book is a purposeful blend of looking back and looking
forward.

My thesis, simply put, is that the revolution in communication technology makes possible a paradigm shift in how human communication is
studied and understood. I argue that this "digital difference" in the current
revolution offers the opportunity for a fundamental rethinking of the foundational concept of "communication effects" and the techniques for systematically measuring their increasingly complex dynamics. I am well
aware that this may be, as they say, a tough sell. There are five fundamental
reasons for a skeptical view of my thesis. First, systematic scholarly research
at the individual and structural levels and in the social science and humanistic traditions of human communication has been under way for many
decades, and central theories and the traditional methodologies are generally well received and not frequently viewed as in need of major revision.
Second, drawing on the importance of digital technologies smacks of
"technological determinism," a crude insensitivity to human agency and
the human role in the design and use of technology. Third, and related to
the second point, is the observation that although communication technologies may have changed abruptly, the evolved human cognitive system
has not, and we are still subject to a variety of systematic distortions in perception whatever technology is used. Fourth, many of the suggestions I
make in the pages ahead will be seen by some as not entirely new as other
researchers have been pursuing these new lines of inquiry already. Fifth
and finally, many scholars take a dubious view of the notion of a paradigm
and paradigm shifts in the tradition of Thomas Kuhn, which strikes them

as simplistic and inappropriately imported into the social sciences and humanities from the physical sciences.

So, all together, there would appear to be a very reasonable case for skepticism, and I welcome it and respond to it in detail. I attempt to make my case as strongly as possible and hope that a little controversy might help draw attention to these issues. One might characterize my project here as a recipe for exasperating friends and colleagues in the field by, in effect, presumptively arguing that they are testing the wrong theories and using the wrong methods. Let me be clear. I have a deep respect for the progress in communication scholarship over the past half century, which should be evident from the abundant citations and recitations of this literature. But I think these theories and their associated methods at this critical juncture offer great promise of reformulation and rejuvenation. This is not at all an abandonment of old theories and methods and related normative concerns but rather something of a pivot in response to the new issues and opportunities associated with the digital difference.

A Revolution in Communication Science?

Over the past century communication scholarship has focused on the power of one-way mass communication to persuade and inform. The received history of this literature posits an initial period focusing on propaganda and strong hypodermic-needle-like effects, followed by a middle period, a reconsideration based on some accumulated finding of so-called minimal effects, and the current period—a return to a conception of strong social and psychological effects of the mass media (Neuman and Guggenheim 2011). In an important article published in 2008 by W. Lance Bennett and Shanto Iyengar, these scholars identified a pressing need to reassess the big-effects versus minimal-effects debate in light of the new media revolution and the increased capacity of the audience member to filter and select from a dramatically expanded media environment. They titled the piece "A New Era of Minimal Effects? The Changing Foundations of Political Communication." They are on solid ground in urging a reexamination of the communication research paradigm, and they are in good company (Rogers 1986; Bryant 1993; Newhagen and Rafaeli 1996; Chaffee and Metzger 2001; Napoli 2010). The article raises the prospect of what might be characterized as a paradigm shift or at least an evolving paradigm in the way we have come to understand the dynamics of the public sphere.

I am an enthusiastic supporter of their call for foundational reassessment but with a twist. Bennett and Iyengar's argument is largely based on a binary conception of big effects versus minimal effects. I think we need to move beyond that paradigmatic and historically grounded notion of communication effects in terms of effect size. *It is not that media effects are either characteristically strong or characteristically minimal—they are characteristically highly variable.* And the source of that variation is deeply and subtlety intertwined in the enduring structure of communication, a primary focus in this book—the routine practices, established institutions, and evolved cultural norms of mass communication. So to posit that the increased opportunity for choice in the digital information cornucopia will lead to more selective attention and thus "minimal effects" misses the opportunity to move away from a notion of a mechanical/persuasive notion of communication effects to one of highly variable resonance between a speaker's message and a listener's interpretation. The dramatic and variable effects of selective attention do not represent a new or newly resurgent phenomenon in the digital age. They are part and parcel of the fundamentally polysemic character of human communication and the abundance of symbolic streams that typically bombard us. Most American adults have heard "The Star-Spangled Banner" many hundreds or even thousands of times. Yet only a third of them can recall the words (Corso 2008). (Among younger Americans it is only one in seven.) Only half of American adults can identify the name of a senator or the representative from their district, again despite many hundreds, sometimes thousands, of media references and numerous billboards, yard signs, and bumper stickers (Delli Carpini and Keeter 1996). At the same time the details of a celebrity divorce, a political scandal, or a dramatic crime may be well known (and frequently discussed) among all but a culturally isolated few. Selective attention was abundantly evident in the industrial age of print and broadcast mass communication.

Accordingly, I propose a strategy of moving away from the alternative celebration and vilification of the potential power of the media to what I believe is a more scientifically focused inquiry into *the conditions under which media effects, both large and small, are or are not in evidence.* As a discipline of communication science evolves it will almost certainly focus on the structural conditions that enhance and inhibit communication effects rather than a defensive posture against those who claim that typical media effects strike them as somehow surprisingly small. Science studies variation, and there is no shortage of that in, for example, the impact of media campaigns—billions are spent each year in promoting products, political candidates, and public service campaigns sometimes with dramatic effect, sometimes with no effect at all, and surprisingly often with demonstrable effects the opposite of those intended (Yzer et al. 2003). The variability and as yet only partially understood *conditionality of communication effects* will be a central theme ahead.

I argue that the key to understanding how the new media will be used and how the media institutions will evolve is to better understand the attentional dynamics of the human mind. *What do individuals want to know? To what do they pay attention? And why?* This fresh paradigmatic notion of communication effects starts with the audience member and listener rather than the media source. It is hard to exaggerate the importance of this shift—it is the question of what people choose to attend to rather than if they are persuaded by what they are shown. This perspective was evident in the "uses and gratifications" tradition of research that was widely acknowledged in the 1970s and 1980s but has lost momentum more recently for a combination of theoretical and methodological issues. Although an attractive notion, researchers never quite figured out how to incorporate such a perspective in a traditional survey or experiment. In the decades ahead, however, digital big data may open new methodological opportunities to follow this enduring theoretical impulse.

Communication scholarship is prone to pronouncements. Numerous books and articles from both the social scientific and cultural studies traditions routinely beseech media executives and regulators to change their evil ways and heed the latest research. Media executives and regulators in turn simply ignore these complaints, utterly ignore them. The economic and political pressures on media business practices and public policy are unambiguous and strong. Most academics have not developed the skills necessary to communicate with media professionals in a way that might resonate with their communities of professional practice. In fact the disconnect between communication scholarship and media industry practice is often celebrated in the academy as the exercise of an independent critical voice unencumbered by an "administrative" perspective. Perhaps so, but it is a voice seldom heard outside the academy.

What makes the current historical stage of media evolution particularly intriguing is a rare opportunity for communication scholarship to be both critical and relevant. Traditional media practices based on push advertising and common denominator cultural fare are in economic jeopardy. When the usual way of doing things in broadcasting and publishing does not work, executives are not only motivated to listen carefully; they may ask some probing and useful questions. At the conclusion of a six-hundred-page examination of the effects of television on human behavior concluded in 1978, the lead author, George Comstock, readily admitted that policy recommendations they had derived were not likely to stir up much attention because the audience was not dissatisfied with what they were getting and the broadcasting industry was economically and politically powerful. Such caveats are no longer warranted.

Perhaps that is the fundamental challenge to commercial research and especially independent scholarship in communications—not just to figure out what audiences will pay for but to understand how public communication can be designed to serve public goals. Collective goals are delicate and fragile to derive in the first place and then, in turn, to sustain in the face of historical developments and institutional frustrations. If the structure of the public sphere is not carefully designed and sustained, polarization and paralysis are likely to arise.

The past century of psychological research has powerfully demonstrated a long list of "hardwired" and systematic biases in the human cognitive system. We tend to interpret information in a way that confirms our preconceptions. We give more weight to negative experiences than positive ones. We attend more to the potential loss of an object than the prospect of gaining an object. We prefer the familiar. We pay a lot of attention to sex and to violence. These and related cognitive biases are addressed in more detail in the pages ahead, particularly Chapters 4 and 5.

Such systematic patterns of cognitive preference and distortion may make sense in terms of the evolutionary survival of individuals in small tribes of hunters and gatherers. However, these systematic cognitive misconceptions now represent a significant threat to our collective survival in an era of instant global communication, airborne terrorism, and missile-based nuclear warheads. So, unsurprisingly, it is not the case that scientific pursuits are independent of and isolated from normative concerns. There is a clear and perhaps even urgent normative challenge to communication science at the dawn of the third millennium—help design technical systems, institutions, and norms of mass communication to counterbalance our bad institutional habits and inevitable cognitive biases as best we can. An epigrammatic quotation from psychological researcher David Huron captures the issue at hand succinctly: "The business of biology is survival and procreation not truth and accuracy" (2006, 105). If our evolving systems of news and entertainment are to sustain a collective movement toward something like truth and accuracy, we cannot expect it to happen "naturally" although clearly we would like it to.

From deep in the soul of twenty-five hundred years of Western culture we would like to believe that a fair-minded jury of our peers will seek justice and that the electorate will deliberate thoughtfully and choose the best candidate. We would like to think that the public would not be persuaded by a trumped-up incident to initiate war. We would like to assert that we are not subject to pressures of groupthink and wishful thinking. But such premises, if not simply naive, are obviously incomplete. The challenge to those who would take up the systematic study of individual and collective communication should be clear. It is a classic Kuhnian paradigm with the requisite puzzle: *How do we design norms and institutions for the public sphere that take into account our individual and collective cognitive patterns that distort our ability to match up means and ends?* The Kuhnian model of what he calls normal science posits that the requisite puzzle is matched to a series of research methodologies designed to "solve" the puzzle. In the case of communication research, the methods are not well matched to the puzzles at hand. In fact, it could be argued that they are matched to the wrong puzzle—one focusing on the urge to demonstrate large media effects and, accordingly, to justify the research enterprise. The resultant mismatch between puzzles and methods is the subject of Chapter 2.

In Chapter 1 I review the historical roots of the dominating focus on large media effects and persuasion. Most modern communication researchers feel that we have moved well beyond the propaganda paradigm of research associated with Lasswell, Lazarsfeld, and Hovland in the mid-twentieth century. An important point. But I try to make the argument that we have not yet moved far enough and that the notion of propaganda still subtly pervades our conception of communication effects.

Some readers will have made note of the repeated use of the phrase *communication science* in these pages and will react skeptically in part because their intellectual home is in a distinctly nonscientific humanistic tradition. I argue, however, in the pages ahead that communications scholarship will benefit from a convergent and consilient contribution from both traditions and that a science of human behavior can and should be humanistic in spirit and well as in purpose and method (Wilson 1998).

Four Puzzles

Science in the Kuhnian tradition is a process of trying to solve puzzles. The issue of whether communication effects are usually massive or minimal is not much of a puzzle and, I argue, not a promising starting point for serious scholarship. Instead I draw attention to four foundational questions.

1. **Profusion of Information:** In an age of abundant information and increasingly complex social and technical information systems, individuals are by definition only able to attend to a shrinking portion of the surrounding information environment and increasingly dependent on technical systems for locating what they seek in the information abundance. Is it inevitable that the ratio between what is knowable and what an individual can realistically know will increase? Will this change evolving social structure?

2. **Polysemy:** Human communication is fundamentally and inevitably ambiguous and polysemic. People tend to underestimate the ambiguity of the messages they send and receive. In most cases individuals overestimate the persuasive effectiveness of information to which they are averse and underestimate the persuasive effectiveness of information about which they are enthusiastic. Can the concept of polysemy be incorporated into a theory of communication effects?

3. **Polarization:** As with many other areas of human endeavor, symbol systems get bound up with various aspects of human identity and centrally held values, and accordingly their interpretation often becomes polarized and contested. Do our institutions of public communication inadvertently nurture polarization?

4. **Pluralism—the Paradox of the Public Sphere:** Given the paradoxical character of evolved human communication behavior and the self-interested and self-reproducing character of most public and private institutions of public communication, is it possible that such institutional systems could be structured to optimally protect an open marketplace of ideas and flexible and responsive public institutions?

Puzzles in the Kuhnian model are matched with a method or set of methods as a community of researchers struggles to find a suitable resolution and then move on to the next question or, as is often the case, the resultant puzzle. Sometimes the puzzle is formulated as a fundamental question, such as what is the nature of light—is it a particle or a wave? Sometimes the puzzle is formulated with a more practical or functional goal associated—for example, we examine the complex dynamics of cancer so we might more successfully treat the disease.

The connection between puzzles, paradigms and methods, as well as the historical zeitgeist, has been illustrated throughout history. Alchemists struggled for centuries with the explicit goal of turning lead into gold by combining it with sulfur and quicksilver in a special oven called an Athanor. They did not succeed, of course, because it is not possible to transform an element with eighty-two protons into one with seventy-nine protons by strictly chemical means; it would require modern particle accelerators and a rather difficult trick of molecular manipulation. But apparently working with an irresolvable paradigm and method did not discourage its practitioners. Active alchemic research thrived from its earliest days in Mesopotamia and Egypt through Greco-Roman times through Islamic culture to the eighteenth century in Europe, where it ultimately evolved into modern chemistry. Turning lead into gold was not motivated by simple greed (although perhaps that helps a bit). It was widely felt that if a technique for purifying metals could be developed, it could be applied to the perfection of the human soul, a noble incentive indeed (Eliade 1979; Lindberg 2007).

Likewise the evolution of our thinking from the geocentric to the heliocentric solar system was not simply a matter of a revised theory and more modern telescopes; our scientific inquiries were deeply intertwined with our understanding of our place in the universe. The scientific process and technical invention are imbedded with our normative concerns and historically grounded collective challenges. Chapter 1 grounds my argument in a detailed historical examination of how concern about the negative impacts of propaganda and brainwashing particularly growing out of the Second World War and the fascist efforts to control and manipulate public opinion in Europe became the basis for systematic social scientific research on media effects and attitude change. The following chapters develop the argument that evolving historical conditions may nudge communication theorizing along. We need not completely abandon the concerns and the evolved research traditions of the last century, but rather expand and refine them in the light of new global conditions and particularly the changing technical infrastructure of interpersonal and mass communication. The allure of alliteration leads me to propose the Four Ps—profusion, polysemy, polarization, and pluralism.

Profusion. The first puzzle and problematique, the explosive profusion of electronic communication is the relatively new phenomenon linked to the expanding capacities of our communication infrastructure. If individuals develop skills with spam filters, search engines, friend lists, voice mail, and DVRs, the dramatically expanded flow of communication need not be problematic as individuals successfully tame the information tide. But the dynamics of informational supply and demand are likely to represent an ongoing issue worthy of attention. At a more clearly normative level, the question becomes will people be able to learn what they need to know—will the capacities for communication be structured for a better informed public? These questions are the focus of Chapter 3, "The Paradox of Profusion."

One element of the exponential growth of information is unambiguous and starkly true—it is a mathematical inevitability that as the amount of accumulated knowledge increases, the ratio of what is potentially knowable to what an individual can reasonably know will increase. This is reflected, of course, in the technical complexity of our environment. When we made our own soap and pumped water at the well, we understood the environment we created and could fix things when they broke as we may have built them in the first place. Not true with GPS navigators and cell phones. We rely on technical specialists for survival in our environment and perhaps spend a reasonable amount of time on hold on a customer service line awaiting their enlightenment. In our professional lives all is specialization. For much of the eighteenth and nineteenth centuries every member of a college faculty was expected to be literate in classical Latin and Greek and each professor could teach virtually any of the math, history, language, or literature classes. The basic elements of the liberal arts curriculum derived from Trivium and Quadrivium of the Middle Ages had changed little in the past eight centuries. But as the sciences flourished, specialization and departmentalization took over and the curriculum was revolutionized and the remarkable new idea of the elective course became institutionalized (Veysey 1965; Rudolph 1993). I review how issues of specialization and the fragmentation of expertise influence the dynamics of collective choice and the public sphere. We have moved from push to pull and from one-way to two-way communication. We have moved from the *Encyclopedia Britannica,* where the anointed few experts would expound, to Wikipedia, were virtually anybody so inclined can expound. And miraculously it seems to work. Assessments of the technical accuracy of *Britannica* and Wikipedia reveal the level of inaccuracy is quite small and about the same in each (Giles 2005).

The Wikipedia case is telling, a hallmark of the new media environment. It may strike us as unlikely that an encyclopedia written by volunteer amateurs and hobbyists would result in a complete and accurate accounting. For a century we became accustomed to relying on official and professional experts, the CBS television network's avuncular Walter Cronkite, who explains with a comforting voice, "That's the way it is." Or *Life* magazine's equally influential and iconic photographs. The fact that these spokespersons of the public sphere and arbiters of public taste were almost exclusively all commercial firms would strike most citizens as utterly natural although the occasional intellectual critic might see it otherwise (Williams 1974; Habermas 1989; Chomsky 2004). There were brief flirtations with two-way communication, including citizens band radio and public access channels on cable television, but none of these experiments seemed to be able to sustain itself with an enduring identity and audience. The industrial character of publishing and the limited spectrum for broadcasting seemed to dictate that the number of voices would be few and that they would be commercial. The vaguely acknowledged notion that commercial entities just give audiences "what they want" because of the audience-size-based economics seemed to reinforce this natural order.

But the one-way commercial push media are not a natural order; they represent a stage in technical evolution. A long stage, perhaps—about one and a half centuries. The Internet and new media revolution changes everything. The blogosphere is immense; it is overwhelmingly nonprofessional; it is growing; and it is self-sustaining. Estimates vary but have recently pegged the size of the blogosphere at between 150 and 200 million reasonably active blogs with more than 100,000 new blogs introduced each day and several million essay-style posts per day (not including the many more millions of short 140-character Twitter postings) (Winn 2008). Importantly, if you have a particular issue of concern and want to find out what the bloggers are saying, you can—there are extensive blog-focused search engines and trend summaries available. The numbers are difficult to comprehend and the environment is dynamic, responding sometimes explosively to new information and issues. It is truly an open marketplace of ideas—sometimes unruly and rude, sometimes deliberative and even poetic. How this vibrant domain of opinion, information, and misinformation has and may continue to interact with the mainstream media and the shared public agenda is a critical question I address in the pages ahead.

Polysemy. The fact that a string of words or images can engender more than a single meaning is hardly a new concept or one that requires a digital media revolution to make its significance evident. But polysemy is at the core of the process of communication, and evolved traditions of communication scholarship continue to struggle with it. The humanist tradition in communication research takes polysemy very seriously indeed, which is a good start. But unfortunately taking polysemy seriously seems to result in skepticism about the prospects of a systematic scientific assessment of communication processes. Perhaps this results from disappointment at how simplistic most experimental and survey designs tend to be. They routinely ignore the polysemic character of personal and mass communication behaviors. The empirical research traditions characteristically treat persuasive messages (vote for candidate X, buy product Y) or entertainment programming (perhaps modeling violent behavior or consuming behavior) as having a singular message and semantic interpretations outside of the officially identified "message as sent" is simply treated as random error and ignored. For example, in measuring political campaign effects, analysts routinely contrast campaign advertising expenditures and corresponding proportions of election voting even though it is widely acknowledged that some political ad campaigns are profoundly effective, others worthless, and many, ironically, counterproductive (West 1997; Kaid 2004). It is an understandable strategy for conducting research. Polysemy is inconvenient. It makes systematic assessment troublingly complex. I address this issue in Chapter 2 on methodological strategies and in Chapter 4 on the dynamics of polysemy itself.

Working polysemy explicitly back into scientific modeling of communication processes is central to the issue of communication conditionality— it is an acknowledgment that a potential resonance in meaning between message as sent and as received is a variable rather than a constant, a variable, I argue, at the theoretical core of communicative practice. Claude Shannon's fundamental and seminal mathematical model of communication formally models the level of errors in transmission between senders and receivers (Shannon and Weaver 1949). And for a decade and a half, social scientific communication researchers toyed with the idea of building from Shannon's starting point but eventually abandoned the enterprise (Smith 1966; Rogers 1994). The likely reason for this is that Shannon had developed a specific and technically limited definition of noise as interference, one that made perfect sense for a Bell Labs engineer— random noise, electronic static. But human interaction and institutionalized communication systems display a much more interesting pattern of interference which causes miscommunication and noncommunication. *The noise and interference in human communication is seldom random; it is primarily systematically structured interference.* It is culturally reinforced bias and prejudgment. The challenge of communication in the public sphere is not that the listener cannot hear the speaker above arbitrary noise in the environment, but that listeners systematically interpret, reinterpret, ignore, and reframe the speaker's utterance.

To make matters more complex, there is deeply ingrained human propensity to presume an absence of misperception. I label this phenomenon the *semantic fallacy*. It is reviewed in some detail in Chapter 4 on the dynamics of polysemy. The fallacy is well captured in Charles Ogden and Ivor Richards's epigram: "Normally, whenever we hear anything said we spring spontaneously to an immediate conclusion, namely, that the speaker is referring to what we should be referring to were we speaking the words ourselves" (1923, 15).

Polysemy and ambiguity are central to the communication process. As Leeds-Hurwitz notes, "Codes are by their very nature full of gaps, inconsistencies and are subject to constant change" (1993, 66). Human beings appear to be hardwired to pay close attention to miscommunication— it is at the core of what humans consider to be comedic. I address this also in more detail later, but briefly—audience members relish being in on what is intended while characters in the narrative misunderstand. Some exaggerated misunderstanding is usually at the core of the situation comedy or theatrical farce, and the classic formulation of a joke—the punch line reveals that what the listener had assumed from the beginning is now turned on its head (Zillmann 2000).

Polarization. In 1957 three psychologists, Charles E. Osgood, George J. Suci, and Percy Tannenbaum, published a monograph entitled *The Measurement of Meaning*. It attracted a fair amount of attention at the time as a creative and novel approach to designing survey questions using pairs of polar adjectives, a technique they labeled the *semantic differential*. They discovered when they matched randomly selected paired adjectives up with randomly selected nouns and surveyed an English-speaking population that an underlying and fundamental semantic structure emerged. They selected the word *father*, for example, and asked respondents to identify the meaning of the word on a variety of randomly selected polar scales fast . . . slow, hard . . . soft, large . . . small, and so on. As with this example, the pairing often did not make particular sense. Nonetheless, a clearly defined structure of three underlying dimensions emerged. Later they confirmed that the same structure evolved in other languages and cultures around the world—a fundamental structure of meaning in human communication. This was an important discovery. The first dimension is labeled *evaluation*. Is the object or concept at hand good or bad, beneficial or harmful, safe or dangerous? It is the most frequently employed dimension of semantic evaluation. The second dimension is *potency*. Is the object strong or weak, tenacious or yielding, large or small? The third dimension is *activity*. Is the object active or passive, dynamic or static, vibrant or still?

The curious history of this little corner of communication scholarship is that the book was interpreted as a contribution to survey item design but not particularly relevant to communication and psychology. That may be in part a result of the detailed technical style in which the original manuscript was written. The survey technique after a brief period of popularity is no longer frequently used, and the fundamental findings on how the evolved human cognitive system structures the world it perceives is only infrequently noted. Apparently it was not immediately clear to Osgood and his associates and their early readers how dramatic these findings are and how they could contribute to a refined paradigm of communication science.

Returning to these findings from a perspective of evolutionary psychology, what dimensions of evaluation and attribution would one expect small tribes of hunting and gathering early humans to use as they invent and refine language for communication and for survival? It may seem rather crude, but the first question at hand in confronting a new object is—is this a potential resource or a threat? Can I eat it, or will it eat me? Friend or foe? Good or bad? Not an unreasonable starting point for sharing communication with your family and tribe. If the object is a threat, is it strong or weak, is it active or passive? All important strategic dimensions of meaning unambiguously connected to the prospects of survival. If the object is friendly and a potential sexual partner, is it strong, healthy, and active?

Humans continue to be fundamentally tribal in their way of viewing the world around them. The key concept here is identity—ethnicity, religion, gender, age, and regional or national identity. The first question asked, in effect, is the evaluative dimension—good or bad, for my team or for the other team, that is, the other tribe, or the other species. People polarize. Groups polarize. It is human nature, actually animal nature, as demonstrated, for example, by warring ant colonies whose territorial borders are characterized by dramatic and distinctly violent combat (Wilson 1975).

Tribal competition, academic competition, athletic competition, commercial competition—perhaps a reasonable practice for sustaining the capacity for survival. Global nuclear war, ethnic genocide, religiously inspired terrorism among competing theological sects—perhaps not so good. The fundamental challenge to interpersonal and particularly to collective communication and the structure of public sphere itself is confronting the deeply ingrained propensity for polarization.

Pluralism. While polarization focuses on the interaction of identity and opinion and the resultant propensity for conflict at the individual level, the issue of pluralism focuses on these dynamics at the collective level. Pluralism in the public sphere posits the challenge—what institutional arrangements work best to promote an open marketplace of ideas, a healthy competition of alternative policy perspectives that is not paralyzed by conflict or offer the awkward stability resulting from the dominance of a particular ideology or dominant identity group. A truly pluralistic politics is difficult to maintain. When the system tips to one side, the pressures often increase toward imbalance rather than self-correction. If a majority is to be tolerant and perhaps even supportive of the expressed views of a minority, it requires a great deal of structural discipline. As various groups within or among societies attain economic and military power and, in turn, political power, it would be naive to expect that the dominant groups would not use their accumulated resources to bias the structure of the public sphere to their advantage and the advantage of their heirs. A century ago Max Weber's student Robert Michels formalized this slippery slope problem of elite control of public institutions as the "iron law of oligarchy" (Michels 1962, 342). Michels studied how union leaders in Europe initially elected from the membership to represent and execute the membership's views would in time increasingly privilege their roles as institutional leaders. In the characteristic pattern the elites stressed the needs of institutional self-reproduction over membership's direct interests. Ironically it was those direct interests that motivated the creation of the union in the first place. Michels's iron law is a natural enough phenomenon in human history evident in virtually all domains of collective human behavior from religious to political, cultural, and military institutions.

In Chapter 6 I review these issues in some detail. I rely on the concept of an open marketplace of ideas as the ideal type of public sphere. Much like the field of economics from which the market metaphor is drawn, I review the many complexities of how markets (in this case, markets of ideas) may be biased, inefficient, and unable to maintain open competition among participants. For example, there is no practical limitation to the number of physical products in an economic market. But just as the individual human mind has a practical limit to the number objects or ideas it can consider at a single time (it turns out the number is approximately seven things) there is a limit to the capacity of the public agenda. New issues push out old ones (Miller 1956; Neuman 1990). Publics tire of problems that resist solution (Downs 1972). Clearly the dynamics of public opinion in aggregate have different properties that the dynamics of an individual's thought and opinion about public issues, and I review both the parallels and the disjunctures. Given the paradox of profusion and the explosive growth of specialized expertise, a centrally important question becomes the dynamics of "issue publics"—groups of concerned and informed members of the general public who are tracking issues that are otherwise "below the radar" of public and media attention.

Partial to Paradox

This book represents something of an aggregation of five decades of professional work. I have been interviewing and experimenting, tracking history and technology, puzzling over policy and politics since the 1960s. My first book, based on my dissertation, was entitled *The Paradox of Mass Politics* and attempted to understand how the democratic process works as well as it does given the celebrated inattentiveness of the average citizen who not unsurprisingly prefers the sports pages and gossip columns to details on the latest legislation proposing financial reform. *The Future of the Mass Audience* followed and with support from a half-dozen media companies, attempted to understand the new media environment from an institutional and policy perspective. *Common Knowledge* and *The Gordian Knot* (with colleagues) developed the themes of public opinion and news and further speculations on the changing policy environment. In the past decade (again with colleagues) the work on *Affective Intelligence* focuses on survey and experimental research to better understand the dynamics of public attention and politics with particular attention to how emotions steer attentiveness. The work spans psychology, sociology, history, economics, political science, policy science, and communication, and accordingly may be appropriately characterized as undisciplined, but not, I would argue, as unfocused. In fact, my reason for raising this brief history now is to make the case that despite their distinct differences in methodology and in level of analysis, they are all about the same fundamental puzzle, or set of intertwined puzzles—how do individuals wrapped up in the very specific challenges of their personal lives take time to observe the world around them as history unfolds in the public sphere? What do they know? What should they know? How will this change in the evolving institutional and policy environment of the communication revolution?

Communication Research in the New Millennium

Scholars in the humanist tradition are sometimes frustrated with the obsessions of behavioral social science and are inclined to question the value of the assorted surveys and experiments that dominate the literature. Experiments in chemistry are fine, but the complexities of the human condition resist analysis by test tube and Bunsen burner. The problem, they might protest, is that this behavioral work is ahistorical, naively seeking some universal truths about human behavior while ignoring the pressing normative issues of our age.

They make an important point. But I challenge the premise. I make the case in Chapters 1 and 4 that, on the contrary, behavioral work on communications effects is as historically and normatively situated as the textual and critical work of the humanists. The focus here is on the study of human communication, a field of study, if perhaps not yet a discipline. It is interesting to note that the modern fields of economics, psychology, anthropology, political science, and sociology were all institutionally formalized at the turn of the twentieth century. But this was not true of the academic study of communication. Departments, journals, and associations of communication would begin to appear only in midcentury, in part in response to the growth of radio and television and the concern about propaganda stemming from World War II.

If we look at the origins of sociology, for example, in the late 1800s, it was clear that urbanization, industrialization, and migration to the big cities led to growing concern about social control. In the rural communities, citizens knew each other, and the town meeting or the cracker barrel at the general store were viable venues for sustaining the public sphere. The challenge of the industrial city was social fragmentation, mass society, and an absent institutional center of identity in church and community. Those are the founding concerns of the field at the turn of the century.

By the middle of the century the thematic has reversed itself and the concern is not too little but too much social control and influence from the media—propaganda, brainwashing, hypercommercialization, influential depictions of violence, political agenda setting by news media. It is at this point that the field of communication starts to establish its own identity, growing from but independent of psychology and sociology.

But now we have come full circle. At the end of the twentieth century with the explosion of the new media revolution the concern again turns to fragmentation, polarization, and the loss of a cultural center point.

Although the historic grounding of these issues may have changed directions in the digital age, the underlying question, the central and paradigmatic question of research on human communication remains the same although sometimes obscured. It is not whether communications effects minimal or not so minimal. It is—*under what conditions do humans successfully communicate (especially among those with diverse perspectives) and how can practices, institutions and norms that develop to structure the flow of information in society promote an open marketplace of ideas and rather than one oriented to protect the status quo and the powerful?* Such a starting point sounds a bit wishful and perhaps naive. Such a notion echoes Jürgen Habermas's (1979) highly idealistic concept of a universal pragmatic—the necessary conditions for reaching an understanding through communication. But the question need not be reserved for philosophical speculation. It could succumb to systematic empirical inquiry. In an age of global communication and global violence, the motivation to move research inquiry in this direction should be strong and, well, pragmatic.

Communication science and the humanistic field of cultural studies both study the structure of human communication. Most researchers in each of these communities largely view the scholarship of the other vantage point as some sort of indecipherable gobbledygook and avoid it like the plague so there is little cross-fertilization. It is unfortunate and, I suppose, just what you would expect from a field that bears the name *communication*.

1

The Propaganda Problem

The thought police would get him just the same. He had committed—would
have committed, even if he had never set pen to paper—the essential crime
that contained all others in itself. Thoughtcrime, they called it. Thoughtcrime
was not a thing that could be concealed forever. You might dodge successfully
for a while, even for years, but sooner or later they were bound to get you.

—GEORGE ORWELL (1949)

Communication research . . . is about effect. It could have been otherwise,
consider the study of art, for example—but it is not. . . . The underlying aim,
not always acknowledged, is to account for the power of the media.

—ELIHU KATZ (2001)

We must have theories which easily embrace new media, rather than calling
for new theory every time there is a new medium.

—ANNIE LANG (2013)

THIS IS A BOOK about communication in the twenty-first century. To
set the historical stage for a better understanding of the dramatic challenges and opportunities that lie ahead, I devote this chapter to a review
of the evolution of communication scholarship in the twentieth century.
The thesis is that there is an important disjuncture between the research
paradigm developed a half century ago and the opportunity to understand

20

the fast-changing dynamics of human communication in the digital age. We start with a puzzle as captured in the quotation above from Elihu Katz. The puzzle, simply put, is this: *How did communication scholarship come to develop the currently dominant research paradigm focusing on the "strength" of media effects?* Research paradigms traditionally proffer not just a puzzle but also a method for addressing the puzzle. In this case the received methodology is to correlate some measure of media exposure with various professed attitudes, changes in attitudes or behaviors. What is the historical origin of this mission to demonstrate that media effects are "not so minimal"? Is there an explanation for the dominant emphasis on negative media effects such as hypercommercialism and violence? Further, why is academic media effects research almost entirely disconnected from the applied research conducted by the various facets of the communication industries? That last element may well turn out to be a positive development, but we could benefit from better understanding its historical origins.

I argue that a large part of the answers to these questions can be found in the activities of a small group of people (mostly men, characteristically in this era, and a few women) in the 1940s and 1950s. The evolving study of mass communication was deeply intertwined with the historical events, normative concerns, and social movements of this period. The zeitgeist drove intellectual inquiry toward certain questions and away from others. Although one could identify perhaps hundreds of diverse projects that contributed in various ways to the developing field of communication study, some of them decades earlier, some later, I focus on one particularly influential group that was organized under the auspices of the Rockefeller Foundation in the late 1930s that continued to guide and fund research into the late 1950s. On September 1, 1939, as the group was meeting, Germany invaded Poland, marking the beginning of the Second World War. The Nazi fascination with manipulation of public opinion and with dramatic propaganda was already well known. The onset of the war generated a distinct culture of urgency as academics, journalists, and public officials began to come to grips with this troubling phenomenon of propaganda.

The term *propaganda,* which so dominated the literature of the era, is no longer in common usage, but this highly normative vocabulary for describing human communication continues to influence the paradigm of research, although, unfortunately, it does so less visibly and is less frequently acknowledged in modern scholarship. *Propaganda* is a shorthand term for false, manipulative, persuasive communication, and its presence in the public sphere necessitates an urgent need for an equally persuasive but "corrective" communication. If the historical origins of this perspective have faded in memory, the etymology is still very much with us.

Could It Have Been Otherwise?

Most scholars in the field of communication research may find it slightly odd to imagine a counterfactual history of research practice. After six decades of conducting surveys and experiments and accumulating findings within an intuitively satisfying paradigm of research practice (Neuman and Guggenheim 2011) there is a certain taken-for-grantedness about how things are done. But, briefly, let us review some arguably quite reasonable alternatives.

The first alternative has already been introduced in the prologue (and will turn up from time to time as a continuing historical trope in this book). Not only could it have been otherwise; communication scholarship could have been built on the polar opposite intellectual starting point. Consider the historical roots of sociology in the late 1800s and early 1900s. Instead of a concern with an overly strong central control via mass media propaganda the central concern of early sociology was a lack of a social glue, a meaningful basis for social organization and coordination that would survive the uprooting transition from smaller rural communities where people knew each other to the anomic crowds of industrialized urban centers (Merton 1968; Coser 1971; Giddens 1976; Jones 1983; Collins 1986). Had the field of communication scholarship started to develop its disciplinary identity fifty years earlier, this might well have been the case. Or communication scholarship, like the beginnings of political science, could have focused not on the psychology of persuasion but on aggregate structures such as communication systems within communities and nation-states. Political science as the comparative study of constitutions and legal systems persisted a half century up to the "behavioral revolution" in the field in the 1950s (Crick 1959). Or the systematic study of mediated communication could have evolved from literary and cultural studies focusing on the text, narrative structure, and cultural resonances of the mass media. That did happen, of course, with abounding energy about twenty-

five years later, but it evolved as a counterpoint to experiments and surveys rather than a precursor and this counterpoint continues to represent
an intellectual fault line in the field (Grossberg 2010). Or one could imagine
the study of the media as evolving as a practical and applied field perhaps
more akin to the curriculum and culture of schools of journalism or cinema
or departments of advertising or public relations. Such departments and
schools exist, but almost always at arm's length from academic communication studies. Also one could imagine an experimentally oriented
field of scholarship focused on learning from the media rather than persuasion as the research paradigm currently characteristic of educational
psychology, educational technology, and some branches of information
science (Chaffee and Berger 1987). Finally, the fields of mediated communication and interpersonal communication scholarship could have shared
common intellectual roots. Some may argue that they do, but since the
publication of Miller and Steinberg (1975) it is generally acknowledged
that these scholarly traditions have largely gone their separate ways.

Hypodermic Effects

Perhaps the iconic image of the pivotal midcentury period is the domi-
nating and demanding face of Big Brother from George Orwell's 1984.
Orwell's book was titled with a year in the near future, but the manuscript
was written just after the Second World War and published originally in
June of 1949. The novel's central thematic of totalitarian propaganda and
mind control was an artful mix of not-so-subtle references to both Hitler's
Nazi Germany and Stalin's Soviet Russia. Orwell's protagonist, Winston
Smith, is a lowly government bureaucrat working at the Ministry of Truth
as an editor, and his primary responsibility is changing the official (in fact,
the only) historical record so that the official past corresponds to the needs
of the current central authority. He routinely rewrites the records and
alters photographs so that those individuals now out of favor become
"unpersons" and inconvenient historical documents are irretrievable
deleted down the memory hole.

A particularly telling sequence early in the book is the required daily
routine of the "two-minutes hate":

> It was nearly eleven hundred, and in the RECORDS DEPARTMENT, they
> were dragging the chairs out of the cubicles and grouping them in the
> centre of the hall opposite the big telescreen, in preparation for the Two

Minutes Hate. . . . The next moment a hideous, grinding speech, as of some monstrous machine running without oil, burst from the big telescreen at the end of the room. It was a noise that set one's teeth on edge and bristled the hair at the back of one's neck. The Hate had started.

As usual, the face of Emmanuel Goldstein, the Enemy of the People, had flashed on to the screen. There were hisses here and there among the audience. Goldstein was the renegade and backslider who once, long ago (how long ago nobody quite remembered), had been one of the leading figures of the Party, almost on a level with BIG BROTHER himself, and then had engaged in counter-revolutionary activities, had been condemned to death and had mysteriously escaped and disappeared. . . . Goldstein was delivering his usual venomous attack upon the doctrines of the Party—an attack so exaggerated and perverse that a child should have been able to see through it, and yet just plausible enough to fill one with an alarmed feeling that other people, less level-headed than oneself, might be taken in by it. . . .

But what was strange was that although Goldstein was hated and despised by everybody, although every day and a thousand times a day, on platforms, on the telescreen, in newspapers, in books, his theories were refuted, smashed, ridiculed, held up to the general gaze for the pitiful rubbish that they were—in spite of all this, his influence never seemed to grow less. Always there were fresh dupes waiting to be seduced by him. A day never passed when spies and saboteurs acting under his directions were not unmasked by the Thought Police. He was the commander of a vast shadowy army, an underground network of conspirators dedicated to the overthrow of the state. . . .

In its second minute the Hate rose to a frenzy. People were leaping up and down in their places and shouting at the tops of their voices in an effort to drown the maddening bleating voice that came from the screen. The little sandy-haired woman had turned bright pink, and her mouth was opening and shutting like that of a landed fish. . . .

The Hate rose to its climax. The voice of Goldstein had become an actual sheep's bleat, and for an instant the face changed into that of a sheep. Then the sheep-face melted into the figure of a Eurasian soldier who seemed to be advancing, huge and terrible, his sub-machine gun roaring, and seeming to spring out of the surface of the screen. But in the same moment, drawing a deep sign of relief from everybody, the hostile figure melted into the face of BIG BROTHER.

Orwell's scenario mixes the anti-Semitic racial propaganda of Nazi Germany and Stalin's vilification of Trotsky (the Goldstein character) as a bête noire and traitor to the cause. Even our protagonist, the reluctant Winston Smith, finds himself caught up in the frenzy of the moment. It may represent rather starkly wrought dramaturgy, but it powerfully captures the concerns of the era about the atomized individual, mass society, the lonely crowd, the individual helpless against the onslaught of psychologically sophisticated propaganda. The extreme case of propaganda, of course, is systematic mind control or "brainwashing" that occupies a special place in the novel's conclusion as Smith's fears and weaknesses are exploited to break down his will to resist.

We continue to use the terms *Orwellian, big brother,* and *thought police* in common parlance today to conjure up these enduring themes. In the decades following its publication *1984* would sell tens of millions of copies and be translated into more than sixty different languages, at that time the greatest number for any novel (Rodden 1989). The hot war had been won, but the cold war of competing ideologies was just beginning. Orwell's novel represents a synecdoche for the very real and not entirely unreasonable dominating mood of fear and concern in the West following the war. It was not just a backward-looking retrospective on the theories and practices of Hitler's propaganda minister Joseph Goebbels (Baird 1975; Herzstein 1978) but a commanding fear that if it could happen in Europe it could happen again there or in the United States or in response to communist propaganda in the developing world. In the early 1950s the fear of communism was briefly but dramatically manipulated by the junior senator from Wisconsin, Joseph McCarthy, with the ironic result that his success in capturing the media spotlight with trumped-up images of a state department overrun with Soviet agents reinforced fears about the threat of propaganda and demagogic manipulation of public opinion, not just from the Left but from the Right, as well (Rogin 1967; Bayley 1981; Hamilton 1982; Gibson 1988; Gary 1996). The war in Korea generated new images of the extreme form of manipulation—Chinese thought reform and brainwashing (Lifton 1961). Richard Condon's 1959 novel *The Manchurian Candidate* and the equally influential 1962 MGM movie version describing former army officers secretly brainwashed in China as sleeper agents to assassinate politicians and take control of the government resonated powerfully with the public mood.

Vance Packard's presumptively nonfiction expose of commercial subliminal persuasion in the *Hidden Persuaders,* published in 1957, warned that imperceptibly brief flashed images of Coca-Cola on a motion picture screen generated sudden and inexplicable thirst among audience members who thronged to the concession stand for an extralarge drink. (Actually the study was a fake, created as a publicity stunt; see Pratkanis 1992.) It initiated a tradition of critical scholarship on manipulative commercial and political advertising still active today (Key 1974; Baker 1994; Goldstein and Ridout 1994; Turow 1997; Chomsky 2004).

We review these prominent themes from the popular and political culture of the midcentury because they are central to the discussion of the propaganda problem. As noted at the conclusion of the prologue, systematic communication research and the institutionalization of departments of communication in the academy began not as virtually all of the other social science disciplines did at the end of the nineteenth century but rather at the middle of the twentieth. Coming to a disciplinary identity at the conclusion of the Second World War, the field of communication research adopted a paradigm designed to address the propaganda problem, the concern about the hypodermic injection of manipulative images and arguments in the perusable public mind. It turns out that the seminal researchers never really used simplistic stimulus–response notions of hypodermic needles and magic bullets of influence that their successors dismissively attributed to them; they were much more sophisticated analysts than that (Bineham 1988; Chaffee 1988; Lubken 2008). But such vivid imagery does capture the zeitgeist of this period.

The Other Marshall Plan

John Marshall was a middle-level program officer in the humanities division of the Rockefeller Foundation. A scholar of English literature and medieval history, he became particularly concerned about the evident power of European fascist propaganda as a result of his travels in Europe on a foundation library project through the 1930s (Buxton 2003). He was looking to make his mark in the foundation and among the intellectual elite of his era so he developed relationships with Edward R. Murrow in England; with sociologist and social activist Robert S. Lynd at Columbia; with political scientist and propaganda specialist Harold Lasswell at Yale; with sociologist and methodologist Paul F. Lazarsfeld, who was soon to establish a highly influential research program at Columbia; and with the influential British literary critic I. A. Richards. Marshall became a man with a plan—a plan to fund research centers around the United States and develop a new systematic, quantitative social science field of research on communication and attitude change. As historian Brett Gary describes this era:

> Beginning in 1939 Marshall drew the principals of the communication projects into a Foundation-sponsored discussion group, interchangeably called the Communications Group or the Communications Seminar, whose primary purposes included achieving consensus on a paradigm for the field.... Through his influence on this group, he self-consciously helped impose theoretical coherence and a "scientific" research paradigm on this inchoate area of interdisciplinary inquiry. The goal, before the war, was to shape an empiricist blueprint for the incipient field, aimed at addressing what Marshall saw as the chief problem for mass communication research: "Mass communication in reaching millions becomes mass influence, for better or for worse." (1996, 126)

The initial focus of research and funding was radio broadcasting, which had become an increasingly dominant medium of mass culture and the public sphere in the 1930s. Marshall persuaded a reluctant Paul Lazarsfeld to accept directorship of the Rockefeller Radio Research project started originally by Hadley Cantril at Princeton, then at Newark University and soon to move again to become an increasingly influential research center at Columbia. But as noted above, the onset of the Second World War in September of 1939 shifted and energized the group's sense of urgency. Marshall reported later, "We saw in this rather disastrous situation of the war an unhappy opportunity to conceptualize this whole field of research. The war as it were put all the factors into sharp focus" (Morrison 1978, 349).

Interestingly Marshall made numerous efforts to persuade the newly influential and increasingly wealthy broadcasting companies to join the foundation in supporting these research efforts financially and intellectually. Marshall argued the results of this research would be useful to the industry as well as the struggling nation. Historians Morrison and Gary trace these developments with particular interest and conclude that the industry was wary of supporting research whose results it could not control, results that could potentially be used to criticize or justify regulation of their business practices (Morrison 1978; Gary 1996). A special assistant to the president of CBS told Marshall candidly, "It's perfectly simple, we don't want to rock the boat" (Morrison 1978, 353). Commercial broadcasting research was highly attuned to maximizing the attractiveness of the medium to advertisers. Marshall ultimately gave up trying to include the media industry in the evolving research effort. This was an important development, and it set a precedent of independence between commercial and academic research that remains dominant today.

As a result of this particular confluence of events and institutional developments, three defining characteristics of the communication research paradigm were firmly established:

1. The research is tied to a pressing potential social problem—the manipulation of public opinion, the propaganda problem. The premise is that the key exemplars of mass communication are not neutral or informational but rather highly valenced. The motivation for research was the pressing need to counter such messages that are intentionally false, misleading, and manipulative with an appropriate successful corrective communication to protect public awareness and a working democratic system.
2. There is a recognition that the field of communication and public opinion is a new domain of inquiry requiring its own coherent and systematic theoretical corpus and new experimental survey research and content analysis methodologies.
3. The burgeoning communication industries were wary of becoming involved with a research enterprise that had the potential to raise questions about their evolved business practices and increasingly profitable business model, so they quietly opted out leaving the research and research funding to government, foundations, and the academy.

There was no particular impetus reflected in the reports, articles, and books of this era to demonstrate a not-so-minimal effect of the media. The power of the media was a given and unquestioned. The influence of this seminal paradigm is still so strong and it has such a taken-for-granted character that it is hard for students of the field to imagine it could have been otherwise (Lasswell 1927; Lee and Lee 1939; Wirth 1948; Hovland, Janis, and Kelley 1953; Katz et al. 1954). I. A. Richards, the distinguished literary scholar from Magdalene College, Cambridge, was an active participant in the Rockefeller planning exercises. One could have predicted that as a father of close-reading methodologies associated with the humanistic tradition of the new criticism, Richards would have developed an approach to the study of content quite different from the quantitative frequency counts that Lasswell, Berelson, and Pool formalized in what came to be known as content analysis (Berelson 1952; Pool 1959). There were dramatic differences in the structure of communication flows in Nazi Germany and later Soviet Russia and communist China compared with the West. Accordingly one could have predicted an approach based on a comparison of media systems perhaps like the dominant approach in political science at the time that focused on comparative constitutions and contrasting legal systems. One could have predicted an energized research project on education, learning, public knowledge, or journalistic institutions as addressed in the well-known work of John Dewey (1927), Walter Lippmann (1922), and even Max Weber (1910). One could imagine an effort to connect our understanding of mass communication with interpersonal and small group communication, but that would not reappear in the literature until the mid-1950s and then as a freestanding discovery of the two-step flow of communication, something quite independent in conceptualization from the other research traditions.

So the serious study of mass communication became institutionalized as the analysis of the effects of "valenced communication"—the battle between totalitarian propaganda (first the fascists and then later the communists) versus the liberal democratic corrective. This model of the atomized and anomic individual seeking to latch on to the reassurance of an authoritarian father figure combined a theory of society and a theory of human psychology.

The Founding Fathers

The various researchers that John Marshall had drawn together in 1940 were already in sporadic contact working at various institutes and universities around the country. It would be another eight years before a full-fledged graduate program in communication research would be established by Wilbur Schramm at the University of Illinois. Still at it in 1948, John Marshall at Rockefeller provided the funding to draw together a distinguished research group with a few new additions to take stock of developments and contribute to an edited volume entitled *Communications in Modern Society* (Schramm 1948), which would become the very first textbook for advanced study in the field. The key principals knew one another well by this time; they had been working together as part of the wartime mobilization of propaganda researchers. Paul Lazarsfeld, an Austrian émigré who had fled in the shadow of Hitler's rise in Germany, was not an American citizen, so he served as a consultant. Wilbur Schramm joined the Library of Congress's Office of Facts and Figures. Harold Lasswell was active in a number of projects, including the Office of Facts and Figures and its successor organization, the Office of War Information; the Office of Strategic Services; the Foreign Broadcast Monitoring Service of the Federal Communications Commission; and the army's Psychological Warfare Branch. Carl Hovland from Yale joined the Information and Education Division of the War Department to evaluate the impact of persuasive films on soldiers (Schramm 1980; Rogers 1992; Cmiel 1996; Sproul 1997; Gary 1999; Glander 2000; Katz et al. 2003; Peters and Simonson 2004; Wahl-Jorgensen 2004; Park and Pooley 2008; Pooley 2008; Simonson 2010). Chaffee and Rogers described this period:

> During World War II, Washington was the place to be for a social scientist. America's enemies seemed to represent such an unmitigated evil that few if any social scientists opposed the war. America's war aims united these scholars in a common cause and brought them together in one place where they formed a network of relationships that endured for the rest of their careers. The war effort demanded an interdisciplinary approach to problems, often closely related to communication study because in so many ways it was seen as "a war of words." Communication was also viewed as the basic tool for mobilization of the American people to volunteer, conserve, and in other ways aid in concentrating the nation's resources on winning the war. In important ways, then, World War II created the conditions for the founding of the communication field, a point that Schramm often stressed to his students in later years. (Schramm 1997, 134–135)

The era was vividly Manichean. The public imagination was dominated by a fear of the power of hidden subliminal messages in the mass media, thought reform and brainwashing and more generally McCarthy-style demagogic manipulation of public opinion (Packard 1957; Shils 1957; Kornhauser 1959). Slowly, careful research revealed that the German propaganda machine was actually not successful in sustaining falsehoods about nonexistent successes on the Russian front especially after the disaster at Stalingrad in 1942–1943, subliminal advertising did not really work, the Chinese were using standard intimidation techniques with prisoners, and McCarthy's inventive falsehoods about a State Department overrun by Soviet agents could not be sustained over time in the public eye (Schein 1971; Baird 1975; Schrecker 1998). It is widely acknowledged that corrections and retractions seldom attract the attention that the original accusation or proposition do. *The Manchurian Candidate* was much more engrossing than dry military histories of what actually happened in Korea and China in the 1950s. I am not aware of any research on the question, but I would wager that among those today who have heard of the subliminal Coke advertisement test, most continue to believe it really happened and are simply unaware that it was a hoax, actually a publicity stunt and that serious attempts to demonstrate the effect were unsuccessful (Pratkanis 1992).

The Waning of the Concept of Propaganda

A central theme of this chapter is that the seminal work in the systematic study of communication was infused with the notion of propaganda. The threat of propaganda motivated the research enterprise. One needed to better understand the power and character of these evolving mass media because the anomic mass audiences were subject to manipulation by forces that were unambiguously evil and duplicitous. It is also posited that although the term itself has faded into near obscurity (except for occasional use in an effort to be rhetorically provocative), the logic of the propaganda concept continues to pervade theories of the communication process. Further, because this historical and intellectual connection is seldom acknowledged, theoretical progress may be inadvertently constrained. My thesis is this: the notion of propaganda is unambiguously and unapologetically asymmetric. The user of the term purports to convey the truth in contradistinction to the untruths, half-truths, distortions, and omissions of the "other party." As modern communication scholarship addresses political communication, health communication, and commercial mass media entertainment, a subtle asymmetry persists—although it is unacknowledged or at least seldom acknowledged. If it were actually true that pernicious and untruthful ideas are more easily propagated by the mass media than virtuous and truthful ones are, that would indeed be an interesting research finding. I suspect that no researcher would be inclined to acknowledge such a premise or paradigm, but, I argue, it is identifiable in research practice, and it pervades the scholarly obsession with negative communication, from the depiction of sexuality and violence through the study of political bias and stereotyping and detrimental health behaviors. After reviewing the intriguing evolution of the propaganda paradigm, I attempt to address some possible responses to this central element of the field's intellectual history.

The use of the term *propaganda* to denote purposefully deceitful and manipulative communication dates back to the First World War. Ironically before the twentieth century the historical origins of the term are derived from the Latin root *propagare* "to propagate," and the term was used only in the entirely positive sense in the Christian tradition of spreading the "good word" of scripture and the propagation of the faith. After the First and Second World Wars, the concept was reenergized in the growing concern about communist propaganda and communist subversion in Western democracies. Textbooks reviewing the findings of recently developed random-sample survey research, for example, were not titled just "public opinion" but *Public Opinion and Propaganda* (Doob 1948). Talcott Parsons's essay from this period captures this conceptual centrality in his opening paragraph: "What is rather vaguely classified as propaganda has come to be one of the most conspicuous social phenomena of the contemporary world, of absorbing interest both from scientific and practical points of view" (1942, 551).

The central argument of Parsons's paper, interestingly, was that the phenomenon of propaganda, although generally seen as a fundamentally psychological dynamic, should be seen as central to the study of social structure and social control, the domain of sociology. One does not fight enemy propaganda with propaganda of one's own; one fights it with "truth." So, when the United States founded the United States Information Agency in the early 1950s to counter the communists, the term *public diplomacy* was coined, and that sounded like a much better enterprise for the propagation of "truth, justice, and the American way" than "counter-propaganda" (Tuch 1990).

There is no evidence of a dramatic intellectual rejection or critique of the propaganda concept; it just gradually fell into disuse as attention moved

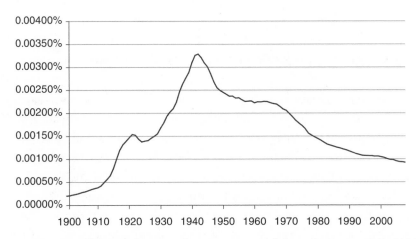

Figure 1.1.   Use of the Word *Propaganda* in American English Books Corpus 1900–2008 (as percent of all words thus controlling for increased number of publications)
Source: Generated via Google Books Ngram Viewer.

on to other concerns and vocabularies. The decline is traced in Figure 1.1 reflecting the use of the word *propaganda* in American book publications by year.

The Legacy of the Propaganda Era. It would be a mistake to assign an overly deterministic role to the influence of the postwar-era thinking on modern communication research. But it may be useful, nonetheless, to explore some of the possible linkages and perhaps to stimulate some fresh thinking about how the received paradigm of media effects research is struggling to accommodate the new media revolution. My thesis is this—*the focal point of media effects research gradually morphed from the study of propaganda and persuasion to the study of a diffuse collection of "negative" effects of communication.* If such a proposition is true or even partially true, it could have important effects on how research studies are designed, how research is funded, and how the paradigm of research practice evolves. A focus on the negative and problematic is certainly understandable—economists often focus on economic crises, medical scientists on biological pathologies. But in such cases there exists an analytic model of a healthy functioning economy and a healthy biological system against which the pathology can be contrasted. That does not appear to be the case in communication research. One of the continuing themes of this book is that it could be and should be. Since at first blush communication would seem to be a relatively positive human phenomenon—the sharing of ideas, communication as an alternative to conflict, the uniquely human achievements of art and culture—this singular focus on the negative and problematic may merit further attention. The study of propaganda and persuasion had two characteristics not carried forward in present-day research. First there was, especially at the outset, a practical application and an urgency to research as scientists gathered in Washington to help the military effort and thereafter the Cold War's war of words. Second, there was an obvious counterpoint to the falsehoods of fascist and later communist propaganda, and that was simply getting out the truth, at the time primarily through dropping leaflets and shortwave broadcasts. It could be said that such notions represented the communication theoretic equivalent of the healthy economy and healthy organism. My argument, however, is that both the sense of practical application and historical exigency and the counterpoint of successful positive communication have largely been lost. Unfortunately, the "critical stance" of much communication scholarship that draws attention to violence, consumerism, stereotypes, and uninformative newscasts seldom leads to changes in the commercial communication industries outside of an occasional congressional slapping of hands or a largely symbolic but ineffective regulatory ruling (Neuman, McKnight, and Solomon 1998).

One of the most effective demonstrations of the near obsession with negative effects is the historical review of media effects on children published by Ellen Wartella and Byron Reeves in 1975. In retrospect the dominating concerns of each era with the new media of its time appear to be a bit quaint. The progression can be characterized as follows—movies appear to lead to juvenile delinquency, comic books to declining reading skills, television to aggressive behavior. Did you know how parents reacted when radio entered their homes? Wartella and Reeves quote a 1936 study on the topic: "This new invader of the privacy of the home [radio] has brought many a disturbing influence in its wake. Parents have become aware of a puzzling change in the behavior of their children. They are bewildered by a host of new problems, and find themselves unprepared, frightened, resentful, helpless" (129–130). The pattern continues today as scholars and journalists review anecdotal and some experimental evidence that "Google is making us Stoopid" (Carr 2008, 2010; Sparrow, Liu, and Wegner 2011).

Frequently in science and sometimes other domains of scholarly en-deavor including the humanities there is sense of discovery, a fresh start, some sense of resolution of some old problems and a shared sense of the importance and promise of some new ones. That was distinctly not the case as concern about propaganda more or less quietly faded away and re-search shifted almost randomly to a diffuse and disorganized set of other issues associated with the general domain of "media effects." There are three difficulties with this curious and almost inadvertent transition from propaganda effects to media effects. First and perhaps most impor-tant, research becomes fragmented into a plethora of increasingly narrow and surprisingly intellectually insolated research subspecialties. Thus, for example, work in health communication would cite and build on theory or findings in political communication only rarely. Second, the sense of ur-gency and connection to real-world practice that was characteristic of the propaganda era is diffused and dissipated. Scholarship is written for and judged by a small circle of fellow specialists. There are exceptions, but again only relatively rarely. And third, the normative grounding is still there, in part associated with the focus on negative or problematic characteristics of communication process, but it has become increasingly diffuse and indi-rect. For example, in the propaganda era the alternative to propagandistic misinformation is accurate information. The alternative to the depiction of violent acts in narratives and news or consumerism is less clear. Regula-tion of the arts and of journalism to limit depictions of violence or con-sumerism runs against the strong tradition of the first amendment and probably as well against strong commercial incentives to attract audience interest. Labeling content so parents and others concerned about violent or commercial content can make more informed choices is a step in the right direction, but that too turns out be highly problematic in practice.

Stepping back from the focal points of modern media effects research, a pattern emerges. My summary of this pattern is incomplete and may miss some important strands of research, but the structure that is evident may help in more integrative and coherent theory building and theory testing. I have come to conclude that the original propaganda problem has morphed into four related lines of research inquiry:

1. Norm violations—particularly violence and sexuality
2. Stereotyping—particularly racial and gender stereotypes

3. Political bias—usually contradictory critiques from the political Left and Right
4. Health communication—usually contrasting the effects of unhealthy and health-oriented behaviors in media content

As the Second World War faded in memory and the Cold War receded in prominence, attention in the United States and most of the industrialized West turned from European fascism and Soviet communism to internal struggles over social and political values and ethnic identities. Albert Bandura's (1977) social learning model and Henri Tajfel's (1982) analysis of in-group/out-group identity dynamics provided a paradigmatic platform for research on how the depiction of violence, sexual behavior, and exaggerated racial, ethnic, and gender stereotypes may reinforce role-modeling of norm violations, desensitization, and prejudice. As Table 1.1 illustrates, it would be difficult to find a publisher if your media text did not cover each of these components. There are, of course, other foci of attention, notably on media and children, advertising effects, video gaming subcultures, music lyrics, educational media, sports, and emotional responses to media, but the "Big Four" are canonic and in time have become inevitable topics for instruction.

All these lines of research are entirely reasonable subjects of inquiry if a slightly odd historically evolved jumble of topics. But the structure of scholarship leaves enormous room for improvement and intellectual integration. I identify four problematic elements in these newly canonic traditions. First of all, each is almost entirely insulated from the other lines of inquiry. Subtopical fragmentation and stovepipe separation are the hallmarks of the structure of journal publication, professional meetings, and the organization of tenure reviews and of "job slot" definitions in academic hiring. Scientific specialization is not the problem; quite the contrary, advanced specialized research is a sign of scientific progress. But specialized work in chemistry and biology, for example, is built on a commonly accepted paradigmatic core that is applied, in turn, on advanced topics. In the structure of communication research there is only passing and partial acknowledgment of a common core and almost no contribution from the subspecialties to an aggregated and integrated theoretical core. Individual scholars should not be blamed, of course, the incentive structures of publication and promotion are not very forgiving, and the evolved tribal

Table 1.1   Central Themes in Media Effects Texts

	Text Title	Norm Violation Chapters	Stereotyping Chapters	Political Bias Chapters	Health Communication Chapters
Perse 2001	Media Effects and Society	"Effects of Violent Media Context" "Effects of Sexually Explicit Media Content"		"Shaping Public Opinion"	
Harris 2004	A Cognitive Psychology of Mass Communication	"Violence: Watching All That Mayhem Really Matters" "Sex: Pornography, Innuendo, and Rape as a Turn-On"	"Media Portrayals of Groups: Distorted Social Mirrors"	"Politics: Using News and Advertising to Win Elections"	"Teaching Values and Health"
Preiss et al. 2007	Mass Media Effects Research	"Effects of Media Violence on Viewers' Aggression in Unconstrained Social Interaction" "Effects of Sexually Explicit Media"	"Effects of Gender Stereotyping on Socialization"	"On the Role of Newspaper Ownership on Bias in Presidential Campaign Coverage by Newspapers"	"Meta-Analyses of Mediated Health Campaigns" "An Analysis of Media Health Campaigns for Children and Adolescents"

(continued)

Table 1.1 (continued)

Text Title	Norm Violation Chapters	Stereotyping Chapters	Political Bias Chapters	Health Communication Chapters
Bryant and Oliver 2009	Media Effects: Advances in Theory and Research	"Media Violence" "Effects of Sex in the Media"	"Effects of Racial and Ethnic Stereotyping" "Content Patterns and Effects Surrounding Sex-Role Stereotyping on Television and Film"	"How the News Shapes Our Civic Agenda"	"Effects of Media on Personal and Public Health" "Effects of Media on Eating Disorders and Body Image"
Sparks 2010	Media Effects Research: A Basic Overview	"Effects of Media Violence" "Sexual Content in the Media"	"The Effects of Media Stereotypes"	"The Effects of News and Political Content"	"Potential Medium Effects on Health" (partial chapter)

identities of subspecialty communities have their own social and professional rewards.

Second, the focus on negative communication, although perfectly understandable, leads to a rather asymmetric structure of research. Audiences undoubtedly pick up a vast variety of positive and negative cues from the media. There are a few proud, brave, and savvy law enforcement role models as well as a few evil and violence-prone criminals in typical media depictions. There are examples of cooperation and collaboration, as well as violence and mayhem. There is a blend of political wisdom as well as self-interested demagogic distortions in the media mix. It seems a bit odd to ignore the vast variety of authentic depictions of ethnic character or complex gender roles in the haste to critique yet another example of institutional racism or sexism. If social learning and identification is stronger for the depiction of violent and intolerant role models in the media than for their more socially desirable counterparts, that would indeed be a finding of fundamental import, but a meaningful analysis, it would seem, would require a comparison and a broader focus of research than is currently in fashion.

Third, understanding such media effects requires a theory of audience selection and preference. It is routinely assumed that media institutions proffer mayhem, blatant sexuality, and cultural stereotypes because it is profitable to do so. Serious study of creative decision making in the news and entertainment media reveals that the process is extraordinarily complex, multilayered, and ill served by simplistic models of pandering and profit seeking (Cantor 1971; Auletta 1992; Waterman 2005; Lotz 2007). Like those who cannot resist the outlines of mysterious figures on grassy knolls, it is a bit convenient to simply assign the shortcomings of the modern media to the fundamental evils of capitalism and the egregious profit seeking of its senior executives. A normatively grounded critique of the institutional structure of a mix of public and commercial media is a laudable enterprise and certainly should be included in the communication research agenda. But the conceit of being utterly independent and critical is ultimately self-defeating if a critical perspective is intended to make a difference. Careful research has revealed that increased violence (gratuitous or otherwise) is surprisingly *not* correlated with increased audience size (Hamilton 1998) even though most simply assume that is the case. The way in which audiences diversely and polysemically interpret ethnic, violent, and sexual depictions, and alternatively avoid them or seek them out, has not yet seriously entered the repertoire of the collective research agenda.

Fourth and finally, the underlying normative basis for these critical perspectives has now become so ingrained in the research process that it is no longer acknowledged or analytically examined. Perhaps an exemplar of "dogmatic slumbers." Such a view is not meant to represent a defense of violence or racism or some sort of postmodern celebration of fundamental relativism. Mine is a call for a careful reexamination of the normative grounding and scientific prospects for studying the structure of human communication in advanced industrial societies. In the chapters ahead I outline some suggestions for just such a grounding.

The focus on propaganda and the related notion of mass society, however, have almost entirely receded from view in the scholarly literature of communication. James Beniger has a theory of what happened. It may not be a particularly elegant or dramatic theory, but it probably captures these intellectual trends most accurately. Beniger posits that the term simply "petered out" and was subtly and at times inadvertently replaced by more narrowly focused theory of media effects. Instead of the organizing concept focusing on the interaction of communication structure and social structure, the new mission became to sustain the conclusion that media effects were not so minimal. In his review of fifty years of public opinion research in the special 1987 edition of *Public Opinion Quarterly*, he argues that the propaganda/mass society tradition was a fundamental stimulus to the evolution of public opinion research. He notes a passage from the editorial forward of the very first edition of *POQ* that reveals a strong normative focus indeed: "A new situation has arisen throughout the world, created by the spread of literacy among the people and the miraculous improvement of the means of communication. Always the opinions of relatively small publics have been a prime force in political life, but now, for the first time in history, we are confronted everywhere by mass opinion as the final determination of political and economic action. Today public opinion operates in quite new dimensions and with new intensities; its surging impact upon events becomes the characteristic of the current age—and its ruin or salvation" (S46).

Beniger argues that central figures such as Lasswell, Lazarsfeld, and Berelson left the field of communication and moved on to other issues and rubrics. But he argues Converse's model of limited issue voting, McCombs and Shaw's agenda-setting paradigm, Noelle-Neumann's spiral of silence model, and Gerbner and colleagues' cultivation model are actually all derived from the notions of propaganda and mass society although those roots are seldom acknowledged. He goes on to characterize a "New Mass Paradigm of Pattern and Process" that moves away from a singular dependence on survey methods to a more methodological pluralistic cognitive perspective that "transcends cognitive psychology and social psychology to include political information processing, macrosociology and much of the traditional subject matter of the humanities" (S53). His upbeat appraisal labels the new work as an "oldnew" paradigm that ought to play a central role in the next fifty years of POQ's intellectual history. He does not seem terribly concerned about the topical and terminological pluralism or the diffuseness of the underlying theory.

But Dennis K. Davis (1990), reviewing these issues a few years later, adds an important caveat. The Columbia School (particularly the work of Lazarsfeld, Katz, Merton, and Klapper) appeared to have challenged the Chicago School by demonstrating that the media effects were actually surprisingly minimal. He continues, "[The] Columbia School views and the data on which they were based quickly came to be interpreted within academia as offering incontrovertible proof that the Chicago School's fears of mass society were groundless" (1990, 151).

This turns out be an important development, especially in that the research paradigm shifts from trying to understand the social and cultural conditions of various media effects and morphs into a campaign to defend the strong effects theory from the challenge of the minimal effects school. Davis argues this spirited defense of strong effects has a compelling normative element. He characterizes the minimal effects tradition as "elite pluralism" and describes the position as asserting that "there is no need for broad citizen participation in government . . . broad involvement is impractical or even dangerous" (1990, 153). The Columbia scholars would reject such a characterization out of hand, of course, but such fighting words help explain the energy behind the celebrated scholarly debate. To heighten the tensions, now approaching a level of sectarian conflict, the Columbia crowd has been characterized as corrupted by their ties to the media industries (Rowland 1983; Chaffee and Hochheimer 1982; Delia 1987). These reviews trace the growth of what they term the *marketing perspective* of Paul Lazarsfeld and his associates. They find that the marketing perspective tends to equate the selling of political candidates to the selling of soap flakes and, as a result, eviscerating the normative core of media effects theory. Delia (1987) asserts, for example, that intimate connection with corporate marketing research made the connection between the commercial and political interests that spawned the original models of communication in society invisible. The state of empirical findings, Delia argues, replaced the original concern with the state of society. There is, in his view, a natural tension between being scientific and being historically relevant. The capture of the media effects paradigm results, he asserts, in "the hegemony of the quantitative social science" (71).

The Puzzle: Communication Research at the Crossroads

This review of the present-day critical tradition in communication research has itself been somewhat critical. I have characterized the transition from the propaganda era to the modern era as overly haphazard and having resulted in a corpus of theory and research that is contested, fragmented, and diffuse. Theory has not progressed as it should through consensual refinement, the integration of findings, the resolution of puzzles, and the collective and convergent rejection of faulty or inadequate hypotheses. It has just sort of accumulated, perhaps not unlike the overflowing piles of seldom used possessions that soon come to so overwhelm the garage that the car's only option is to park in the driveway. I am not arguing that there has been a complete absence of progress. My point is that the literature is so fragmented and diffuse and its authors so anxious to distance themselves from their hypodermic or minimal-effects predecessors, that what progress there has been is hard to identify. To pursue the garage metaphor just a little further—one may come to abandon much hope of finding an older object among the piles in the garage and accordingly find it simply easier to obtain a new one. Accordingly those who have forgotten or ignored extant research are unconstrained from rediscovering well-known patterns and simply giving them new names. Such practices may have been prevalent enough to have led to Robert Craig's skeptical conclusion about the state of the field: "Anderson (1996) analyzed the contents of seven communication theory textbooks and identified 249 distinct 'theories,' 195 of which appeared in only one of the seven books. That is, just 22% of the theories appeared in more than one of the seven books, and only 18 of the 249 theories (7%) were included in more than three books. If communication theory were really a field, it seems likely that more than half of the introductory textbooks would agree on something more than 7% of the field's essential contents. The conclusion that communication theory is not yet a coherent field of study seems inescapable" (1999, 120).

It is probably the case that most active scholars in the field even if they acknowledge Craig's thesis are not terribly troubled by it. Because of the field's disconnect from professional practice and unlike the disciplines of, say, medicine or law, pressing real-world problems are not generating demands for the field to demonstrate its worth in helping resolve such challenges. There is also little internal pressure within the field because the incentives for professional advancement encourage specialization and inadvertently encourage fragmentation. So one might speculate that without some dramatic external intervention, such a state of affairs is not likely to change.

This is where the story gets interesting. There is development of significantly dramatic consequence. Perhaps it is obvious. *It is the digital revolution—the shift from push to pull, a fundamental challenge to the viability of a century-old model of advertising-based industrial commercial mass media of news and entertainment and the selling of books and recordings. Further there is a global intrusion into what were once independent and virtually unchallenged nation-state controlled or regulated systems of telephonic and mass communication.* These circumstances lead to labeling this section "Communication Research at the Crossroads." Indeed, this entire book is organized as an attempt to address the question of how a revolution in communication practice makes possible a revolution in communication scholarship. Centrally important, in my view, is Annie Lang's call to arms, included among the quotations at the beginning of the chapter: "We must have theories which easily embrace new media, rather than calling for new theory every time there is a new medium" (2013, 23).

And the theme of this chapter is that rather than bolting away from the historical origins of communications research in the era of propaganda, we should embrace them. Not embrace in the sense of studying Nazi propaganda posters. Not embrace in the sense of ignoring the past six decades of scholarship. But embrace as part of an effort to rediscover the roots of research, exemplified in the commitment and energy among those working with John Marshall. Of special interest is the sense of urgency and practical real-world significance of needing to understand how the dynamics of the public sphere actually work in order to protect it from the propagandists. The puzzle continues—how would this be accomplished?

I offer three suggestions—three potential keys to the puzzle, if you will. The first is to return to the concept of propaganda to see if can be reworked, reimagined, and scientifically generalized as an analytic tool for understanding the modern dynamics of media and audiences. My candidate for that exercise is to abandon the obvious asymmetry of the original propaganda concept (their views are manipulative, self-interested propaganda and our views are simply truthful and unbiased observations) and examine the simple prospect of *valenced communication*—communication processes deeply imbued with the identities and interests of different social groups and resulting in the likely if not inevitable polysemic conflict.

The second suggestion is to reground the practice of communication research in the historical moment. The historical thematic of the 1950s was the mass society—the anomic and atomized individual subject to the manipulative powers of the growing industrial and governmental media. Such a thematic may have indulged in some effective exaggerations and emphases. But that is central to the character of the intellectual moments and fashions that energize the academy and the public sphere. My suggestion is to explicitly explore the transition from the mass society to the information society, carefully noting continuities of the human condition and the historical disjunctures, as well.

The third suggestion is to reframe and reground the research enterprise as a continuing effort to understand the conditions under which media effects are alternatively large or small rather than attempting to supersede a misguided era of minimal effects with some shiny new analytic model in which media effects are proudly and inevitably "not so minimal."

The First Key to the Puzzle: Valenced Communication

Before directly addressing the prospect of how to refine older theories of propaganda effects in terms of the effects of valenced communication, it might be useful to step back and contrast the subject matter of human communication with the behavioral foci of several of the other social sciences. Many of the paradigmatic models are based on a fundamental game theoretic notion of zero-sum dynamics—that is, that the benefits for one individual or social group are likely to be associated with a cost or loss for another. The economic model of the marketplace, of course, assumes a willing buyer and a willing seller, which sounds cooperative rather than competitive. But clearly a market exchange is based on a differential valuation of the item of exchange; that is, each participant is cheerfully taking advantage of the other's error in valuation and profiting therefrom. In political studies, the zero-sum dynamic involves power rather than wealth, but a gain in power for one political entity is likely to be associated with a loss of power for another entity. In evolutionary biology the zero-sum dynamics of the survival of the fittest is clear-cut. One might ask, then, is such a dynamic typically present in communication? As noted above, historically communication has been defined in terms of a commons, a sharing of information, a coming together, a communion. So as with the willing seller and willing buyer in the abstracted marketplace, at first examination, the communication exchange like the market exchange has the earmarks of a cooperative and collaborative process, a positive-sum rather than a zero-sum process. In mass communication the willing audience member attends to the content and the ads, or pays for admission seeking entertainment and information from authors and journalists who are duly compensated, often well compensated, for success in an open and competitive information marketplace. Such a perspective seems reasonable enough (Napoli 2001).

So let us introduce the demagogue and propagandist into the mix. A self-motivated cultural and political actor purposely manipulates the gullible audience and distorts what would otherwise be an open marketplace of ideas. The normative posture of the propaganda era is to counter the self-motivated falsehoods with the unvarnished and un-self-motivated truth and if possible remove the demagogue or defeat the evil empire or current axis of evil. This reflects, of course, the inherent asymmetry of the propaganda paradigm.

Alternatively, we could introduce a notion of valenced communication—the view that virtually all human communication is at least in part self-interested behavior. Humans seek reinforcement of their identities and their ideals in the news and entertainment of the public sphere. It could be noted again, as in the prologue, that each of the three fundamental dimensions of linguistic meaning as derived from the semantic differential analysis (and replicated in multiple languages and cultures) speak to the question of self-interest and identity (Osgood, Suci, and Tannen-baum 1957). The first dimension is "evaluation" (good versus bad), the second is "activity" (active versus inactive), and the third is "potency" (strong versus weak).

So one imagines the individual enters a new situation; the first question asked is friend or foe? Further (particularly if foe) is this new entity active or passive and strong or weak? It would seem to make evolutionary sense to look at the world that way. Perhaps it could be said to be part of the human condition. We are hardwired that way. It shows up in the patterns of the languages we invent and use. It is not unnatural or inappropriate to be inherently self-interested in entering into communication behavior. We interpret complex and polysemic messages in ways that make sense to us and reinforce our identities. We speak in ways that highlight our virtues and values. *Human communication, especially in the public sphere, tends to be valenced communication.* And because our identities, our geographies, and our economic interests vary, communication in the public sphere tends to be polarized. Polarization is not inherently a bad thing. It is the natural condition. It is not like propaganda, something to be countered and stamped out if possible, although it is often easiest to perceive the valenced views of others in such terms. The appropriate response to bad ideas is not censorship but the entry of good ideas in an open market-place of ideas, pluralism on a level playing field.

The structure of communication, simply put, implies the analytic and normative question—is the playing field actually level? Because communication is valenced, most participants in the public sphere are thoroughly convinced that the playing field is not level; it is biased against them and fundamentally inhospitable to their perspectives. In the American context, conservatives are convinced that the mainstream press is in the hands of a biased liberal journalistic establishment. Liberals, of course, are convinced of just the opposite—the fat-cat capitalist press leans to the right and is in the pocket of advertisers, big corporations, and influential lobbying groups. This mutual perception that the playing field is not level is a good start, and may be taken as evidence that it is roughly level from the perspective of Left and Right. Although it may seem to be a problematic condition for the public sphere, it is just as it should be in a healthy democracy. Communication tends to be valenced. I return to these questions at some length in Chapters 5 and 6.

The Second Key to the Puzzle: Historically Grounded Research

There is nothing like a world war to help get things organized. And, in the absence of military conflict, a cold war with an evil empire is a pretty good substitute. The history of modern communication research in the West, of course, has taken place under just such historical conditions. As the Cold War dissipated, the focus of communication research became more fragmented and diffuse and certainly lacking in a disciplinary culture of urgency and connection to professional practice that was characteristic of the Marshall era. Is it possible to rekindle the energies and intellectual focus of the early days of the propaganda era?

Probably not. And although it is romantic to imagine returning to the thrilling days of yesteryear, it is not necessarily a step in the right direction. But the prospect of rekindling the connection to modern historical conditions is a more promising option.

I have argued that the concept of propaganda and the related notion of mass society, critically important in their day, are no longer appropriate conceptual models for historical connectedness. I have just attempted to make the case that a concept of valenced communication may represent a promising replacement for the bluntly asymmetric notion of propaganda. At this point I try to make the case that a reenergized notion of the information society could serve as a highly appropriate and hopefully provocative means of grounding the research agenda both historically and normatively.

The defining problematique of the mass society was the manipulative singularity of the media message. In Orwell's powerful prose it was the exaggerated caricature of Big Brother representing an appropriately terrifying admixture of Adolf Hitler and Joseph Stalin. In later critiques the consumer was the helpless pawn of the mass society dominated by powerful commercial interests promoting material consumption and credit card indebtedness. Although both notions continue to have entirely appropriate resonance today as forewarnings about the power of big government and big business, it may be useful to emphasize what is new about the marketplace of ideas in our times—the digital revolution. Throughout most of the twentieth century big governments had dramatically powerful control of the industrial media through censorship of print media and largely direct control of spectrum-based broadcast media. Forbidden tape recordings and carbon-copy, samizdat-style underground pamphlets and newspapers circulated revolutionary ideas beyond the government's reach, but only to a tiny minority of scattered and primarily intellectual protesters and only under conditions of severe sanction, usually extensive imprisonment for any public displays of independence (Pool 1973). And likewise in the democratic West, it was difficult to imagine life outside a constant bombardment of commercial messages unless one was living self-sufficiently in a remote mountain cabin (Baker 1994).

But in the age of the information society, information is abundant. There are still plenty of governmental and commercial messages, but satellites and the Internet frustrate the ability of nation-states to insulate their citizens from ideas emanating from outside their borders. And digital video recorders, pop-up blockers, and sophisticated online filters allow an individual with sufficient motivation and ingenuity to squelch the commercial stream. No mountain cabin required.

Like the notion of the mass society, the information society literature has stimulated a diverse set of critics who point out weaknesses and ambiguities—particularly that the information society concept minimizes the many continuities in social structure and cultural dynamics between this century and the last. But such criticism may miss the point. Highlighting the particularly unique way enduring issues of social life are represented in our time is motivation for the periodization of zeitgeists, evolving schools of thought, and cultural and political fashions. If an era is declared to be modern, there must be a postmodern and soon enough a post-postmodern until the cycle starts afresh.

The explosive profusion of information and communication and the rich literatures on the mass society and the information society is the primary subject matter of Chapter 3. The terminologies are introduced here to bring some closure to the era of propaganda and to highlight the many if seldom acknowledged continuities between the era of John Marshall's community of discipline builders and our own.

The Third Key to the Puzzle: From Effect Size to Conditions of Effect

In the era of propaganda there was no question about the strength and effectiveness of media messages—that was the founding essence of both public and scholarly concern. The early round of survey and even some experimental studies in the 1940s and 1950s, however, threw out some surprisingly counterintuitive findings and anecdotes. Most audience members are far from gullible and isolated pawns. Audience members discuss what they hear and read with others—the famous two-step flow. Some values and perspectives are deeply ingrained and strongly felt and thus not subject to media manipulation. Some media messages in the mix reinforce existing views rather than changing opinions. And some modicum of self-selective behavior permits many to seek out agreeable content and avoid messages with which they disagree. These initial findings were ably summarized by Joseph Klapper, initially in 1949 as his dissertation at Columbia and updated and published as *The Effects of Mass Communication* in 1960. Klapper's review is actually quite nuanced and reasonable. Not once in the three-hundred-page book does Klapper use the phrase the *minimal effects* of the media, but like "Play It Again, Sam" in *Casablanca*, the nonexistent phrase caught on in the literature as his central thematic and the poor Dr. Klapper became the much demeaned villain of the field as the false prophet of the ill-considered "minimal effects" school. And what a convenient villain. Klapper worked for CBS, which made the minimal-effects conclusion look like a self-interested apologia for the industry. As an industry-based professional he was not publishing actively and he was afflicted with rheumatoid arthritis and died in 1984, so he was not in a position to mount a proper defense or update his initial conclusions. So in the gamesmanship of academic publishing it became de rigueur to organize articles and even books as refutation of Klapper and the minimal-effects school through the demonstration of not-so-minimal effects using more refined theory and more sophisticated measurement techniques. The argument here has been elaborated in more detail in (Neuman and Guggenheim 2011) but can be paraphrased as follows—despite its pedagogical allure, the minimal-effects–significant-effects polarity could function as an impediment to theorizing. There are three elements to the argument.

First, the minimal-effects–significant-effects polarity conflates the empirical effect size of media impacts and their theoretical and practical importance. A mathematically tiny effect can accumulate over time to play a decisive role. Frequently, as in many election campaigns, a tiny fraction of the electorate becomes a pivotal swing vote. In the practical terms of electoral outcomes, the fact that the large majority of voters do not appear to be swayed by political ads and bumper stickers is simply beside the point. Numerically small but scientifically important results require no apology.

Second, the narrative unduly simplifies the history of communication research, and by diminishing earlier scholarship, it awkwardly puts younger scholars in the position of needlessly reinventing ideas and repeating research in a manner that is less constructive and accumulative. Lasswell's (1935) ideas about the interaction of psychopathology of national identity, for example, have new resonance in a post-9/11 world. Lazarsfeld and Merton's (1948) nuanced theorizing about conformity and status conferral provides abundant grist for modern-day critical theory and analysis (Katz 1987; Simonson 1999; Simonson and Weimann 2003; Chadwick 2013). In addition, even Klapper's (1960) much derided compendium and analysis offers thoughtful discussion of the conditions under which media effects tend to be the strongest and advice on how further research might clarify our understanding of those conditionalities. A close reading of Klapper reveals that he called for further research on: (a) the psychological predispositions of audience members; (b) the situated social context of message reception; (c) the broader social, societal, and cultural context of message reception; and (d) the structure of beliefs among audience members, not just the direction of beliefs. Each of these four represents a critically important condition of the communication process, and each has served as a foundation for theoretical advancement and refinement.

Third, the minimal-effects–significant-effects polarity is a demonstrable impediment to the design and interpretation of media effects research and the evolution of an accumulative agreed-on set of findings about the conditions that impede and facilitate those effects at the individual and aggregate level. For example, as late as 1999, Emmers-Sommer and Allen, in their overview of the field, conclude: "Taken together, these findings can be used to lend insight for future research directions. Overall, we can conclude that the media do, indeed, have effects" (1999, 492). It would appear that even after fifty years, simply to demonstrate a statistically significant effect in the ongoing battle against the vestiges of Klapper's evil empire is sufficient justification for celebration and publication. The fact of the matter is that the research corpus in media effects documents an impressive range of effects from no effect at all to very large effects. The challenge to progress in communication research is to systematically theorize and test the conditions that may facilitate or impede such effects and not simply to celebrate that the mean measure of effect size is larger now than the effect sizes typically assessed by preceding generations of researchers.

This chapter started with a bit of a historical puzzle—how did communication scholarship come to develop the currently dominant research paradigm focusing on the relative strength or weakness of media effects? The proposed answer to this puzzle has been a bit circuitous in its development. In the era of propaganda most analysts took strong and significant effects simply as a given. The practical question was how to effectively counter the propagandist. Further, the asymmetry of the propaganda concept was associated with an asymmetry in the assumption of strong media effects—that it is hard to counter psychologically sophisticated propaganda with the simple and unvarnished truth. This presumption is captured in the widely cited epigram—"a lie will get halfway around the world before the truth gets its boots on"—variously attributed to Mark Twain and Winston Churchill. The expectation that seductively presented negative communication is inherently more powerful than its positive counterpoint continues to resonate in communication scholarship and may explain in part the curiously asymmetric attentiveness to negatively valenced communication within the discipline today.

There is a famous finding in communication scholarship called the third-person effect (Davison 1983). It is a frequently demonstrated empirical phenomenon among audience members. It turns out most audience members report that persuasive propaganda may affect others but certainly not themselves. Researchers disparage this cognitive bias in the mass audience. But it appears that a variation of this propensity is actually characteristic of the researchers themselves. Their paradigm is built on the assumption that messages they "disagree with" are more effective than those they might find more resonant with their beliefs. I return to these perceptual asymmetries among audiences and researchers in the chapters ahead. But first in Chapter 2, a look at the paradigmatic method of measurement that came with the paradigmatic model of not-so-minimal media effects.

2

The Prospect of Precision

It is the mark of an educated man to look for precision in each class of things just so far as the nature of the subject admits.

—ARISTOTLE

To speak with precision about public opinion is a task not unlike coming to grips with the Holy Ghost.

—V. O. KEY JR. (1961)

The state of research on media effects is one of the most notable embarrassments of modern social science.

—LARRY BARTELS (1993)

CHAPTER 1 MADE the case that the field of communication research has been organized around a particular Kuhnian puzzle—the issue of media effects historically framed by a deeply ingrained concern about propaganda effects following the Second World War. Kuhn's model of scientific paradigms, of course, posits the existence of not just a puzzle, but a puzzle paired with a methodology designed to "solve" the puzzle. This chapter focuses on that evolved methodology that has, like the puzzle, dominated the field for the past half century. And, as noted above, given the shift from push to pull media, the methodology may also benefit from refinement in consideration of the new media environment. The central theme—the prospect of precision in measuring what the media do—is based on the premise that the phenomenon of human communication is particularly resistant to reliable measurement and the use of traditional social scientific quantitative analysis. It is resistant for two reasons: (1) profusion, the incredible abundance of words and images in the individual's daily environment, increasing even more in quantity and diversity in the digital age, and (2) polysemy, the fact that each of these words and images is subject to dramatically variant interpretation by different individuals.

Economics deals with profusion and large quantities, but the monetary units studied are clearly defined and their value, at any point in time, not subject to divergent interpretation possibly based, for example, on deeply ingrained values and social identities as is the case in communication. Political science deals with large quantities, as well, but, like economics, voting statistics, for example, are quantitatively clear-cut; a vote is a vote. Sociology, a more diverse field of study deals with issues of inequity in class, status, and power, and, again, like economics, measures of such phenomena as socioeconomic status are no longer controversial or problematic. The measurement of meaning, however, is fraught with additional levels of complexity. Trying to document whether a persuasive message has had an identifiable effect on an audience member within the churning message flows of modern society is like trying to count all the stars in the heavens.

Counting Stars

Counting stars is difficult because there are so many of them, because they are in constant motion, because they are frequently obscured, and because they are, in fact, inaccessible to direct examination. The faint, flickering patterns of light are ephemeral. Human thinking about heavenly bodies resonates with the most famous and one of the most historically controversial of scientific paradigm shifts—the transition in human perception of a geocentric to a solar-centric solar system. It is difficult for the human being looking up at the nighttime sky to make sense of the distance of and volume of individual stars, let alone to understand their structure and profusion. Typically we can see several thousand stars with the naked eye. Astronomers explain that what we see at night is really an infinitesimal fraction of the trillion stars in our own galaxy. From other astrophysical measurements and theory, scientists estimate that there are about a tril-lion galaxies. So when we look up, although faint beyond human percep-tion and obscured in various ways, we are looking at a trillion trillion stars (Comins and Kaufmann 2011). Indeed, we have been looking at the eve-ning sky and struggling to make sense of it since the beginning of human life on earth. For most cultures the patterns of the stars are interpreted as figures of humans and animals and, of course, divinities of various sorts. We look up and see the hunter Orion with a club and a sword belt and household objects like dippers, big and small. We exhibit some remarkable creativity in perceiving patterns among the stars and attributing meaning and causation to them. But it would appear to be foolish to try to actually count them, to enumerate each one for a precise quantitative assessment. For some critics such an enterprise would be an exercise in hubris; for others, worse than hubris—such an ill-advised venture into the nature of the heavens would be blasphemous. For the purposes of the extended metaphor concerning a search for precision, the lesson is a little less harsh: precise measurement of the expansive phenomena of human communi-cation is extraordinarily difficult and will require both the perseverance and analytic creativity in interpreting limited and error-filled data. Perse-verance and analytic creativity have served modern astronomers well. The splintered enterprise of communication research should take heart.

The Fundamental Paradigm of Media Effects Research

The evolved research paradigm for media effects research is remarkably simple—it consists largely of the analysis of correlations between varia-tion in media exposure and variation in behavioral response. At this level of description it might strike one as an entirely reasonable model for organ-izing and testing hypotheses and, it would appear, relatively easy to im-plement. The seminal experimental design was pioneered by Carl Hovland and his associates at Yale as they systematically varied the character of persuasive messages and assessed the relative degree of evident attitude change (Hovland, Lumsdaine, and Sheffield 1949; Hovland, Janis, and Kelley 1953). The corresponding survey design was developed by Paul Lazarsfeld and colleagues at Columbia, who examined the changes in vote intention for voters of different backgrounds and for those routinely dif-ferentially exposed to newspapers and radio (no television yet) (Lazarsfeld,

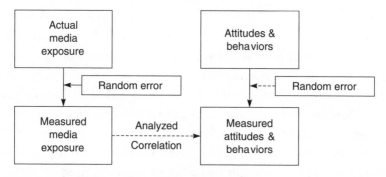

Figure 2.1. Fundamental Analytic Model of Media Effects

Berelson, and Gaudet 1944). The systematic quantitative assessment of mass media content known as content analysis was explored by Harold Lasswell and his team of propaganda researchers in Washington during World War II and later at Stanford and Yale (Lasswell 1935). Lasswell's dream study design developed back in the 1930s (but never fully implemented) was the World Attention Survey, which proposed to correlate newspaper content with different predominant attitudes and behaviors of different nations (Lasswell 1941). The following sections of this chapter review in some detail why, unfortunately, what appears at first glance to be a reasonably specified and straightforwardly implemented analytic causal model turns out to be anything but. The fundamental model has been diagrammatically expressed numerous times (McQuail and Windahl 1993; Jensen 2011) albeit seldom with reference to the prospect of measurement error, but at its core has the fundamental structural properties as expressed in Figure 2.1. This model is elaborated as the chapter proceeds.

Each chapter section explores examples of research and various calls to action for refinement of research that have been articulated over the years as scholars have struggled with the theories and with the data. But two themes, as noted above, will dominate the review, the two challenges that are characteristic of although not entirely unique to the study of human communication: the issues of profusion and polysemy.

Briefly, the profusion of common messages over virtually all media and accumulatively over many years for any adult subject results in the practical impossibility of finding a defensible control group, that is, an otherwise socially and psychologically equivalent sample that is not exposed. If the same fundamental messages and narratives are also present in books, motion pictures, broadcasting, and print media, then measuring the use of different media (as typically done in the uses and gratifications tradition) is not particularly helpful. If one systematically varies exposure and non-exposure to, say, three persuasive messages or three particularly violent narratives in an experiment and then looks for differential responses in otherwise equivalent samples, one can explore message effects only *at the margin*—that is the difference between a subject who has seen 100,000 such messages and one who has seen 100,003. If one examines comparisons across nations and cultures who are exposed to systematically varied messages, it is not practically possible to parse out differences in communication flows from other differences in culture, social structure, economics, and historical context.

Correspondingly—polysemy results in a broad and complex distribution of potential behavioral responses rather than a single measurable "effect." Thus a persuasive message carefully designed to, say, demonstrate the "benefits of healthy eating" may for a number of those exposed simply remind them of their hunger pangs as they make a mental note to head out for a generous helping of sugary or salty fast food as soon as the study is finished.

Profusion's Challenge to Measurement

Dealing thoughtfully, precisely, and realistically with these massive flows of mediated text, audio, and images is the challenge at hand, and an attempt to respond to this challenge represents one of the central themes of this book. Chapter 3 is entitled "The Paradox of Profusion" and addresses these new trends in the tidal quantities of communication flows further. For now attention is drawn to questions of appropriate methodology and the prospect of some reasonable precision and reliability of measurement.

One might approach the question of exposure to, for example, depictions of violence with a new and novel methodological approach. What if instead of systematically varying levels of exposure to violence experimentally, a researcher simply sat down with a few young respondents and asked them to remember as many as possible of the presumptive 100,000 mediated violent acts they have seen or read about that they could recall. The researcher could assure them, "Take your time—we have all day." It is likely that that the respondents' memory for specific depictions would be exhausted in a few minutes after recalling several dozen shootings, stabbings, stranglings, and perhaps a poisoning or two. The subjects would be reminded that social scientists usually also include "verbal threats of potential violence" in their counts of depictions of violence, and that might stimulate another dozen or so recollections of threatened violence. This is an essential characteristic of the causal dynamics of media effects. The ratio of actual exposures to those exposures that can be recalled or otherwise identified is many thousands of actual exposures to every recalled exposure, something in the order of 10,000 to 1. There may well be important subliminal effects of exposure to violence without explicit recall, but subliminal processes are even more elusive and resistant to assessment.

Take any number of theoretically prominent phenomena in the tradition of communication research—stereotypes about young African American males, differentiated gender roles, the trustworthiness of the federal government, the conspicuous consumption of consumer goods, sexual behavior (Jeffres 1997; Perse 2001; Preiss et al. 2007; Bryant and Oliver 2009; Sparks 2010). How many relevant messages has, say, a healthy, active, and engaged twenty-year-old been exposed to accumulatively in his or her life thus far—messages in school, media messages observed directly or passed on by parents or peers? Ten thousand messages for topics this broadly defined is probably too low. One million messages—perhaps too high. It is difficult to tell, but perhaps something in the neighborhood of 100,000. Only a tiny percentage of relevant messages can be explicitly remembered, but there is no reason to believe various forms of potential "effects" would require explicit and current conscious awareness.

The scientific understanding of human behavior is based on the careful analysis of variation—thus inevitably the first paragraphs of any traditional social science study report identifies the "variables" of primary interest and a few hypotheses about the structural relations among them. And most studies in mass communication are no exception. Communication researchers routinely rely on such variables as self-reported exposure to television or newspapers in survey studies or systematically varied exposure to several minutes of persuasive messages or narratives in experimental studies. What is typically taken as nonproblematic, and what clearly should not be, is the character of the underlying distribution of those variables.

Communication at the Margin.   Communication researchers, not surprisingly, tend to focus on the potential role of media depictions and persuasive messages in domains of social life that are currently attracting interest and controversy. Thus, for example, the depiction of racial or gender stereotypes, news reports and political ads, the depiction of violence and sexuality have each inspired large components of the accumulated research literature (Berger, Roloff, and Roskos-Ewoldsen 2009). This may be as it should be, but it creates a fundamental challenge to research design because the quantities of media content that deal with race, gender, public affairs, and the like is massive and utterly pervasive. Take as a case study the depiction of interpersonal and social violence in the media.

The question of whether violent media content raises the propensity for violence in the modern world is a staple of communication theorizing. It has roots in the earliest prototypical effects studies in the first half of the twentieth century (Wartella and Reeves 1985). It spawned several thriving academic journals devoted to this specific topic and other research reports on media and violence continue to represent a dominating presence in numerous other journals in the field of media psychology. One review study succeeded in tracking down 3,500 studies on this media violence published since 1950 (Grossman and DeGaetano 1999).

Experimentalists in this tradition will routinely assign random subsamples of subjects (usually youthful subjects who are of obvious particular interest) to different conditions defined by the presence and corresponding absence of depicted violence in film or video narratives, text-based scenarios, or video games. Typically differences in self-reported propensity for or actually observed antisocial behavior immediately following exposure are found to be in evidence. But it is difficult to interpret these associations as causal in the traditional scientific sense of a meaningful and potentially longer-term change of state. The reason is that the only variance available to be scientifically explored is *variance at the margin*. By the time they finish elementary school, the average American youngster has witnessed 100,000 violent acts and about 8,000 murders on television alone (Huston et al. 1992). At an average of fourteen violent acts per hour (Strasburger and Wilson 2002) even in the programming aimed at children (aside from programming for adults children routinely watch), how is it possible not to be working with slivers of difference in exposure at the tip of a massive distribution of exposure? The few children from intensely religious or other ideologically motivated households who have little or no exposure to this component of popular culture represent an interesting curiosity but embody an entirely noncomparable subsample for purposes of research. It is, of course, true that accumulative exposure over time could represent a highly significant effect. But in the real world with longitudinal field data researchers are forced to confront the causal difference between exposure to typically something of the order of 85,000 violent acts as opposed to 115,000. Because of the fundamental fact of communicative profusion, such empirical and analytic challenges are simply inherent in the phenomenon at hand.

Systematic Inattention and Forgetting. It is unrealistic to imagine that even some rare individual with something approaching a photographic memory could possibly resurrect anything more than a few percent of such media depictions of violence, sexuality, or persuasive political arguments. It is the nature of the beast—or in this case the nature of the causal processes under study. The fact of the matter is that our cognitive systems are thankfully designed to forget large portions of our visual and auditory perceptions. The human brain functions with entirely separate systems for long-term and short-term memory (Cowan 1998; Klingberg 2009). We recall our hotel room number while we are at the hotel, but a week later can typically no longer resurrect the specific number as it is no longer relevant. Like many of the sensory perceptions of our existence, to keep such context-specific details in memory would simply clutter our minds. The way humans (in fact all organisms) deal with immense flows of sensory data is that they filter out, ignore, and forget all but a few fragments of perception. We attend to the salient elements of our environment that our relevant for survival and reproduction. We may remember our first kiss, our first day at school, and when we tried to dance like John Travolta in *Saturday Night Fever*. Systematic inattention and selective attention are simply inevitable responses to tidal quantities of mediated messages in our environment. We do recall having seen John Travolta in another movie of that era, *Urban Cowboy*, but we can't recall anything else about it except that he rode on some sort of mechanical bull in a honky-tonk bar trying to impress the girl. (Who was that actress?)

Reinforcement. The communication research paradigm typically identifies the key dependent variable to be "attitude change." The implication, of course, would appear to be that if a fixed belief or opinion was not subject to evident change there was no "effect." But given that there are competing flows of pro and con messages on most contested issues and certainly as well in the domain of product marketing, a potentially important communication "effect" may be the reinforcement of a given belief or opinion in the face of competitive messages. If it can be demonstrated that in the absence of a reinforcing pattern of messages, beliefs and attitudes would indeed change in response to a communication environment, that would unambiguously represent an "effect." As a result of Joseph Klapper's (1960) famous discussion of reinforcement and the troubling association of reinforcement with the dreaded evidence of "minimal effects," the paradigm of accepted research design has awkwardly and unfortunately avoided the systematic study of patterns of reinforcement. Given that one of the most widely acknowledged factors in selective attention and selective recollection is familiarity, this one-sided methodological focus on attitude change has constrained theoretical and empirical progress.

Opinion Change versus Opinion Creation. There have been several references to the carefully constructed and cumulative research designs of the Hovland team at Yale in the 1950s. As psychologists with a background in learning theory, it was not surprising that their attention was drawn to cognitive dynamics rather than the embedded social character of the topics they picked for their experimentally manipulated attitude change experiments. In marked contrast to the sociologists at Columbia who found very little evidence of change over an election period in the socially embedded beliefs concerning policy, party identification, and candidate preference, the Yale group members were relatively casual about the topics they picked. In fact they made an unapologetic point about picking topics that subjects were not likely to be familiar with to increase the likelihood of demonstrating "attitude change." Some topics were notoriously strange or obscure—one persuasive message focused on the consumption of chocolate-covered grasshoppers and another on the esoteric issue of requiring dentists to carry liability insurance. In retrospect, many analysts of the field remarked that what was described in the original studies as evidence of attitude change was in many cases more appropriately characterized as attitude creation (McGuire 1985). Accordingly, as the hard work of methodological refinement continues and paradigmatic goals evolve from celebrating not-so-minimal effects to systematically assessing the conditions of effects, careful attention to the embeddedness, strength, or level of commitment to opinions and beliefs subject to change or reinforcement will become more clearly identified.

Cultivation Analysis: A Small Step. One of the most creative and influential approaches to dealing with the profusion of media messages and the more subtle and potentially subliminal influences over time of mediated themes was developed by George Gerbner and his colleagues at the University of Pennsylvania under the banner of cultivation analysis. Gerbner was not a fan of traditional effects research and explicitly avoided the term *effects* in opting for *cultivation*, which captured the prospect of the long-term envelopment of individuals in a particularized message environment (Gerbner 1956, 1967, 1969). The research team focused on broadcast television and pioneered a survey-based approach to matching beliefs about such issues as the threat of crime and the prominence of the medical and law enforcement professions with self-reported levels of daily television viewing (Gerbner et al. 1976, 1978, 1979, 1980; Gerbner and Gross 1976). Despite Gerbner's critical bent, he ended up adopting a classic variation of the fundamental communication research paradigm. The surveys revealed that those who reported higher viewing levels were more fearful of crime and exaggerated the percentage of law enforcement and medical professionals (as emphasized, of course, in television's crime and medical genres). The difficulty here is that those of less education and lower social status (who have every reason to be more fearful of street crime and may be less familiar with the distribution of professional careers) watch television at much higher levels and unlike the college-educated respondents, do not seem to feel any embarrassment in describing their viewing levels, which further exaggerates the class–viewing level correlation. As a result it is difficult to adequately control for or parcel out the "TV effect" from the "social class effect" and the "selective exposure effect" (Hirsch 1980, 1981a, 1981b; Hughes 1980; Gerbner et al. 1981a, 1981b; Wober and Gunter 1982; Rubin, Perse, and Taylor 1988; Potter 1994). The attempt to address the issue of long-term and accumulative effects is certainly a step in the right direction and the notion of an "emersion" in a message flow continues to be intuitively appealing (Shrum 2007), but the simplistic exposure-attitude linkage of the fundamental paradigm breaks down. As Rubin and colleagues conclude—the heavy-viewing–scary-world correlation is largely spurious: "methodology may explain cultivation effects that have been attributed to television exposure . . . it is fallacious to believe that television viewing can have only negative effects . . . other antecedent and intervening variables accounted for more of the variance in the social attitude indices than did exposure levels . . . television [may] affect personal perceptions, not from inordinate exposure levels, but from content selectivity tempered by individual differences and [pre-existing] audience attitudes and activities . . . people actively and differentially evaluate television content before integrating it into social perceptions" (1988, 123–126).

The Measurement of Media Exposure. As noted above, the survey methodology is based on self-reports of behavior. Self-reported behavior, it is widely recognized in survey research, is notoriously distorted by multiple sources of systematic and random error. People have a hard time remembering past behavior, are notoriously bad at estimated quantities such as the number of hours of activity, and are subject to severely underreporting behaviors they believe to be socially undesirable. In one classic study, the analyst compared what people said they watched on television in a survey study with actual set-top measurements of their viewing behavior. Lower-class respondents said they watched a lot of quiz shows, dramas, and situation comedies, and that is indeed what they watched. College-educated professionals claimed to watch only sports, news, and public television while the set-top boxes revealed that their diet was pretty much the same as their less educated counterparts—a lot of quiz shows, dramas, and situation comedies (Wilensky 1964).

A central problem for the assessment of communication behavior is that the media activity is often a secondary or tertiary activity—taking place while cooking, cleaning, or talking on the phone (Robinson and Godbey 1997). As a result when researchers measure television or radio listening using the standard twenty-four-hour diary method of time-use assessment, they get exposure rates half that or even less than half of estimates from the ratings services based on set-top and people meter technologies (Robinson 1971).

Content Differences across Media. One frequently used methodological tool for the past half century of communication research has been to assess differential exposure to various media. One classic survey-based approach is to compare those who report getting most of their news from a newspaper as opposed to television news. It turns out that those who report depending more on the print media are systematically better informed politically and more politically active in terms of voting and contributing to campaigns (Robinson and Levy 1986). It is widely noted that a network television newscast is only twenty-two minutes long and the text from the newscast would not fill even the front page of the *New York Times.* It is noted that newspapers have much more extensive coverage, which allows those interested to read in greater detail. And it is summarily concluded that this is evidence of a media effect drawn from the difference between television and newspaper news content. The relative preference for different media— notably such phenomena as a preference for books over movies—became a staple of the "uses and gratifications" tradition of research that attempted, appropriately, to acknowledge that active audience motivations and motivated selectivity represent an important part of the paradigm rather than passive bullet-like effects (Blumler and Katz 1974; Rubin 1986). Unfortunately the use of the physical mass medium (movies versus television and the like) as the analytic variable ignored some of the most interesting differences in content and symbolic emphasis. Consider the contrasts in the character of coverage in the *New York Times* versus the *Daily News,* the PBS *News Hour* versus *Entertainment Tonight.*

Interestingly, the past decade of research on the "impact" of the Internet has adopted this medium-is-the-message model of analysis contrasting those who report relying on the Internet versus newspapers and the like. As media convergence continues, what now is simplistically identified as newspaper, radio, or television content will all be equally available on the web. As a result of these developments, analysts will have to be more focused in their matching of differential exposure and differential attitudes and not rely on the more-exposure-to-medium-x-equals-unique-beliefs trope (DiMaggio et al. 2001).

Hovland's Paradox: Experiments versus Surveys. A few years before his untimely death, Carl Hovland, a towering figure of the first generation of

Table 2.1   Hovland's Comparison of Experimental and Survey Methodologies

Factor	Experiment	Survey
1) Exposure	Random assignment	Motivated self-selection
2) Message complexity	Single message	Message campaign
3) Message length	Short	Long
4) Communication context	Authoritative source/ lab	Diverse sources/natural settings
5) Typical sample	Students	Representative adult samples
6) Typical issue	Susceptible to modification	Socially significant, strongly held

communication researchers, published an influential paper in the *American Psychologist*. The title of this 1959 essay explained its purpose with the crispness and clarity that was characteristic of Hovland: "Reconciling Conflicting Results Derived from Experimental and Survey Studies of Attitude Change." Hovland was an experimentalist by training and personal proclivity, but he recognized that divergent findings from surveys and experiments resulted from the fact that each methodology had unique virtues and in the interest of scientific progress; he believed that the most sophisticated theory testing would be derived from integrating the two (see Table 2.1). He mentions, as noted above, that experimental psychologists purposefully select attitudes that are "subject to modification" rather than those that are strongly felt and/or connected with an individual's identity. As a result their work borders on the boundary between attitude creation and meaningful attitude change. Hovland goes on to note that there is no reason experimentalists could not include more socially connected and deeply held attitudes (in addition to novel ones) in their research akin to the work on political party affiliation and candidate evaluation that characterized survey research. So Hovland concludes that there really is not a "divergence" of findings between surveys and experiments, but rather that, for a variety of reasons, the two scholarly traditions ended up studying different kinds of communication situations and different kinds of messages with different characteristic samples. There is no hint in his language proclaiming the superiority of his own research tradition, but rather a clear call for the benefits of multimethod integration structured by common theoretical concerns.

The paradox of this paper is that writing the year before Klapper published his influential review *The Effects of Mass Communication,* which came to be seen as the principal canonic document of the "minimal effects perspective," Hovland too reflected on the spirit of the time and appeared to emphasize the difference between minimal and significant effects. As Hovland puts it: "The picture of mass communication effects which emerges from correlational studies is one in which few individuals are seen as being affected by communications. . . . Research using experimental procedures, on the other hand, indicates the possibility of considerable modifiability of attitudes through exposure to communication. . . . The discrepancy between the results derived from these two methodologies raises some fascinating problems for analysis" (1959, 496–497).

However, in due course in his analysis he actually acknowledges that it is not a discrepancy after all, but rather a perfectly reasonable finding that different messages and contexts generate effects of characteristically different magnitudes.

Unfortunately, the challenge for multimethodological approaches to integrated theory Hovland issued in 1959 did not translate into effective practice for active researchers in subsequent decades who in large part continued to specialize in particular message types, contexts, and corresponding methodologies of convenience. This is perhaps not surprising since the academic incentives in communication like other fields continue to reward specialization and identification with smaller, self-identified peer review communities. An academic career, for example, might easily focus on the study of political campaign ads or antismoking public service announcements. The good news is that with the transition from typical one-way push analog media to digital media a variety of field experiments that permit monitoring and manipulating communication flow in natural settings give the field a second chance to take Hovland's exhortations seriously.

Polysemy's Challenge to Measurement

When Hovland reviewed the challenges to communication research methodology, he duly noted the difficulty that the phenomenon of selective exposure presents to research design and theory building. Reflecting his own intellectual trajectory from educational and learning psychology to media effects and perhaps also reflecting the spirit of the times in midcentury he came to characterize the problem quite narrowly. He did this in two ways. First, he more or less identified the situation of motivated avoidance of exposure to a message as exogenous—"outside" the domain of communication research. If the message was not actually conveyed to an audience member, it is not communication, and accordingly outside the paradigm and no longer subject to theorizing. Clearly, and it is increasingly true in a digital world, the motivations for, the habitual patterns of media behavior, and the technical capacities for opting out of or even fast-forwarding through exposure has to be endogenous and subject to measurement, manipulation, and theory testing. Second, again given his tradition of work with authoritative persuasive messages, he treats the phenomenon of exposure to a message with which an audience member disagrees as subject to distortion rather subject to interpretation. When an audience member does not get the message "as intended," it is characterized as a "reception" problem. Thus, according to this view, ambiguity and variegated polysemy of the message and the equally diverse perspectives and evaluative dimensions of an audience member's perception are simply exogenous to the model. But, perhaps, they need not be.

The Complexity of Selective Exposure and Selective Interpretation. The widely acknowledged phenomenon of audience self-selection has been a fundamental and centrally troubling challenge to communication research methodology from the outset. It is beyond dispute that partisans are attracted to like-minded content in the media. This phenomenon was addressed by the first generation of quantitative communication researchers and the notion of reinforcement became associated with the notion of "minimal" effects as characterized by Klapper, which in turn became an irresistible bugaboo for researchers to disavow and disprove. (Ironically the phenomenon of reinforcement, as noted above, need not be characterized as the absence of an effect; reinforcement *is* an effect. Unfortunately, however, as the field of research has evolved, most researchers appear to treat reinforcement as an uninteresting noneffect or dreaded minimal effect.) The key causal challenge, however, is making sense of a correlation between a measure of exposure to a particular type of content and a corresponding measure of an attitude or behavior assessed at one point in time. Recalling the case example of exposure to the depiction of violence in the media—the delicate challenge continues to be differentiating the fact that those with antisocial tendencies may be drawn to higher levels of exposure to violent genres from the fact that exposure may engender antisocial behavior. (The prospect of a spiral process whereby more exposure leads to higher levels of self-selection will be addressed in the following section.) The classic technique in psychological research to overcome the self-selection problem is random assignment of subjects in the experimental design. It is an effective tool in some instances, but, as noted above, a brief interval of exposure measured in minutes does not come close to capturing the potential causal mechanisms involved in exposure over years to hundreds and thousands of hours of mediated content.

As noted above, the issue of selective exposure has been characterized by some as exogenous because in the traditional "push environment" of headlines and prime-time network television, the capacity to exercise choice was limited. This challenging problem becomes more serious as the high-choice environment of abundant content and sophisticated algorithms of choice and search become available. Audience members so motivated can locate content corresponding to their specialized interests (not just violence but, say, a particular kind of violence), and, again if so inclined, they can view only those parts of a narrative that feature that content (perhaps skipping conversational sequences and going straight to the action scenes). The increased significance of self-selection in the new media environment has stimulated increased attention and concern in the research community, which has adopted the term *endogeneity* (meaning within the model) to identify the problem (Clarke and Kline 1974; Chaffee and Metzger 2001; Bennett and Iyengar 2008).

Selective Exposure and Spiral Mechanisms. The fundamental paradigm of communication research with its historic concern about the atomized citizens of mass society posits a fairly straightforward relationship between "cause," variation in exposure to a persuasive message, and "effect," variation in attitudes or behaviors. Setting aside the ambiguities of self-report measures for the moment, the model appears to make sense intuitively. The individual viewer or reader cannot influence media content; the character of such content is determined by complex decision processes in large industrial institutions. So any correlation between message exposure and individual outcomes is interpreted as a one-way causal process of "media effects." Some individuals may ignore or systematically misinterpret a message, of course, but within this paradigm it is typically defined as random noise or measurement error rather than a causal process in the reverse direction.

But upon reflection it would seem a perfectly straightforward strategy to model the nature of media effects as reciprocal and interactive with some media content potentially resonating with an audience member, which, in turn, over time, effects attitudes and further selective attention and selective perception of further media messages. Such a model would seem to be even more appropriate in an environment of pull media with abundant diversity, instant gratification, and on-screen tabs and buttons that lure individuals to further content (Chaffee and Metzger 2001; Bennett and Iyengar 2008; Brosius 2008). In the American case it appears that the expansion of the media flow is coincident with a more clearly labeled, more partisan, and more intensely opinionated set of media options (Mutz 2006; Nivola and Brady 2006; Harwood 2009; Stroud 2011). Studies of world media have not yet established whether this is a global phenomenon of our age (Esser and Pfetsch 2004).

Given the promise of such a model, one might wonder why it is only rarely invoked. The answer in this case is strikingly clear and powerfully influential. Such research designs are extremely difficult and expensive to implement. Most scholars and students have limited time and limited financial resources and are drawn to the practical simplicity of the single-shot survey or experiment. As noted above, self-reports of attending to a lot of (violent, conservative, sexual) media and self-reports of associated attitudes and behaviors have multiple sources of covariance such as perceived cultural acceptability and style of speech. Attributing all such covariance to "media influence" seductively strengthens the chances that any such hypothesis put forward will be sustained by the data collected. Most researchers are keenly aware of such limitations and engage sophisticated statistical controls to try to distinguish actual causation from mere correlation as best as possible. But without time as an analytic variable built into the research design, a clear distinction eludes even the most advanced statistical manipulations.

Besides the additional cost and complexity, over-time studies raise other vexing problems. Inevitably, some subjects will be missing from successive waves of inquiry, and the likelihood of encountering missing data may be associated with key analytic variables of interest. Over-time research by definition requires the researcher to identify specific time intervals for measurement, a design task more difficult than it may first appear to be. In the real world, media effects may only become evident over years or even decades, obviously not a practical time interval for a typical research design. Furthermore, subjects may remember responses made earlier in a questionnaire and attempt to replicate earlier responses to exhibit consistency of view rather than potentially evolving views. Over time research is not an easy undertaking, but for the reasons outlined above, it emerges as an extremely important additional methodological tool for the field to supplement single-time surveys and experiments.

Increased interest in reinforcing spiral mechanisms has been stimulated by an influential article in *Communication Theory* by Michael Slater at Ohio State University. Slater's work on health communication and violence in the media had convinced him that attention to the dynamics of the reciprocal interaction between persuasive media messages and audience member selectivity was critical to understanding how media effects work. His overview paper was entitled, "Reinforcing Spirals: The Mutual Influence of Media Selectivity and Media Effects and Their Impact on Individual Behavior and Social Identity." As a methodologist, Slater devoted a number of pages to a detailed discussion of the mathematical fine points of analyzing over-time data. But the core of the paper resonates strongly with the argument being made here. He posits, for example, that reinforcing spiral dynamics are widely acknowledged but seldom explicitly theorized: "Surprisingly, however, there has been limited systematic effort to synthesize the process of media selection and media effects into a more comprehensive model. . . . This notion that together media selectivity and media effects form a reciprocal, mutually influencing process is noted or implied, though not extensively developed, in a variety of classic sources" (2007, 281–283).

He reviews the central role of spiral dynamics in the classic studies of spirals of silence by Noelle-Neumann (1984), selective exposure in the uses and gratifications of entertainment media by Zillmann and Bryant (1985), cultivation processes in the work of Gerbner and associates (2002), and extensively the over-time reinforcement of social identity in the work of Tajfel and Turner (1986), another subject that resonates strongly with the argument here. Such methods are difficult and time-consuming, he notes, but the critical requirement is assessment at multiple points in time as

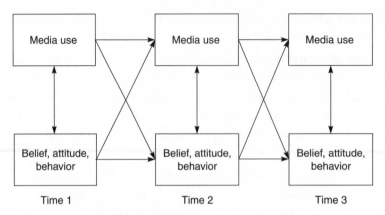

Figure 2.2.   Analytic Model for Over-Time Assessment of Potential Reinforcing Spiral Effects
Source: Slater (2007), Figure 1, p. 284.

noted in this classic three-wave model of analysis, reproduced here as Figure 2.2, an extension of the previous model focusing on over-time dynamics rather than the complexities of self-report measures.

He mentions the interaction of new media and new forms of data collection only in passing, but it is clearly a fertile area of methodological exploration. It is interesting to speculate on how these techniques may give us a better understanding of the ritualized and routinized character of media behavior and how such routines are changing as digital media become both mobile and ubiquitous. Commercial media services such as A. C. Nielsen have been tracking television viewing and radio listening for decades—the "drive time" radio audience peaks reflecting commuter habits and the 9 p.m. evening peak of television viewing reflecting household daily routines have changed little so far (Nielsen 1986, 2010). Newspaper audience research refers to the "newspaper habit" and remorsefully traces its dramatic decline among the young (Bogart 1989). The predictable scheduling of broadcast programming had reinforced the routinization of audience behavior as a sizable proportion of the audience for *The Tonight Show* over the years watched Johnny Carson and then Jay Leno through their toes in bed as a comforting nightly ritual. It is not yet clear how these rituals will evolve, although Nielsen (2010) reports that the bedroom TV has new competition as iPads and e-books crawl into bed with their owners. Ritualized media behavior becomes associated with the reinforcement of social identity as social pressures especially among younger media users dictates behavior (Anderson and Subramanyam 2011). The evolving linkages between habituated media patterns and social identity and the framing of social issues may well be subtle and complex in the sense of the concept of *habitus* elaborated by Pierre Bourdieu (1991, 1993). These issues will be addressed again in later chapters.

Having reviewed a frustratingly long list of impediments to over-time research, it is fitting to turn to a more encouraging development. As noted above, the digital age provides an abundance of digital footprints. As more and more media messages are digitally mediated through online search, and associated platforms such as iTunes, Amazon, and Netflix, we confront a resource much more accurate than human memory and self-reporting. With appropriate protections of individual privacy, those patterns of content exposure can be associated with later exposure, online behaviors, and even attitudes and beliefs in natural communication flows and, of course, online surveys. Researchers, especially younger researchers, are increasingly intrigued by these possibilities. (Thomas Kuhn would be proud.) This evolving research tradition is generally associated with the term *big data*, and a brief review of its promise and perils is advanced at the conclusion of this chapter. It should be clear that the argument here points strongly and enthusiastically in the direction of this new development (Wu et al. 2011; boyd and Crawford 2012; Choi and Varian 2012).

The Puzzle of Selective Interpretation. Setting aside the causal complexities of selective exposure for the moment, the question of what happens when, for whatever reason, an individual is exposed to a persuasive or informational message can be addressed. When in the course of habituated media exposure the sender and receiver come from closely aligned cultural backgrounds and share beliefs and values and the message is relatively straightforward, the odds are enhanced that the message received will coincide roughly with intended meanings of the message as sent. But when the cultural distances and beliefs are more distant and the message richly complex it really gets interesting. Note, for example, the epigram from *New York Times* editor Bill Keller in an essay on the prevalence of conspiracy theories in the public sphere: "Conspiracy theories erupt whenever unfathomable news collides with unshakable beliefs." He was paraphrasing political historian Robert Goldberg's (2001) views when he composed that sentence, but it does capture the dynamic of selective interpretation colorfully, and the phrase was picked up by numerous bloggers. This is, at its core, simply an acknowledgment of the phenomenon of polysemy. Consider the response of public opinion in the Arab world to the events of 9/11. The striking "success" of a small group of Arab terrorists to challenge the American behemoth seemed so incongruous to many Arab observers that they looked elsewhere to explain what happened. It appears that fewer than one in four of Muslims surveyed in recent surveys believes al Qaeda was responsible for the September 11 attacks. Countries bordering on Israel blame Israel (43 percent in Egypt, for example). And in Mexico 30 percent of the respondents were convinced the whole event was staged by the United States (Kull et al. 2009). These eye-of-the-beholder dynamics are widely recognized, of course, but not yet systematically incorporated into communication research practice.

Analysts can expect any complex message to generate not a single interpretation but a distribution of interpretations. And, again interestingly, under some conditions more of those exposed interpret the message to be at odds with, or even in a completely opposite vein from, the intentions of the message as sent. This is, of course, not a recently discovered insight. It resonates with the powerful imagery of Plato's fleeting shadows on the wall of the cave or, more recently, Walter Lippmann's famous discussion of the power of stereotypes in his 1922 manuscript *Public Opinion*. Lippmann takes some care to describe an experiment conducted a few years before in Germany in which a group of psychologists, presumably well-trained observers, unbeknown to them, witnessed a brief and carefully choreographed scuffle among a group of actors who rushed in and then out of the room. The forty observers were immediately asked to write down in detail what they had just witnessed. Lippmann proceeds to describe the analysis of their "eyewitness" reports:

> Only one had less than 20% of mistakes in regard to the principal facts; fourteen had 20% to 40% of mistakes; twelve from 40% to 50%; thirteen more than 50%. Moreover in twenty-four accounts 10% of the details were pure inventions and this proportion was exceeded in ten accounts and diminished in six. Briefly a quarter of the accounts were false. . . . The ten false reports may then be relegated to the category of tales and legends; twenty four accounts are half legendary, and six have a value approximating to exact evidence. Thus out of forty trained observers writing a responsible account of a scene that had just happened before

their eyes, more than a majority saw a scene that had not taken place. What then did they see? One would suppose it was easier to tell what had occurred, than to invent something which had not occurred. They saw their stereotype of such a brawl. All of them had in the course of their lives acquired a series of images of brawls, and these images flickered before their eyes. In one man these images displaced less than 20% of the actual scene, in thirteen men more than half. In thirty-four out of the forty observers the stereotypes preempted at least one-tenth of the scene. (55)

Thus Lippmann provides us with a particularly dramatic description of the vagaries of eyewitness accounts and the prominence of the Rashomon effect in human perception. For our purposes this example makes clear the danger in research design of treating a persuasive or informational message as a single object with a single meaning, or as being persuasive in an unambiguous or singular direction.

Agenda Setting, Framing, and Priming. There is a particular characteristic of polysemic selective interpretation that has drawn the attention of communication researchers. One could have predicted such a concern given the historically grounded roots of systematic media effects, persuasion, and attitude change research in the shadow of the Second World War as discussed in Chapter 1 and in a recent review of the accumulative character of the media effects literature (Neuman and Guggenheim 2011). Because the emphasis in this research tradition is on "effects," especially the Holy Grail of finding evidence of "strong effects," researchers came to grips with repeated cases of relatively small fractions of an audience changing opinions by seizing on the prospect that if an actual opinion position on an issue had not changed, perhaps an interpretation of an issue had. This perspective introduces the influential concepts of agenda setting, framing, and priming to the media effects research paradigm. Each of these research traditions posits that a media depiction may influence audience members to emphasize one or another attribute of a complex issues and events in their thinking as described in Table 2.2.

Although the language of the early studies in each of these three traditions makes it increasingly clear that the researchers viewed these more nuanced concepts and methods as a way to salvage media effects analysis from the discouraging prospect of minimal effects, in retrospect these developments can be seen as an unambiguous step forward as empirical

Table 2.2   Models of Polysemic Effects

	Agenda Setting	Framing	Priming
Analytic emphasis	Media emphasis on some public issues over others potentially raises the salience of those issues in public opinion over time.	Media emphasis on some interpretations and attributes of a complex issue rather than others potentially raises the salience of those interpretations and attributes in public opinion over time.	Media emphasis on some interpretations or attributes of a complex issue over others potentially raises the salience of those interpretations or attributes in short-term memory and thus the cognitive accessibility of those issues in concurrent thinking.
Prominent methodologies	Compare issue salience in media agenda with issue salience in public opinion.	Compare issue interpretations in media with those in public opinion; manipulate emphasis experimentally.	Manipulate emphasis experimentally.

researchers begin to struggle with the polysemic character of the phenomenon at hand rather than ignoring it.

McCombs and Shaw (1972) prominently introduced their notion of media agenda-setting effects by quoting Bernard Cohen's (1963) now famous epigram: "The press may not be successful much of the time in telling people what to think, but it is stunningly successful in telling its readers what to think about." "What to think" clearly refers to the old paradigm of traditional attitude change research and the counterintuitive pattern of low correlations between media exposure measures and attitude change. But now a new wrinkle—the analysis of the relative prominence of some issues over others in public concern, a more subtle dynamic perhaps, but the prospect of a "stunningly successful" effect. The agenda-setting literature is immense. In a review published in 2004, Max McCombs notes the existence of more than four hundred agenda-setting studies worldwide. The basic causal model shifts from the individual level to the aggregate level with the proposition that the variation in the quantity of aggregate media coverage of a set of issues and events in the news should be correlated with the proportion of population who rate that issue as an "important problem" in public opinion surveys. Subsequent studies refined the model by examining the agenda-setting correlation for different types of issues, different types of media, different types of audiences, and different time lags between media coverage and audience response (McCombs and Shaw 1993; Dearing and Rogers 1996; McCombs, Shaw, and Weaver 1997; McCombs 2004; Wanta and Ghanem 2007). A few in this tradition tackled the difficult prospect of measuring both media agendas and public agendas over time to better sort out causal directions (Fan 1988; Neuman 1990), but the number of such attempts is so few that they were eliminated from a meta-analysis with the notation: "Because the majority of agenda-setting studies have used Pearson correlations, eliminating these few time series studies did not substantially reduce the number of studies included in our analysis" (Wanta and Ghanem 2007, 43). It turns out that despite the strong theoretical start and confident language provided by Professor McCombs and colleagues, the accumulated findings assessing the correlation of media and public agendas ranges widely, so widely it would appear that whether an effect is evident is heavily dependent on how the key variables are operationalized. Take the aforementioned meta-analysis of agenda-setting studies, for example (Wanta and Ghanem 2007). The authors' final sample of studies included forty-five publications and a total of ninety independent tests of the agenda-setting hypothesized correlation between the media agenda and public agenda. The results are reproduced here at Table 2.3. Even an informal perusal of the results reveals that they are inexplicably inconsistent, ranging from a low of a trivial .05 correlation in the McLeod and colleagues (1974) study to the high correlation of .967 from the original McCombs and Shaw (1972) study. The standard meta-analytic technique is simply to average the results found to get a best estimate of the underlying causal pattern, and Wanta and Ghanem calculate that to be a correlation of .53, which would correspond to approximately 25 percent of the variance in public issue salience being associated with media issue salience. But such a technique represents the statistical equivalent of averaging apples and oranges. Until it can be explained why some studies find no meaningful correlation at all while others find a near-unitary correlation, prudence requires caution in drawing theoretical conclusions. Wanta and Ghanem appropriately explored subsets of studies to see if average correlations varied by method but the analysis was inconclusive.

Table 2.3 Meta-Analysis of Agenda-Setting Studies Correlations between Media and Public Agendas

Author	Date	r	N
Atwater	1985	.64–.46	304
Atwood	1978	.44	150
Behr	1985	.73–.37	MIP
Benton	1976	.81–.62	111
Brosius	1992a	.62	1,000
Brosius	1992b	.12	1,000
Demers	1989	.21–.77	MIP
Eaton	1989	.48	MIP
Einsledel	1984	.45	488
Erbring	1980	.10–.11	MIP
Funkhouser	1973	.78	MIP
Heeter	1989	.96	193
Hill	1985	.19	1,204
Hubbard	1975	.24	150
Iyengar	1979	.35–.47	MIP
Iyengar	1993	.85	1,500
Jablonski	1996	.19	1,324
Kaid	1977	.64	166
Lasorsa	1990	.57	624
McCombs	1972	.967	100
McLeod	1974	.05–.16	389
Miller	1996	.59	577
Palmgreen	1977	.50–.70	400
Salwen	1988	.54–.98	304
Salwen	1992	.56	629
Siune	1975	.91	1,302
Smith	1987	.65	400
Smith	1988	.71	471
Sohn	1978	.24	150
Stone	1981	.47–.55	302
Swanson	1978	.45	83
Tipton	1975	.75–.88	42–303
Wanta	1994a	.54	MIP
Wanta	1994b	.29	341

(*continued*)

Table 2.3   Meta-Analysis of Agenda-Setting Studies Correlations between Media and
            Public Agendas

Author	Date	r	N
Wanta	1994c	.60–.92	341
Wanta	1992	.31	341
Watt	1981	.35–.69	MIP
Weaver	1980	.27–.31	339
Weaver	1975	.21–.33	421
Williams	1977	.49–.83	350
Williams	1978	.11–.24	503
Williams	1983	.22–.78	356
Winter	1981	.71	MIP
Yagade	1990	.79	MIP
Zhu	1992	.52	MIP

Note: The abbreviation "MIP" refers to the standardized *"Most Important Problem* Facing
America" historical survey data set (Gallup.com).

Furthermore, there is an even more troubling critical problem with the
agenda-setting paradigm—how does one distinguish the effects of the nature of an event reported from the effects of the reporting itself? In other
words, if a major military engagement or economic or environmental crisis
is at hand, it may be that, not unreasonably, both the public and the journalists appropriately and independently perceive the event to be politically
significant and newsworthy. The journalists put it on the front page; the
public rate it as an "important issue."

There in not necessarily a causal link between the media emphasis and
public response. It might better be described as the independently assessed
significance of the issue or event by both the media and the public. This
has been acknowledged under the terminologies of *issue obtrusiveness* and
real-world cues (Zucker 1978; Erbring, Goldenberg, and Miller 1980; MacKuen
and Coombs 1981; Behr and Iyengar 1985; Demers et al. 1989). The only
way to tease out causal linkages, such as they are, is to develop an independently assessed measure of event magnitude, which should be possible particularly in matters military and economic, for example, and measure variation in event magnitude, media coverage, and public response over time.
It is a frustratingly complex enterprise, but the only way to get around the
"reality problem" of agenda-setting research. McCombs addressed this

issue thoughtfully in his 2004 book *Setting the Agenda* in a chapter entitled "Reality and the News." Drawing on Lippmann (1922) he notes two things: (1) sometimes the media get in a tizzy about some issue that turns out to be a false alarm but the public is nonetheless alarmed at least for a time, and (2) complex events are simplified in the public consciousness (the world outside and the pictures in our heads), and the media may play an important role in that simplifying process. Both points are well taken, but the first case is relatively rare (McCombs draws on a few historical case studies) and the second is not agenda setting in the sense of issue salience but what has come to be called "second order agenda setting," which is basically the phenomenon of framing.

"Framing" in media effects research refers to the prospect that by emphasizing certain attributes of a polysemic issue, actor, or object in the public sphere, the media can influence how the public perceives and responds to these phenomena. Framing effects are not a new development. One could imagine, for example, that in prehistoric times around the campfire perhaps some speakers would tell the story of the day's hunt, framing the story to emphasize their particular contributions to the collective effort. But in the era of a few predominant mass media reporting the news of the day, such human impulses take on special significance.

Because of the complexities of polysemic communication, there is no straightforward determination of an appropriate hypothesis-testing methodology to explore the conditions under which framing effects may be most evident. Further, most of this work focuses on the framing of a single issue or a few issues and each issue may generate a noncomparable typology of alternative frames of attribute emphasis. Bob Entman's widely cited review of framing research refers to it as a "fractured paradigm": "Despite its omnipresence across the social sciences and humanities, nowhere is there a general statement of framing theory that shows exactly how frames become embedded within and make themselves manifest in a text, or how framing influences thinking" (1993, 51). Dietram Scheufele, in a follow-up review, explains why: "Research on framing is characterized by theoretical and empirical vagueness. This is due, in part, to the lack of a commonly shared theoretical model underlying framing research. Conceptual problems translate into operational problems limiting the comparability of instruments and results" (1999, 103).

Framing research has a natural resonance with persuasion and attitude change research because it is often the case that one frame puts an issue in a more positive light than another and accordingly elicits higher levels of public approval. A classic example of this dynamic from Sniderman, Brody, and Tetlock (1991) is paraphrased in Entman's review: "The effect of framing is to prime values differentially, establishing the salience of the one or the other. [Thus] . . . a majority of the public supports the rights of persons with AIDS when the issue is framed (in a survey question) to accentuate civil liberties considerations and supports . . . mandatory testing when the issue is framed to accentuate public health considerations" (1993, 54).

There is a psychological tradition of framing research that precedes the elaboration of framing effects models in communication research. It was conducted by the Israeli team of Amos Tversky and Daniel Kahneman (Kahneman, Slovic, and Tversky 1982; Kahneman and Tversky 2000). Referred to as *equivalence framing* or *prospect theory*, this pioneering work on human cognitive biases demonstrated dramatic differences in how individuals valued exactly equivalent outcomes depending on whether they were framed as a loss or as a gain. Prospect theory has not yet stimulated a significant level of research in communication. An extensive meta-analysis of 165 studies, for example, of loss-framed and gain-framed persuasive messages found no meaningful differences in their persuasive effect (O'Keefe and Jensen 2006).

There have been, however, identifiable steps forward on another front in framing research in the two decades since Entman pronounced framing research to be a fractured paradigm. Entman's original critique made two points: first, that the hypothesized causal model was unclear, and, second, that each researcher would define the nature of an issue frame anew depending on the issue or issues under study. Important theoretical work on both fronts has been published that allows for more comparable and accumulated findings. The key analytic progress includes the distinction between whether the frame or frames are being analyzed as an independent or dependent variable, attention to level of analysis (media vs. public opinion) and the distinction between a more generalized multi-issue framing mechanism versus a framing mechanism intimately tied to a particular kind of issue such as racial stereotypes or gender roles. An overview of these developments in the literature is summarized in Table 2.4.

Of particular importance is the shift away from trying to demonstrate a not-so-minimal framing effect to identifying consistently important moderating variables—the conditions under which the effects are more or

Table 2.4　Progress in Issue-Framing Research

Causal model	Frame as dependent versus independent variable	Scheufele 1999
Level of analysis	Media versus public opinion frames	Scheufele 1999
Frame type	Single issue versus multiple issues	De Vreese 2005
Moderators of framing effects	Issue importance	Lecheler, de Vreese, and Slothuus 2009
	Knowledge	Krosnick and Brannon 1993
	Strength of prior opinion	Druckman 2001
	Frame strength	Chong and Druckman 2007
	One-sided versus competitive frames	Chong and Druckman 2007
Multiple issue frame typologies	Thematic versus episodic	Iyengar 1991
	Strategic versus issue oriented	Cappella and Jamieson 1997 Neuman, Just, and Crigler 1992
	News story frames	Gamson 1992
	New story frames	O'Keefe and Jensen 2006
	Equivalence frames	

less prominent. Interestingly, in the best sense of a Kuhnian puzzle, results are conflicting among studies on whether the more knowledgeable or less knowledgeable among the citizenry are more susceptible to framing effects (compare Kinder and Sanders 1996 with Nelson, Clawson, and Oxley 1997). The weight of research appears to be moving toward a tentative conclusion that the more knowledgeable are more susceptible because of increased familiarity with the issues and the logical connections between framing elements and policy evaluations (Druckman and Nelson 2003). A general or "all-purpose" model of public issue frames has not yet emerged, but Iyengar's (1991) distinction between episodic and thematic frames and Patterson's (1993) work on strategic frames have been influential.

The Distribution of Responses to Media Messages. A key development in integrating the inherently polysemic character of human communication into models of communication effects is to acknowledge that a typically

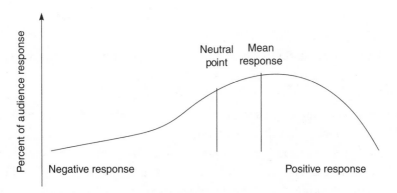

Figure 2.3. Hypothetical Distribution of Responses to a Persuasive Message

complex media message, intended to be persuasive or otherwise, is not likely to stimulate a singular response, but rather a distribution of responses across a population of those who have encountered the message, as depicted in Figure 2.3.

It is well known in the field of health communication where the effects of health-oriented public service announcements are extensively studied, for example, that many messages are misinterpreted by some audience members, and in many cases at least some audience members' attitudes move in the direction opposite of the pro-health direction intended. In one study of public health messages intended to discourage the use of marijuana, for example, the researchers found that those exposed to testimonial messages that smoking pot leads to smoking crack and serious addiction were actually less likely to agree with the marijuana-as-gateway-to-hard-drugs argument than a control group (Yzer et al. 2003). In a parallel antimarijuana study (Hornik et al. 2008), researchers found no decrease in the intention to use marijuana among those exposed to their health message. What they found in an over-time panel was actually a statistically significant increase in reported intention to use marijuana, a finding sufficiently frequently observed as to merit a research jargon terminology: the "boomerang effect." Such null and negative effects are frequently found in assessing traditional commercial ad campaigns, as well, but because such results are somewhat embarrassing to the entities involved (including commercial clients, advertising agencies, and the publishing and broadcasting media institutions) they are labeled "proprietary" and seldom see the light of day (for some rare exceptions to this pattern, see Schudson 1984; Lodish et al. 1995).

The dominant tradition in survey and experimental research that linked variation in message exposure with variation in attitudinal and behavioral outcomes is to focus on the net or aggregate differences in the key dependent variables. Thus in the experimental tradition, the means of the dependent variable for the control and exposure conditions are compared. In the survey tradition the comparison would be typically the same with self-reported exposure as the independent variable and various attitudinal and behavioral reports as the dependent variables. The statistical techniques engaged are typically an analysis of variance or linear regression. So the distribution of responses to message exposure are brought to bear through the statistical procedures that are derived from the ratio of covariance (or "explained variance") to total variance. The problem with this dominant tradition is that variation in attitudinal or behavioral responses to a complex message is treated as a statistical artifact necessary to compute the statistical significance of an aggregate "effect" for the population under study. *Given the inherent polysemic character of complex messages, the variance of response could and should be seen as of central theoretical interest.*

Take the following thought experiment as an example. Imagine that researchers tested two political ads for a presidential candidate among potential voters. Both ads are intended to generate support for the candidate because the candidate is committed to not raising taxes. In each case the researchers find a statistically significant increase of 3 percent in support for the candidate among undecided voters after exposure to the ad. In the dominant tradition, the research task is complete—the two ads are equivalent in their power to stimulate aggregate attitude change. But if the variation of response to the ad is three times greater in the first ad compared with the second, there is evidence that something much more complicated is going on in how diverse audiences are interpreting these messages. In one case, for example, there might be a small positive shift in attitudes toward the candidate in question. In the other case there is a large positive effect and a somewhat smaller boomerang effect, leading to an aggregate positive effect equivalent to the first case. Such differences are important, and if researchers are interested in the question of why such messages are effective, rather than just if they are effective, then an examination of variance of response rather than just net response has to be central to the research design. It may be that some audience members interpret the candidate's position as being thoughtfully responsive to public opinion while others see it as cynically pandering to the public mood of the moment. Serious researchers and serious campaign professionals, as well, should want to know why a message works, not just if it does. Such an orientation burdens researchers with much more complex designs that engage a mixture of qualitative and quantitative assessments of the often dramatic variation in how individuals respond to such messages and message campaigns.

Jackson's Conundrum. Sally Jackson is a communications scholar and methodologist (and more recently the chief information officer of the Urbana-Champaign campus of the University of Illinois). Her academic specialty is the study of rhetoric and persuasive communication. Starting in the early 1980s she initiated a campaign of persuasion among her disciplinary colleagues to convince them to be a little more careful about drawing conclusions from experiments and surveys. Working with several colleagues, including Scott Jacobs, Daniel O'Keefe, and Dale Brashers, she makes the following argument: "To do empirical research on effects of message variables, it is generally necessary to examine responses to actual messages that represent . . . the values of the variable of interest. The adequacy of actual concrete messages as instantiations of variables is central to any assessment of the validity of such an experiment. During the long history of experimental message effects research, virtually no attention has been paid to this issue. The seminal studies of message effects conducted by Hovland and associates during and after World War II set the precedent for how to deal with 'operationalization' of message variables that has been essentially unchallenged within communication and social psychology" (Jackson, O'Keefe, and Brashers 1994, 984).

She and her colleagues describe communication researchers as interested in such message properties as the trustworthiness of the source or particular rhetorical strategies. In her parlance these are "attributes" of a complex message, and her central point is that the results of any single experiment could have resulted from the many other attributes of the selected concrete experimental message. In other words, communication is polysemic. Thus while as researchers one might attribute the power to engage attention or to persuade to the extremity of an argument, it may well have been an incidental use of a colorful metaphor that resulted in an observed effect. As a methodologist, her recommended resolution to the problem is straightforward: treat any selected message that appears to have an attribute as a single sample from the universe of messages that appear to have that attribute and before drawing any conclusions replicate the study design with other sampled messages—the more sampled messages with similar results, the stronger the evidence. If exceptions are found, the puzzle to be addressed is what attribute of the complex polysemic message is actually generating or suppressing the behavioral or attitudinal result of interest to the researchers. Professor Jackson and colleagues have published by my count eight articles and a full-length methodology text on this topic over two decades and provoked at least four sets of respondents who have published commentaries and critiques (Bradac 1983; Jackson and Jacobs 1983; Jackson, O'Keefe, and Jacobs 1988; Morley 1988a, 1988b; O'Keefe, Jackson, and Jacobs 1988; Hunter, Hamilton, and Allen 1989; Jackson et al. 1989; Jackson 1991; Slater 1991; Jackson 1992; Jackson, O'Keefe, and Brashers 1994; Brashers and Jackson 1999). One pauses to ponder how influential this carefully argued and illustrated methodological campaign has been in the field. Unfortunately, the answer appears to be not very influential. The seminal 1983 article by Jackson and Jacobs has been cited more than seventy times, but the citations have been declining from four citations per year in the 1980s to two per year in the 2000s. One critical reviewer characterized "each new missive" in her campaign as "falling on deaf ears" and "the pummeling of a point most methodologists consider obvious" (Harris 1994, 474).

I label this strong argument and the tepid reaction to it "Jackson's Conundrum." The argument she and her colleagues make is critically important. And the fact that the argument appears to have largely fallen on deaf ears may be even more important. In this case her methodological framing of the question appears to have led to a series of persnickety technical debates over fixed versus random effects and crossed versus nested experimental designs in the analysis of variance (see Gelman 2005). The fact of the matter is that different members of an audience may be reacting to entirely different components of a complex message, and, further, that different members of an audience may react to the same component of a message in profoundly different ways. Jackson's recommendation for extensive replication is understandably burdensome, and that may contribute to its limited popularity and infrequent execution. She argues that to better understand these complexities researchers need to replicate the same basic research design multiple times with systematic variation of sampled messages to make sure any single result is not a fluke. Such a practice would contribute powerfully to the accumulative character of the collective research enterprise. And when puzzling anomalies arise, even further replications may be required to tease out what may be causing what. Unfortunately such disciplined comparability of research has not yet become characteristic of the communication field and appears to be surprisingly rare in many fields of science (Zimmer 2011). Editors are reluctant to publish replications, and researchers apparently feel their work would be seen as derivative rather than original. Replication is not a statistical nuance akin to selecting a single- or double-tailed test of significance. The systematic practice of replication is fundamental to better understanding the structure of human communication. The good news is that this issue is getting increased attention across the social sciences (King 2004; Benoit and Holbert 2008; Gerber and Green 2012).

Measurement Error in Media Research. There are many engaging examples of the perils of survey research as it must rely on the faulty memories, partial attentiveness, and biased self-perceptions of respondents. One of my favorites was a surprising finding of a particularly strong interest in "foreign affairs" in the rural American south in election surveys conducted by the University of Michigan during the 1950s. These were, of course, face-to-face in-home interviews conducted by local interviewers who read out the questions and made note of the answers on a clipboard. A little probing revealed that in the characteristic drawl of the American south, the word *foreign* sounds a lot like *farm*, and the respondents were actually expressing a concern about "farm affairs." It is an amusing and possibly apocryphal anecdote, but it speaks to a critically important issue in research design and that is the fact the research process itself engages the polysemy and ambiguity of human communication and thus complicates our ability to draw reliable conclusions from our observations.

The most dramatic evidence of systematic error in self-reports comes from contrasting individuals' reports of particular behaviors with actual recorded evidence of those behaviors. One component of this literature has tracked self-reports of recent communication with friends or coworkers with actual evidence of calls and emails. In one study of teletype exchanges within a deaf community, log records revealed that the person most often communicated with was listed among the self-reported top four communication partners only 52 percent of the time (Bernard et al. 1984). Further research indicated that many respondents interpret the question "who did you communicate with" as "who do you like" and that either question elicited basically the same answers (Bernard et al. 1984). As a result of their review of a large number of similar cases where there was often no meaningful association between reported behavior and measured behavior these analysts reluctantly conclude in a companion paper: "We must therefore recommend unreservedly that any conclusion drawn from the data gathered by the question 'who do you talk to' are of no use in understanding the social structure of communication" (cited in Romney and Weller 1984, 60). Similar conclusions were drawn from analyses in the areas of child care behavior and health-care-seeking behavior where independent records were available. It is not just a matter of faulty recall. Evidence revealed patterns of systematic overreporting and underreporting in different conditions (Bernard et al. 1984).

Survey researchers and experimentalists are well aware of the delicacies of designing items and scales to assess attitudes and behaviors. This literature is also rich and extensive (a small sampling might include Achen 1975; Schuman and Presser 1981; Turner and Martin 1984; Zaller 1991; Zaller and Feldman 1992; Nunnally and Bernstein 1994; Krosnick 1999; Hansen 2009; Alwin 2010; Babbie 2010; Marsden and Wright 2010; Bucy and Holbert 2014). Three phenomena that have attracted particular attention in this domain of research are (1) the dramatic variation in opinion responses to modest variations in question wordings, (2) the variation in answers respondents provide to opinion items when asked again after a short time interval, and (3) the fact that individuals trying to be helpful sometimes invent "opinions" on the spot in response to vague or unfamiliar policy questions.

A classic example of the first phenomenon is survey measurement of attitudes toward abortion. Depending on how the question is framed, the number of those responding to a survey favoring the legal exercise of abortion can move from a substantial majority to a tiny minority (Westoff, Moore, and Ryder 1969; Cook, Jelen, and Wilcox 1992). Another study revealed that the order in which two abortion questions were asked generated a 20 percent difference in levels of approval (Schuman, Presser, and Ludwig 1981). This is not an indication that individuals are careless or arbitrary about their responses, quite the contrary. The context of an abortion is critically important in how people think about it. When the health of the mother is at risk large majorities favor the possibility of abortion. When abortion is simply a matter of parental preference, large majorities oppose. The abortion issue is not unique; most policy issues are subject to framing or interpretive effects of various types, which is, of course, a central topic of communication structure addressed elsewhere in this chapter.

Methodological Fragmentation

Two widely circulated anecdotes characterize the difficult state of the search for precision and validity in the systematic study of the structure of human communication—the stories of the drunkard and of the hammer. In the first oft-told tale there is the drunkard who has lost his keys and is searching without success under the lamppost. In response to the query if had in fact lost his keys there, he replies without missing a beat, that no, he lost them way over there but this is where the light is better. In the second narrative there is a ten-year-old (a boy rather than a girl, as the story is usually told) in possession of his first hammer. The boy looks up smiling from his shiny new possession with the realization that just about everything within sight in the environment around him suddenly needs hammering.

A short survey or experiment, especially utilizing the handy captive audience of college sophomores, is hard to resist. This is where the light *is* better. But the external validity and even the internal validity of our capacity to measure attitudes and behaviors and responses to complex messages with precision is highly constrained for the many reasons outlined in the course of this chapter. The academic incentive structure, however, and the capacity to successfully publish in a field's prestigious journals is likely to pull us back again and again to the familiar lamppost. This is, as Kuhn has pointed out with powerful effect, what should be termed *normal science*—the routine application of established methods to well-accepted puzzles in each field of specialization. Furthermore, as graduate students, most young scholars work effectively as apprentices with senior faculty who have likely perfected a particular variant of an experiment or survey or content analysis that will become the shiny hammer for the newly minted researcher. Such patterns of professionalization are common enough in most fields of endeavor and there is certainly no intention here to derogate the professionals or the practice of professional training. But it represents a particularly troubling problem for progress and accumulation of findings in the study of human communication. Debates in the literatures tend to proceed within methodological "stovepipes" or "silos" and only seldom across them. It is a problem of methodological fragmentation. Findings derived from variant methodologies are simply judged to be incommensurate and are routinely ignored. It is not always the case but it is much too frequently the case.

My primary evidence for this observation is the collection of textbooks that are designed to address the issue of communication research methodology. I may have missed a few, but I have reviewed two dozen commonly cited communication research methods texts and roughly half of them simply devote separate chapters or book sections in turn to individual methodological approaches, typically surveys, experiments, content analysis, and sometimes textual analysis and depth interviewing or participant observation (Stempel and Westley 1989; Berger 2000; Bertrand and Hughes 2005; Weerakkody 2008; Priest 2009; Sparks 2010; Anderson 2011; Zhou and Sloan 2011; Bucy and Holbert 2014). Some texts focus on undergraduate students collecting research information from journals and libraries (Berger 2000; Rubin, Rubin, and Haridakis 2009). Some are straightforwardly edited collections of research reports and reviews (Singletary 1994; Hansen 2009; Bucy and Holbert 2014). Others focus on statistics or specialized qualitative methodologies (Monge and Cappella 1980; Hayes, Slater, and Snyder 2008; Lindlof and Taylor 2010). There is an occasional chapter attempting to draw the research output from these diverse perspectives into an integrated whole, but such efforts are rare. On the whole the message is to each his own hammer, and practitioners of each methodological/epistemological specialty should, as it is said on Broadway, "stick to your own kind."

Rethinking the Fundamental Paradigm of Media Effects Research

This chapter has thus far followed the classic structure of a narrative arc. The protagonist was introduced—in this case not the noble warrior of humble birth but a hopeful idea—the prospect that through thoughtful and structured measurement of collective human communication behavior we could better understand its structure and character and protect ourselves from its potential pathologies. Born of an urgent concern in the mid-twentieth century to defend otherwise stable democracies from the distortions of high-powered electronic propaganda, social scientists forged a new field of inquiry through the use of surveys, experiments, and content analyses of the mass media. The fundamental paradigm of analysis was simple—study the variation in exposure to the mass media messages and its potential correlation with the corresponding attitudes and behaviors these messages promote. Then the plot thickens. It turns out there are a series of methodological challenges that make what at first seems a simple task a nearly impossible one. Our review of these villainous obstacles to our protagonist's quest has been organized under the working concepts of profusion and polysemy to emphasize the unique difficulties of working with the elusive phenomena of collective human communication processes:

1. Profusion
    • Communication at the margin
    • Systematic inattention and forgetting
    • Reinforcement
    • Opinion change versus opinion creation
    • Cultivation analysis: a small step
    • The measurement of media exposure
    • Content differences across media
    • Hovland's paradox: experiments versus surveys
2. Polysemy
    • The complexity of selective exposure and selective interpretation
    • Selective exposure and spiral mechanisms
    • The puzzle of selective interpretation
    • Agenda setting, framing, and priming
    • The distribution of responses to media messages
    • Measurement error in media research

The very phenomenon that attracts our attention to media effects in the first place and resonates so strongly with our intuition—the tidal volumes of professionally crafted messages in which we are immersed as citizens of the modern world turn out to constitute a critical impediment to systematic analysis. Only an isolated and incomparable sliver of the population is not heavily exposed. Furthermore, because audiences are selective in their exposure to media and even more selective and differentiated in how they react to the complex and polysemic messages that swirl about, the notion of a singular persuasive "effect" represents a strikingly simplistic model. And further still, when we ask people to recall their media behaviors and interests, these very attitudes we wish to assess as "dependent variables" may bias their self-reports of what they like to watch and read and provide us only spurious indicators. One step in the right direction is the extended analysis of interaction, contextual, or moderation effects in addition to possible direct effects (McGuire 1968; McLeod et al. 2001; Preacher and Hayes 2008). But the fundamental problems of measurement and inference persist.

This is a precarious position in this narrative arc. A hero is needed, perhaps an entire heroic army to ride to the rescue. And as some readers may suspect, given at least a dozen not-so-subtle clues along the way, I have just such a deus ex machina in mind. The Latin phrase, literally "God from the machine," is derived from the tradition of contrived plot resolution in ancient tragedy and comedy when an actor or puppet representing a God was lowered onto the stage by a crane to resolve the otherwise irresolvable. And the machine I have in mind is the very digital technology that has been contributing to the most recent profusion—the ubiquitous Internet.

Let us return to the fundamental communication research paradigm introduced at the beginning of the chapter. Figure 2.4 replicates the original model and includes notations of how errors or biases or interpretations of survey questions in self-reported exposure measures may result in spurious conclusions about media effects. There is every reason to believe that the error introduced is systematic error because attitudes and social position affect how people describe both their media behavior and their political behavior. To date researchers have been forced to rely almost exclusively on these survey-based self-reports they have been unable to separate possible measurement error from actual behavior and actual attitudes. Further, because of the characteristic reliance on one-shot single survey measures researchers are unable to parse self-selection of exposure from possible effects of exposure—that requires over-time measurement (as depicted in Figure 2.2). Comparisons using the broad media labels of

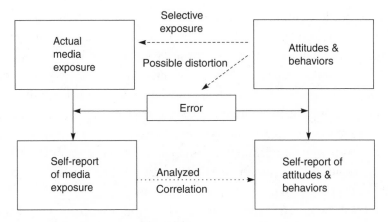

Figure 2.4. Fundamental Analytic Model of Media Effects Revised

television versus newspapers versus cinema, and so on, made it difficult to get a finer-grain picture of the messages involved since there is such dramatic diversity within each traditional broadcast/publishing medium.

This highly constrained and error-filled research paradigm, however, is no longer necessary in the new media environment. The flow of mass communication—virtually all of what is now associated with television, radio, books, movies, magazines, recordings, and newsletters—will ultimately be delivered electronically. This is already largely true for younger audience members. Technology is moving from newsprint and rabbit ears to laptops, e-book readers, iPads, iPods, smartphones, and Internet-enabled video screens. This digital flow leaves detailed digital footprints, so called big data. *The future of communication research will be compelled by analysis of big data* (Chang, Kaufmann, and Kwon 2014; Shah, Cappella, and Neuman 2015). Some prefer the term *computational social science* rather than big data, but the reference is the same. Our emphasis here is not on the size of the data set but rather on the fact that the data are naturally occurring rather than artificially situated or self-reported. This turns out to be critically important in the domain of research on human communication.

The Promise and Perils of Big Data

At the beginning of the digital era many (usually somewhat older) critics muttered that they would never trade in their treasured ink-on-paper medium to sit in front of an uncomfortable flickering screen. But the distinctions between traditional and digital media are declining, and many of the remaining distinctions favor the ease of use and flexible interactivity of digital storage and display. The era of email, texting, tweeting, posting, on-demand online video, on-demand streaming audio, and on-demand e-books provides a treasure trove of opportunity for the communication scholar. One can imagine that in the near future the overwhelming preponderance of human mass and interpersonal communication will be digitally mediated. Skeptical? When was the last time someone took pen to paper to write you a letter?

The following list presents several of the top-level attributes of conducting media research online making use of the same technologies the audience members would normally use to read an editorial, to Google a question of interest, to write a blog essay, to share a favorite news clipping or to watch a movie:

1. Actual fine-grained media exposure and selectivity can be assessed directly.
2. Measurement is possible continuously over time and for extended periods.
3. Many "real-world" behaviors online can also be assessed directly.
4. Individual perceptions and interpretations can be unobtrusively assessed in context and in real time.

All of these measurement techniques, of course, need to be undertaken with the fully informed consent of the participants, with their right to withdraw made clear and with careful attention to the protection of personal privacy. All of these elements of privacy and personal control of personal information are routinely addressed in present-day experiments, surveys, and observational studies and enforced by independent institutional review boards federally required of academic and federally funded research institutions in the American context since the passage of the National Research Act of 1974.

Early research on "new media" typically contrasted those who reported using the Internet frequently with those who did not as researchers examined possible differences in knowledge, attitudes, and behavior, and experimentalists contrasted paired comparisons of online and traditional media representations of content (DiMaggio et al. 2001). The research question was McLuhanesque in its focus on the medium of transmission. Such research designs were arguably appropriate, because in the earlier days of Internet diffusion, now dubbed Web 1.0, online newspapers and other media were essentially a re-creation of their original content and format on a video screen. Two things about the relationship between the medium and the audience member, however, have changed. First, increasingly large majorities are online and ubiquitously so, and as a result the online-offline contrast makes less sense as an analytic approach. Second, the nature of how individuals retrieve information and entertainment has changed, as noted above, characterized as a shift from push to pull as daily headlines and scheduled prime-time television offerings are replaced by an unadorned search box under the control of the individual audience member (Doyle 2013). This prompts a third development: the content audience members seek out and how they react to it is now increasingly digitally accessible to communication researchers. Some, especially more youthful audience members, may get their news about the world around them from social media as often, or even more often, than from traditional media. Two-step and multistep flows of communication have always been with us (Katz and Lazarsfeld 1955), but the velocity, intensity, diversity, and complexity of socially mediated flows of information is growing (Pew Project for Excellence in Journalism 2010) and, importantly, it is increasingly accessible to direct analysis.

I argue this development can be understood to represent significantly more than just a shift in strategy of data collection; it may represent a theoretical shift in how we understand media effects themselves (Neuman et al. 2014). In experimental research researchers assess the potential effect of randomized message exposure contrasted with its absence. In survey studies, researchers typically assess the correlation of self-reported message or medium exposure with attitudes, knowledge or behavior. When statistically significant differences are found, they are understood to be potential "effects" with appropriate caveats about confounds and problems of spuriousness. These designs have the distinct advantage of permitting the study of individual differences among audience members and of the conditions of exposure and systematic factors of mediation (Shah, Cappella, and Neuman 2015).

Real-world data collected from extant digital social and traditional media draw the analyst's attention to a more holistic media ecology as various media and messages rise and fall, representing continuous variation of public attention over time. The diversity and complexity of messages in social context leads the analyst to shift from message "x" potentially affecting attitude "y," to a question of which of the multitude of messages appear to resonate with public consciousness, which attract attention or comment, which are passed on, and which are sustained over time? It complements rather than replaces models of traditional media effects. Real-world big data are particularly well attuned to analyzing agenda dynamics, self-reinforcing mechanisms, damping effects, thresholds, and spiral phenomena in historical context (McCombs and Shaw 1972; Noelle-Neumann 1974; Neuman 1990; Page and Shapiro 1991; Erikson, MacKuen, and Stimson 2002; Slater 2007). Importantly, however, it lends itself to only a modest capacity for researchers studying individual differences and fine-grained contextual moderators.

The emphasis on exposure/persuasion effects may have made historical sense in the early days of communication research in the mid-twentieth century in part because of the limited number of published and broadcast sources of information for the average audience member. But in an age for which each broadband connection to the web is equally technically empowered to speak as well as to listen, our theoretical lens shifts from the dynamics of attitude change to the dynamics of attention—of the many voices in the digital cacophony, what attracts attention and which framings of public issues seem to resonate most strongly in the ongoing public discussion?

Because most early big data research was commercial, attention has been given to social media reputation and brand monitoring rather than building an abstract theory of public attention. One recent review analyzed the social media monitoring services of eighty companies competing for commercial business in this space (Moffitt 2011). As a result of this commercial activity there are ongoing advancements in analytics and data processing of which academics can take advantage. The commercial folks, of course, are primarily only interested in what happened this week concerning their brand or brand category. But the aggregated data are available to academics who are more likely to be assessing trends over months or years. Entering the search phrase "Obama," for example, for mentions in the last year in the popular Sysomos Map analysis system reveals more than ninety-three million hits in the social media, blogs, online discussion fora, and in the online versions of the traditional broadcast and print media. One can trace the ups and downs of traditional media commentary on Obama and, for example, the specific issue of Obamacare tracking official news and public commentary over time. It is an amazingly rich and as yet hardly tapped resource for social scientists interested in opinion dynamics.

Using social media big data as an indicator of public sentiment upsets some survey research public opinion traditionalists (cited in Hargittai 2015). Big data is not a representative sample of the public who have been asked to address an issue of potential public concern. Indeed, it is something quite different—a sample of those who were moved to comment on a public issue without being asked. It is arguably as reasonable an indicator as a survey especially now when percentage response rates to traditional mail and telephone surveys has fallen to the low twenties and teens (Ansolabehere and Schaffner 2014).

The terminology of "big data" may strike some as a bit self-aggrandizing, but it seems to be catching on as a generic label for data and analyses of this general type (boyd and Crawford 2012). Early promoters of these new directions, although full of enthusiasm and perhaps a bit of missionary zeal, are generally well aware of the many limitations and biases of these methodologies and acknowledge that these new research opportunities will complement and expand, rather than replace, more traditional methods (Bollier 2010). One notable problem in this domain is that big data systems effortlessly generate large numbers of colorful visualizations of text patterns and over-time trend graphs so as a result analysts may be inadvertently seduced into a reliance on description rather than theory testing (Borrero and Gualda 2013).

Imagine the following scenarios. Immediately following the downloading of a provocative news story, a citizen fires off an email to their congressional representative and immediately thereafter makes a financial contribution to the representative's rival in an upcoming election. Or: following exposure to an online ad for a new beauty product, the viewer orders the product online. Or: over a two-year period an individual becomes increasingly dependent on Fox News, drops subscriptions to less conservative news sources, and reports increasingly conservative views and voting intentions. Or: in a researcher-managed ongoing online political discussion group, it is demonstrated that the injection of verifiable factual information leads to more moderate views and deliberative discussion while parallel untreated control groups tend to spiral toward polarization. Or: it is demonstrated that after adventuring into heavily liberal or conservative news/talk environments for an extended periods, certain types of citizens tire of the advocacy and emotionality and find themselves returning to familiar mainstream media for the bulk of their media diet.

The use of online survey panels and online behavioral assessment do not proceed without serious difficulties of inference, representativeness, panel wear out, and, whenever self-reports are involved, difficulties of re-call and interpretation and occasional intentional misrepresentation. But because the medium of research is the same as the basic medium of com-munication in natural environments, the benefits are significant. Manip-ulation of exposure in the experimental tradition is still possible. With the consent of the subjects involved, numerous techniques are possible to systematically steer and filter the flow of information they are exposed to over time. For obvious reasons, most volunteers would not want to have what they see and hear in the media manipulated for decades, but it may well be possible for weeks and months, and that is a significant step ahead of the standard thirty- or forty-minute experimental lab study—the cor-nerstone, indeed, the gold standard of the experimental tradition.

It may be noted that the humanistic and critical traditions of commu-nication scholarship have occasionally ventured out in the field to assess how typical viewers or readers were making sense of the mainstream media fare available to them. One of the most prominent and widely cited of these exercises was the *Nationwide* study series conducted by David Morley and Charlotte Brunsdon in the late 1970s. *Nationwide* was a popular current affairs magazine program broadcast weekdays from 6:00 to 7:00 p.m. on BBC1, and the research team managed to arrange showings of two broadcasts in mostly adult educational settings with twenty-nine small viewing groups (three to thirteen participants) in London and the Midlands who would discuss the program in an unstructured format following the viewing (Morley and Brunsdon 1978, 1999; Morley 1992). Because of the costs and difficulty of the enterprise such empirical explorations of what viewers are thinking as they are viewing are extremely rare (see also Neuman 1982). Nowadays, however, most popular programming has multiple naturally occurring online discussion groups full of speculation about plot twists, character motives, and, of course, "what it all means" (Wohn and Na 2011). Such a rich source of audience reaction can be supplemented with researcher-driven and more representative samples of viewers (rather than the high-energy fan base that dominates the discussion boards) and include specific queries on how individuals are reacting to the complex flow of narrative twists and turns as well as the framing of news and public affairs (Jenkins 1992).

For the past fifty years researchers have been watching TV and reading newspapers and magazines with a clipboard in hand and a detailed codebook for a rigorous content analysis of media messages for the purposes of description and of characterizing the themes of the dominant media fare. Analyzing text quantitatively and qualitatively with an interpretive flair has been an area where humanistic and social scientific traditions have overlapped somewhat. It is not yet clear, but at some point sophisticated automated content analysis may approach human-coded analysis in its richness, reliability, and validity. One group that systematically codes the flow of news around the world (mediatenor.com) continues to rely exclusively on a small army of extensively trained human coders rather than computer-based coding, so this automation may take some time yet. This research group has every incentive to automate the process. But since their financial bottom line depends in large measure on the accuracy of the subtle evaluative story framing over time, they have not yet abandoned their human coders with clipboards. In any case, with the extensive archiving of online content under way, it is increasingly practical to review and rereview content for a mix of automated and human-based exercises in tracking message trends and the linkages of those trends to attitudinal and behavioral responses among those actually exposed (Tufekci and Wilson 2012).

Clearly a great deal remains to be done. But the prospects for a fundamentally fresh approach to the systematic measurement of the flows and structure of electronically mediated communication are clearly promising. If Thomas Kuhn's skeptical views about the inertia of scientific research practices among older researchers is sound, it may be up to a new generation of researchers to fully explore these options, hopefully with the encouragement and hard-earned experience of their elders. So we cannot conclude our narrative arc with our protagonist, sword raised triumphantly and standing astride the vanquished evildoer. This is a book on evolving research in practice. Our narrative is perhaps more akin to structure of episodic television. The conclusion is simply—stay tuned.

3

The Paradox of Profusion

A wealth of information creates a poverty of attention.

—HERBERT SIMON (1971)

For a term that sounds as if so technically exact, information overload
is a strangely elusive concept.

—JAY BLUMLER (1980)

In our ironic twentieth-century version of Gresham's law, information tends
to drive knowledge out of circulation

—DANIEL BOORSTIN (1989)

IN THIS CHAPTER I confront the issue of information abundance. Information is generally a good thing. Abundance would seem to be a good
thing. Is there a problem, a paradox?

I argue that there are several paradoxes associated with the electronically mediated profusion of information (and, of course, also an abundance
of misinformation) that characterizes the modern era and that both the
working conceptions of information profusion and information overload
generally used in scholarship and in journalistic and policy debate are strikingly simplistic, unquestioned, and undertheorized. In Chapter 2 we reviewed the special difficulties that the sheer volume of communication

generates for measurement and inference from data collected "at the margin." Here we address a set of issues and arguments about profusion that may strike the reader as self-contradictory and accordingly paradoxical. So, at the outset, let me outline the path ahead.

Problem number one: the profusion of information and the explosive expansion of its movement among humans and machines have drawn surprisingly little attention within the field of communication research. Profusion is a given—a nonproblematic, taken-for-granted, and self-evident fact. Although communication researchers have frequently obsessed over the impact of a particular communication medium such as radio, television, video games, and networked computers, the broader issue of the overall quantity of the symbolic flow has only infrequently attracted the systematic attention of the research community. My view is that that is an important oversight. As a result, an initial task is to try to draw attention to, to theorize, and to problematize the broader issue of the quantities of symbolic flows.

Problem number two: when scholars have addressed the profusion issue, the guiding theoretical trope is typically the notion of "information overload" most often framed as some sort of psychological pathology. So my task is complicated. While attempting to draw attention to an issue I'm arguing that the way the issue is typically framed is actually something of a red herring. This is especially difficult because the notion of being overloaded by information is journalistically powerful, intuitively satisfying, and in strong resonance with many personal experiences.

Problem number three: the normative issues of maintaining a free and open public sphere without censorship in the industrial era of push media were relatively straightforward. Elite censorship can be documented and condemned relatively straightforwardly. But the more subtle forms of hegemonic, commercial, or institutional influence on the attentional agenda that may submerge and minimize critical views in the age of information abundance are much more difficult to identify and address. If a presidential aide persuades the head of a television network to deep-six a story, we have a scandal. If Google's complex attentional algorithms bury a story and effectively hide it from public view, we have a problem Google's engineers may not fully understand, let alone the public at large.

Throughout this volume I return to the notion of a fundamental historical transition in the relationship between the "media" and the "audience"—a movement characterized as from push to pull. Bruce Bimber (personal communication) reminds us that the evolving digital environment for the foreseeable future will likely represent not just pull but a lively mixture of push and pull. He makes an excellent point. The opportunity to "pull" may be exercised only some of the time and under some conditions, an important topic for research. As one focus group participant put it when asked about confronting the digital cornucopia—"It's impossible to keep up with it . . . I've kind of got to the point where I let it wash over me" (Hargittai, Neuman, and Curry 2012, 169).

We will turn first to the available data that characterize the phenomenon of profusion, the underlying historical trends, and what we know thus far about the human response to an overwhelmingly rich information environment.

Throughout this chapter I move back and forth between the individual level of the cognitive psychology of information processing and the social/cultural/organizational level of collective structures designed to deal with these tidal information flows. This chapter has a singular conclusion in responding to these various puzzles associated with the real information explosion, an answer, a punch line. In most formats the punch line is saved until the end, the answer to "who done it," or the punch line of the joke. But in the organization of effective scholarly prose, to which this chapter aspires, the logic is reversed. One starts with presenting the proposition or hypothesis to the reader and then reviews the relevant evidence and remaining unresolved mysteries for the reader's due consideration. So here goes.

Is information profusion a significant issue meriting sustained attention and research? Yes, indeed. The trouble is that the issue has largely been left on a theoretically isolated island of research focusing at the individual level on the potential for information overload. Further, the psychopathology framing is largely a red herring. Most people treat profusion as choice rather than overload and are much more delighted than frustrated. The fundamental connection to theories of accessible information citizens need in the public sphere has been obscured or simply neglected. The question of how selective attention works with massive information flows requires new energy, new methods engaging big data, refined theories in the spirit of culturation.

The phenomenon of losing information as captured in the familiar phrase "a needle in a haystack" becomes better described as losing a specific needle in a haystack-size mountain of needles. *The solution to the problem is not less information; it is more structure.* The power of the haystack analogy is that the proverbial haystack is an unstructured and unorganized pile and the needle could be anywhere. If the hay were neatly arrayed and organized, laid out in parallel, stalk by stalk from shortest to the longest, one could probably spot the needle right away over at the short end of the array.

This talk of haystacks may strike the reader as simplistic and unserious, but I am going to argue otherwise. My reasoning is as follows. Information is rarely neutral. What one person would like everybody to know, others may find threatening, unsettling, libelous, or blasphemous. In simpler times, especially before the expansion of the printing press, unwanted information was censored and unwanted speakers were hanged or burned at the stake. In the modern age, unwanted information is buried in an informational haystack of irrelevancies or contradictions. As polymath Herbert Simon is quoted in the beginning of this chapter, "A wealth of information creates a poverty of attention" (Simon 1971, 40). What is more likely in simply "burying" information in the battle for attention among opposing interested parties is a systematic bias in the labeling and structuring of information. This becomes the battleground in a pull-media public sphere—moving potentially significant information from the first page in response to a Google search to the one hundredth page means it will in all likelihood not be seen, a powerful haystack indeed. In an attempt to maintain an open marketplace of ideas, the conflict will center not on the existence of information but its structure and accessibility in the profusion.

The Harvard Business School is particularly noted for its "case method" of instruction. The students are challenged by casebooks of notorious length and complexity about a particular problem typically faced by business executives. With rare exception the casebooks contain extensive material that is irrelevant and often material that is intentionally incorrect or misleading. The reasoning, of course, is that that is what business executives face in the real world when colleagues and competitors have strong motivations to mislead and misrepresent. The "A"s go to the students who most quickly and successfully filter out the irrelevancies and get to the heart of the matter. Perhaps more of higher education in the information age should pursue similar techniques of pedagogy.

There is an engaging anecdote that may or may not have been scientifically verified, but makes an important point nonetheless. It goes like this. If you put a frog in boiling water, it immediately reacts and jumps out. If you put a frog in lukewarm water and very gradually warm it to boiling the frog just sits there until its ultimate and unfortunate demise. The point, of course, is that gradual change is hard to detect, and appropriate reaction may be delayed well beyond the point that is too late.

This insight has been highlighted powerfully by Daniel Bell. In his introduction to the English translation of Nora and Minc's seminal 1980 review of the impact of computers on society he notes:

> It is a rare moment in cultural history when we can self-consciously witness a large-scale social transformation (as distinct from a revolution). Few persons realized, when the industrial revolution was beginning, the import of what was taking place. The term itself was coined only a hundred years after the start of the process, by Arnold Toynbee, in 1884, when he gave a set of lectures retrospectively viewing the era that he called "The Industrial Revolution." Today, with our greater sensitivity to social consequences and to the future—indeed, with a readiness to hail any new gadget (conditioned as we are by science fiction) as the magic wand of social change—we are more alert to the possible imports of technological and organizational change; and this is all to the good, for to the extent that we are that sensitive, we can try to estimate the consequences and decide which policies we should choose, consonant with the values we have, in order to shape, accept, or even reject the alternative futures that are available to us. (x–xi)

The observation is poignant for at least two reasons. First, in this volume we are examining the implications of the information revolution and it appears to be increasingly likely that the social transformations are fully as significant as the Industrial Revolution that preceded it. How ironic that a comprehensive concept like the "Industrial Revolution" only came into common use a full century into its historical progression. We are prone to proclaim that such a lag in recognition certainly could not be the case this time. Books and articles on informatization (or as Nora and Minc put it more felicitously, *telematique*) number in the thousands. I argue here, however, that we may not yet fully understand the magnitude of this technical revolution. Second, Bell makes the important point that the better we understand these developments, the better we can derive policy and architect information structures to serve collective goals. Our strategy here, as throughout most of this book, is to explore these dynamics at both the individual and the social level.

Information Overload

The notion of information overload draws our attention to the individual and psychological level of dealing with informational abundance. It would appear that we all have memories of frustrating experiences we have interpreted as some sort of information overload, so we are suckers for journalistic accounts that string anecdotes together with a selective summary of a few laboratory experiments that explain what is afoot and, importantly, that it is not our fault. The newspaper headlines and book titles are revealing. The *New York Times* 2010 series entitled "Your Brain on Computers" drew your distracted attention with the following headlines:

Growing Up Digital, Wired for Distraction

Digital Devices Deprive Brain of Needed Downtime

An Ugly Toll of Technology: Impatience and Forgetfulness

More Americans Sense a Downside to an Always Plugged-In Existence

Hooked on Gadgets, and Paying a Mental Price

While at your local bookstore the following book titles beckon:

The Shallows: What the Internet Is Doing to Our Brains

Media Unlimited: How the Torrent of Images and Sounds Overwhelms Our Lives

iBrain: Surviving the Technological Alteration of the Modern Mind

Data Smog: Surviving the Information Glut

Distracted: The Erosion of Attention and the Coming Dark Age

Yes, the Coming Dark Age. One wonders whether these book titles are the work of the authors or whether the publishers of such popular and self-help-style volumes have a special backroom office of title-writing stylists who dream up these dire threats. As a result of an abundance of activity in this little corner of the world of journalism and book publishing, I find myself in the curious position of trying to sort, filter, interpret, and make sense of this overload of commentary on overload. I'm with Jay Blumler, who three decades ago expressed some frustration with the dire predictions and selective exaggerations.

> For a term that sounds as if so technically exact, "information overload" is a strangely elusive concept, to which my own reactions are quite mixed and ambivalent. On the one hand, I applaud its broadly humanitarian thrust. References to "overload" are welcome cries of sympathy for the supposedly beleaguered, overburdened and bewildered receiver of too much unsatisfying data. It is a curious fact, however, that we really do not know whether many audience members see themselves in this light and therefore truly deserve our expressions of sympathy . . . perhaps we should be trying to put that hypothesis to empirical test by developing ways of monitoring the more negative reactions of audiences to major information sources in the modern world. On the other hand, the concept of "information overload" is open to much abuse. . . . Some writers on this topic vastly over-extend its focus of concern, treating "overload" as rather like nuclear reactors that have got out of control, pouring out masses of data that pollute every corner of the environment and causing or exacerbating a very wide range of social problems . . . when "information overload" is so indiscriminately caught up in such a mood of weltschmerz, all possibility of using the term precisely is hopelessly lost. (1980, 229)

It turns out there are three distinct subgenres in this domain. One variety focuses on the information environment itself, documenting, as I do here, the exponential growth of information. Todd Gitlin's *Media Unlimited* (2002) and David Shenk's *Data Smog: Surviving the Information Glut* (1998) fall into this category. The second variety focuses on the psychological burdens of extended opportunities for choice in daily life, noting, for example, the burdens of negotiating a supermarket isle featuring 85 different brands of crackers and 285 varieties of cookies. Sheena Iyengar's *The Art of Choosing* (2010) and Barry Schwartz's *The Paradox of Choice: Why More Is Less* (2004) represent best sellers in that category. The third variant, my personal favorite, focuses on the impact of the information environment—typically characterized as "the Internet" or simply "Google" on the human brain.

Small and Vorgan titled their volume published in 2008, *iBrain: Surviving the Technological Alteration of the Modern Mind*. Nicholas Carr chose *The Shallows: What the Internet Is Doing to Our Brains* (2010) as his book-length follow-up to his 2008 *Atlantic* article "Is Google Making Us Stupid?"

The drill in these books seems to be about two hundred pages of hand-wringing and angst about information overload emphasizing oceanic metaphors of tidal waves and torrents followed by about fifty pages of "how-to" tips on time management and getting organized. My take on the your-brain-on-computers literature is that these popular authors have read the literature on neuroplasticity and basically got it backward. *Neuroplasticity* is a general term for the spectacular adaptability of the human brain. It was originally thought that only certain parts of the cerebral cortex were particularly adaptive and that it was typically limited to children whose brains were still developing. But recent advances in neuroscience reveal that virtually all sections of the brain are highly plastic and that the malleability of the brain to literally change its physiology as a function of experience extends through adulthood (Buonomano and Merzenich 1998). As a result, our intrepid journalists cite studies that demonstrate that extensive use of computers and computer games has measurable effects on brain functioning. For example, Carr, reviewing Gary Small and colleagues (2009), "Your Brain on Google: Patterns of Cerebral Activation during Internet Searching," from the *American Journal of Geriatric Psychiatry* notes:

> The most remarkable part of the experiment came when the tests were repeated six days later. In the interim, the researchers had the novices spend an hour a day online, searching the Net. The new scans revealed that the area in their prefrontal cortex that had been largely dormant now showed extensive activity—just like the activity in the brains of the veteran surfers. "After just five days of practice, the exact same neural circuitry in the front part of the brain became active in the Internet-naïve subjects," reports Small. "Five hours on the Internet, and the naïve subjects had already rewired their brains." He goes on to ask, "If our brains are so sensitive to just an hour a day of computer exposure, what happens when we spend more time [online]?" (2010, 121)

The original study, it turns out, was actually quite positive about the cognitive *benefits* of this form of cognitive stimulation. The press release accompanying the study, for example, used the headline, "UCLA Study

Finds That Searching the Internet Increases Brain Function." The trick in making it sound like information overload leads to permanent brain damage is to ignore the fact that ongoing plasticity of the brain will respond to whatever tasks are next undertaken, so read a little Shakespeare and listen to some Mozart after Googling, and the brain is accordingly modestly rewired anew. For more detail, interested readers should consult the book on this topic by a real neuroscientist, Torkel Klingberg, *The Overflowing Brain: Information Overload and the Limits of Working Memory* (2009), which makes the case clearly and fairly.

For me the real fascination in the brain-on-computers genre is its inherent believability and popularity. *New York Times* reporter Matt Richtel (2010) reports that his series of articles on information overload was especially popular and, ironically, forwarded by many emailers, bloggers, and twitterers. One could argue that this is the inevitable digital extension of the mechanical rhythms of the Industrial Revolution and the frustrations of time management and multiple obligations of modern life make for a sympathetic audience that takes some comfort in being told these are evil forces and such circumstances are not the readers' fault.

My reading of this frustratingly inchoate literature leads to the conclusion that within the technical studies and between the breathless observations of the popularizers there are in fact four underlying thematics as summarized in Table 3.1. Iconic examples of information overload dynamics usually involve a fighter pilot or perhaps a battlefield commander. Somewhat less frequently a surgeon in a high-tech operating room or a bond trader in front of multiple screens might be typified. Such exemplars illustrate the central importance of dimensions 1 and 2 above. Military and medical decisions are often matters of life and death and typically also made under time pressures often measured in seconds rather than hours or days. In the financial case the time pressures are equivalent and the importance of the decision is measured monetarily. So the emphasis on "amount of information" in the typical treatment of this concept may be a bit misleading—the key issues are time constraint and time-constrained decision making. In the light of that observation we can step back and make note that typical web browsing and information seeking online is rarely time constrained, and rarely requires critical decision making. There are exceptions, of course—eBay auction participation, online games and contests, perhaps online gambling. But online reading is more like the

Table 3.1 Dimensions of "Information Overload"

Dimension	Underlying Dynamic
1) Time sensitivity	A key element is the perception of "overload" is the time constraints on reviewing available information.
2) Decision requirement	Related to time sensitivity is the time constrained need to make a decision, especially critical decisions.
3) Structure of information	The "amount" of information may be less critical than the extent to which the information is structured, permitting the observer to retrieve what is judged to be relevant.
4) Quality of information	Many grievances about "information overload" turn out to actually concern the quality of information or the information variate of the engineering concept of signal-to-noise ratio.

reading of paper-based newspapers, magazines, and books or like television viewing. On the issue of a vast variety of websites or video streams from which to choose, as we shall see below, the verdict is in. People value the choice and are not overwhelmed by the quantities involved. Of the five hundred channels on digital cable, the typical viewer has personalized their remote control to display typically about a dozen favorites watched most regularly. When Google reports it has found possible websites responsive to your query measured in the hundreds of thousands, search engine users do not faint at the thought. In fact they rarely search beyond the first page of matches displayed. Not a problem. Thinking back to the industrial age public library with many thousands of volumes and a card catalog, nobody fainted then either at the prospects of more information resources than they could possibly consult. And, of course, the card catalog and the search engine identify the central importance of the third dimension in this domain—the extent to which information is structured and the extent to which the structuring is responsive to the information seekers interests. Card catalogs worked, but not very well as they were limited to author, title, and a simplified topical list—U.S. History—Post–Civil War—not terribly helpful. But the characteristic of the digital age is extensive structuring, labeling, and classifying information. Google Books, for example, famously permits full text searches of tens of millions of volumes. Recommendation systems make it possible for individuals to benefit from the opinions, judgments, and descriptors of like-minded seekers.

Finally there is the issue of information quality. The recurrent fear in this literature is the needle-in-a-haystack notion that the irrelevancies will obscure what is important. Making it easier for the "unqualified" as well as the "qualified" to comment and opine may indeed in some sense lower the signal-to-noise ratio, but on that two comments. First, people coming to an issue anew may provide fresh perspectives. Diversity of perspectives in decision making is a point emphasized by system theorists like Scott Page (2007) and psychologists such as Philip Tetlock (2005). Second, the structuring, evaluating, and labeling of digital information noted above makes a big difference in managing the quantities of information typically available online. So our review of all this ceremony about overload leads us, one more time, to a paradoxical conclusion—the increased quantities of information are indeed significant, but the problem is not "information overload." Such a terminology represents something of a red herring and perhaps a distraction from more important questions. Key issues revolve around accurate and responsive systems for labeling and categorizing information that are responsive to human needs, providing access for the full citizenry to benefit with minimal distortions from marketplace profiteering and manipulation of scarcity and minimal distortions from self-protective state institutions. In later chapters I address questions of intellectual property law, state censorship, and the economics of niche markets and long tails, which I argue are indeed critically important potential gatekeepers that will require ongoing attention and further analysis, rather than the endless hand-wringing over potentially disoriented web surfers.

That said, research at the individual level informed by advanced psychological theory designed to optimize the human capacity to locate and utilize information is certainly appropriately included in the research agenda. Some strands of research address matters of user interface design, navigation, and database structuring (Nielsen 2000; Nielsen and Pernice 2010). Other strands focus on user attitudes and skills (Hargittai 2002).

A few more brief comments on information overload and cognitive capacity issues before turning to the information society literature. In reviewing these literatures I came across a thoughtful review and informal meta-analysis of the overload literature by two Swiss management professors (Eppler and Mengis 2004). Management Information Systems (MIS) is an active and growing field in business education, and these researchers searched the online journals titles and abstracts for the keywords: *informa-*

tion overload, information load, cognitive overload, and *cognitive load,* which resulted in more than five hundred retrieved articles. That struck them as
a bit unwieldly, so they filtered by more recent publication and several
other criteria to reduce the sample to just under one hundred. Then they
proceeded to clearly and thoughtfully summarize the findings in terms of
causes, symptoms, and countermeasures for overload problems. It would
appear that the irony had escaped them. In fact they had themselves confronted a bit of an overload problem and simply filtered and organized the
information at hand to get the job done apparently with such modest effort that they did not even take note.

We actually know a few rather specific facts about the human capacities for cognitive load. I am referring to what is something of a classic paper
in psychology—George Miller's 1956 paper "The Magical Number Seven,
Plus or Minus Two: Some Limits on Our Capacity for Processing Information." Miller finds, remarkably, in study after study that the human brain
has, obviously enough, a finite capacity for holding separate entities in
mind simultaneously. As the paper title proclaims, the number turns out
to be seven or very close to seven. It varies some by individuals and by
domains of psychophysical discrimination. The paper is understandably
famous for celebrating the magic number, something akin to the psychological equivalent of the mathematical constant pi. But, in my view, the
central insight of the paper is less about magic numbers and more about
what humans instinctively do when they find themselves confronting a
number of dimensions of consideration greater than seven.

What do we do? We "chunk." (This is Miller's evocative vocabulary.) We
cluster similar considerations in chunks so that what would have been a
dozen evaluations is meaningfully reduced to seven or fewer. The process
of chunking, of course, involves abstraction—that uniquely and importantly human cognitive trait. It is how we naturally and often without
self-awareness deal with what otherwise would have been information
overload; indeed, as we noted above with the two Swiss management
professors.

One final and, I believe, critically important addition to this literature.
This insight draws on the work of Herbert A. Simon, who pioneered research in such diverse fields as cognitive psychology, computer science,
artificial intelligence, information processing, decision making, attention
economics, organization theory, and complex systems. I am referring to

what might be termed the other evolved human trait for dealing with what would otherwise be information overload—the trait of satisficing (Simon 1956). The term was coined by Herbert Simon in the 1950s, a portmanteau of "satisfy" and "suffice" that refers to the fact that humans only rarely maximize the amount of potentially relevant information in making decisions. They satisfice. They instinctively calculate at the margin, whether collecting and evaluating further information justifies the effort.

Today we are amused by the self-assured proclamations of the nineteenth-century Belgian doctors who warned that humans were simply not capable of viewing the passing scenery now that trains were traveling at the unprecedented speeds of more than twenty miles an hour. What they failed to fully appreciate is that even when an individual is standing still, the human sensory apparatus generates hundreds of thousands of bits of sensory information only a tiny fraction of which reach the level of conscious cognitive processing (Zimmermann 1989). When sensory input does reach the level of conscious processing, Miller's "chunking" and Simon's "satisficing" go a good distance at the individual level in taming the information tide. But humans are social animals and also fabricators. So in the balance of this chapter I shift to the social and cultural levels of social analysis to explore how social and technical networks and mechanisms respond to information profusion.

The Information Society

The concept of "the information society" has several interesting properties. The terminology (and several closely related cognates) has been around for a while. This vocabulary first came into use in the 1960s when Princeton economist Fritz Machlup (1962) published some pathbreaking work demonstrating how the information processing and service sectors of the economy were growing dramatically as agriculture and manufacturing were declining. These notions were picked up and popularized by business guru Peter Drucker (1969) and later futurist Alvin Toffler (1980). The most thoughtful and provoking work of this era was Daniel Bell's *The Coming of Post-Industrial Society*, published in 1973. Related terminologies included "the knowledge society," "the network society," "the information revolution," and "the control revolution." What is interesting and frustrating about this by now vast literature is its shallowness, elusiveness, and the

Table 3.2   The Structure of the Information Society Literature

	Emphasize Change	Emphasize Continuity
Postive emphasis	Drucker 1969; Bell 1973; Masuda 1980; Toffler 1980; Gilder 1989; Negroponte 1995	Cohen and Zysman 1987
Neutral	Machlup 1962; Price 1963; Porat 1977; Pool 1983; Dordick-Wang 1993; van Dijk 1999	Beniger 1986
Critical emphasis	Nora and Minc 1980	Traber 1986; Murdoch and Golding 1989; McChesney 1999; Schiller 2000; Mattelart 2003; Mosco 2005; Fuchs 2008; Mansell 2009

absence of a theory, puzzle to solve, or methodology to guide future research. It is rather curious. Scholars actively publishing in the field are often deeply ambivalent and some even hostile to the notion as an organizing intellectual framework. Take, for example, English sociologist Frank Webster. He has published now three editions (most recently 2008) of *Theories of the Information Society* in which he notes that by use of such a concept "people were marshaling yet another grandiose term to identify the germane features of our time. But simultaneously thinkers were remarkably divergent in their interpretations of what form this information took, why it was central to our present systems, and how it was affecting social, economic and political relationships" (2). Toward the end of this, now his third edition, he concludes, "I am convinced that a focus on information trends is vital to understand the character of the world today, though most information society scenarios are of little help in this exercise" (263).

I have become convinced that, like the related notion of postmodernism—"the information society," although it need not be, is something of a neutral and mostly empty vessel that is used to signal that issues should be addressed historically. Further, and importantly, it has evolved into a conceptual Rorschach test that draws out an analyst's views on two fundamental dimensions of social analysis as illustrated in Table 3.2.

Table 3.2 attempts to array some exemplars in the literature along two dimensions. The horizontal dimension contrasts an emphasis on social and economic change wrought by developments in information technology versus historical continuity particularly in elite-driven politics. The vertical dimension contrasts a normative perspective of enthusiasm and optimism versus pessimism and concern. A quick perusal of the table reveals that the dimensions appear to be correlated as by far the most exemplars can be found in the upper-left and lower-right quadrants. It makes sense. The upper left is populated by technically oriented analysts who are not terribly displeased with the day-to-day workings of the capitalist system and their scenarios of change depict increased efficiencies and productivities. The lower-right quadrant, in contrast, is populated largely by men and women of the political Left who are openly sarcastic about such scenarios and warn gravely of how capitalist elites will utilize these technologies to strengthen their political power and cultural dominance. The neutral category in this typology identifies major works that a primarily descriptive of economic, behavioral, or technical change or focus on technical issues about the measurement of the information economy. The Beniger study is an exception. His widely cited study uses the terminology "the control revolution," and he makes the case that the dramatic growth of information and control technologies was initiated fifty to eighty years earlier than commonly thought and was necessitated by the demands of the Industrial Revolution, notably electronic timing and switching technologies to keep the shiny new trains from crashing into each other, primarily as a result of human error. Typical of the lower-right quadrant critically emphasizing continuity is Michael Traber's opening salvo in *The Myth of the Information Revolution* (1986, 3): "All this has far-reaching consequences, but not in the direction of human emancipation and liberation, or of improving the quality of life for ordinary people. If anything, the communication revolution is turning out to be an exercise in consolidating the military, economic and political powers of the elite."

Contrast such considered pessimism with (from the upper-left quadrant) Yoneji Masuda's heady scenario of what he calls "Computopia," a portmanteau of computer and utopia.

> The information society that will emerge from the computer communications revolution will be a society that actually moves toward a universal society of plenty. The most important point I would make is that the information society will function around the axis of information values rather than material values. . . . Thus, if industrial society is a society in which people have affluent material consumption, the in-

formation society will be a society in which the cognitive creativity of individuals flourishes throughout society. And if the highest stage of industrial society is the high mass consumption society, then the highest stage of the information society will be the globalfuturization society, a vision that greatly expands and develops [Adam] Smith's vision of a universal opulent society; this is what I mean by "Computopia." (1980, 147)

An alert reader may have noted a prominent omission from our informal analysis of exemplars of the information society literature—the multivolume works of the widely cited sociologist Manuel Castells (1996, 1997, 1998, 2004, 2009) focusing on, in his vocabulary, "the network society." The reason I set Castells apart is that, to his enduring credit, his richly theoretical analysis is not easily captured by any of the four quadrants. He emphasizes change and even celebrates the distinctive qualities of the evolving technical environment. His provocative if slightly elusive concepts of "timeless time," "space of flows," "mass self-communication," and the polarity of the "net versus the self" each speak of the unique character of our age. But his analysis is deeply grounded in political continuities of social identity, social movements, the evolving dynamics of the public sphere, and both the strengths and weaknesses of capitalist institutions. In other words, Castells is engaging in an enterprise quite similar to my own aspirations in this volume—to empirically examine the relationships between each of the dimensions of technical change and the corresponding possible social, political, economic, and cultural changes that may result. And the exercise is based on the same fundamental sociological questions about inequality, political engagement, social identity, polarization, fragmentation, and pluralism that have engaged social scientists since Marx, Weber and Durkheim first tried to make sense of the onset of industrial capitalism. Such an enterprise cannot be characterized as either optimistic or pessimistic because the ramifications of technical change, first, will likely result in a mix of positive and negative effects, and, second, will hopefully, when better understood, be subject to appropriate and intelligent collective intervention.

What strikes me as particularly interesting is that the information society literature so consistently ignores its many structural connections and contrasts with the mass society literature that preceded it. It is in my view unfortunate, but it helps explain some of the theoretical thinness of work on the information society. A few of the information society theorists,

Table 3.3   The Mass Society Meets the Information Society

	Mass Society Literature	Information Society Literature
Period of prominence	1890–1960	1960–current
Key theorists	Tönnies, Durkheim, Park, Fromm, Riesman, Arendt, Bell, Lipset, Kornhauser, Putnam	Machlup, Porat, Bell, Masuda, Pool, Beniger, Mansell, Castells, van Djik, Fuchs
Key theses:	Usually framed as highly problematic and dire.	Most often framed as an unsolved puzzle.
Household	Decline of family life—nuclear family replaces the extended family, family members spend less time together, children attend large, centralized, anomic school systems, working mothers may be absent, television watching replaces family conversation.	A central but unanswered research question—will growth of mobile and social media potentially strengthen or weaken bonds among family (and extended family) members?
Occupations	Alienating workplace—mobility from job to job and isolating work conditions in large organizations makes both the workplace and work associates less important to the individual.	Shift from manufacturing to information processing and service sector work, impact on workplace alienation and economic inequities unclear, critical theorists very skeptical of improvements.
Geography	Decline of local community—urbanization and suburbanization replaces small town and gives residents little sense of community.	Increasing significance of globalization, potential weakening of nation-state and growth of networked global corporations and nonprofits, a mix of positive developments and new problems.
Identity	Weakening of religious and ethnic ties—local religious institutions become less important, over time ethnic communities blur into a massified urban landscape.	A central puzzle of information society—will evolving media reinforce polarization, fragmentation; will globalization lead to a clash of civilizations?
Group life	Decline of participation in voluntary associations.	Will new media environment reinforce or further impoverish group life and voluntary associations?
Communication	Direct and unfiltered exposure to propaganda—anomic individuals may find comfort in the pseudo-authority and pseudo-community of the mass media.	Focus on overly powerful central media institutions shifts to expanded diversity and possibly fragmented, chaotic, and polarized information environment.

notably Bell, Beniger, Rogers, and Castells, make a point of connecting and contrasting these two traditions. My sense of the mass society literature is that the central questions were never really resolved successfully and that the attention of social scientists just gradually turned to a variety of other questions which had a fresher character and engaged new methodologies and data. As Table 3.3 suggests, analysts of the new media and the information society could, and in my view, should understand these dynamics as a critically important continuation of an older intellectual tradition with some new wrinkles, new questions, and new methods.

Illustrative of the shift from the hand-wringing angst and self-assuredness of the mass society tradition to the puzzled curiosity and intermixed hopefulness and skepticism of analysts of the information society is Robert Putnam's (2000) influential *Bowling Alone* argument. His voice is self-assured and powerfully supported by numerous charts and graphs of declining social capital of voluntary association participation largely due, he argues, to anomic and isolating television viewing and associated generational differences in sociability. But writing in 2000, it was already clear that the Internet and new media were challenging television's dominance, so he includes a brief section on whether the Internet and new media might reverse these trends. It is an articulate and thoughtful review of such prospects, but in the end he basically admits he we don't yet have a handle on these developments, a welcome admission.

Putnam's concern with citizen participation in public organizations, however, raises an issue not yet fully addressed in these pages. Citizens do more than vote and speak as individuals; they organize. How does the new information environment impact the prospect of collective behavior, mobilization, protest, and/or rallying around the flag? Clearly important questions for theory and research, but for practical purposes, it remains outside the scope of this volume. Fortunately, there is an active tradition of scholarship on this issue (Bimber 2003; Castells 2012; Bimber, Flanagin, and Stohl 2012).

The Profusion

If human existence on earth were compressed into twenty-four hours, the emergence of writing would come at eight minutes before midnight, Gutenberg's popularization of movable type and mass printing at forty-six seconds before midnight, and finally the telephone, radio, television, and Internet all within seconds of midnight. It is difficult to fully grasp how recently and briefly in human history we have confronted the stunning proliferation of media and messages. So thinking in terms of the familiar twenty-four hours helps to provide prospective. A few seconds is indeed a tiny fraction of a day. It would seem that ten thousand generations of highly selective survival in a tribal existence of hunting and gathering would have produced an evolved cognitive system ill designed to process the flood of electronic images and sound that swirls about us. But we seem to manage pretty well. How is that possible?

Again, starting with the bottom line first and then developing the supporting argument—we manage the information profusion pretty well because the notable trait of the evolved human cognitive system is not its finely attuned hardwired responsiveness to consistent environmental challenges but rather its plasticity, its reliance on cognitive appraisal and strategizing rather than on fixed capacities of speed, strength, or agility. The challenges of survival through hunting and gathering required purposeful and selective attention to acquire sufficient food and to avoid abundant and diverse potential predators (Bettinger 1991; D'Andrade 1995). Charles Jonscher tells the story with an effective scenario.

> The year is 8000 B.C. A hunter-gatherer is standing amid the plains of Northern Europe, scanning the horizon for animals—animals which might be food for him, or which represent danger. Although he wouldn't think of it in these terms, the panorama before him is generating immense number of photons. Every leaf, stone and passing bird is sending out beams of reflected light at varying levels of frequency and intensity in accordance with the source's shape, colour and texture. These streams of photons are data—raw signals—and they number in the millions upon millions. The pattern of leaves on the trees alone contains more data than all the pages of the Encyclopedia Britannica. But as the light rays pass through the air they are not yet information. Nobody has yet seen them; nobody has been informed. Some of these photons reach the eyes of the hunter. They are focused onto the rods and cones of his retinae, causing signals to be sent via the optic nerves to the visual cortex in the brain. A great deal of processing happens along the route. What the mind registers is not raw data but a distillation of it which he can digest and make sense of. He sees patterns and regularities in the light

signals and these he interprets as trees, rocks and birds. Only now has the data become information. (1999, 35–36)

The Long View. It turns out that the amazing capacities of the human brain for language and for sophisticated forms of selective attention and selective forgetting is a well-researched field of inquiry. Ethologists have studied animal communication and social behavior in laboratory and field settings with particular attention to our primate ancestors (de Waal 1982; Hauser 1996; Deacon 1997). Archaeologists and anthropologists have modeled the social structure and likely quotidian practices of the tribal existence that comprise 90–99 percent of our history as humans (Chance 1976; Barkow, Cosmides, and Tooby 1992). Because there is no surviving record of symbolic communication among humans until the development of the first written language systems approximately six thousand years ago, we have only fragmentary information. Verbal language through these early years was probably a roughly equal mix of gestures, grunts, and verbalizations, probably primarily oriented to the coordination of the hunting of larger game with small groups and some strategic division of labor (Kenneally 2007).

There are rather large technical and popular literatures about human evolution and prehistory. The complex debates of the literatures concerning human prehistory are a bit of a digression from our ultimate purposes, but a brief and disciplined look is rewarding and reveals some important insights for interpreting the digital revolution of our age, especially concerning the ongoing interaction of human psychology and evolving technology.

Four questions arise from the literature:

1. How different is human communication from that of our animal ancestors?
2. Why is the written transcription of human speech such a recent phenomenon?
3. Why have we witnessed such an explosive rate of recent change?
4. Why are we not utterly overwhelmed by the information explosion?

Are Humans So Different? Numerous observers have pointed out that genetically speaking we are not so different from our prehominid ancestors.

Roughly 98.8 percent of the DNA of chimpanzees and humans is identical (although details of the decimals are subject to technical debate) (Diamond 1992; Grehan and Schwartz 2011). Such an estimate roughly corresponds to the genetic similarities of lions and tigers and of horses and zebras. So part of the answer to the first question above is that although there are only tiny physiological differences between humans and our near ancestors, how do we explain the rather dramatic differences in behavior and cognition? Well, it turns out those small physiological differences in brain function make a very big difference in behavior and perception. The basic difference is that animal communication is largely (though not entirely) hardwired. Although there is some learning by listening, birdsong is largely innate physiologically and functionally fixed. When vervet monkeys are raised without any exposure to conspecifics they signal with their species-specific warning vocalizations. If raised within a colony of primates that has a different hardwired signaling system, they will come to understand the meaning of the systems, but themselves only signal in their innate system of warning and attention signals (Hauser 1996).

Marc Hauser, a specialist in animal communication, has become well known for arguing that human communication skills are not all that unique. Various nonhuman species have in varying degrees the capacity to imitate and invent vocalizations, to equate vocalizations with meaning, to discriminate the elements of continuous speech, and to make limited attribution of intention to speakers (Hauser, Chomsky, and Fitch 2002). There are fascinating parallels in animal communication learning and human acquisition in childhood. Birds not exposed to birdsong of their species in youth turn out to have a difficult time mastering it later in their life cycle, a well-known phenomenon of human language learning and the skills of native speakers. So although there are numerous technical debates in the literature of biolinguistics, we can conclude that most cognitive and neuromotor capacities that permit human communication through speech are actually shared with much of the animal kingdom, and that most differences that result in the unique richness and complexity of human speech are matters of degree of the capacity to process grammatical complexity and symbolic diversity that evolved gradually in the six million years since we last shared common ancestors with our modern primate cousins. Researchers are particularly careful and tactful in reporting these findings because most of us find lack of an evident evolutionary saltation or "giant step" hard to accept.

So we have another puzzle. It is a puzzle that has become a near obsession for American psychologist and primatologist Michael Tomasello, now heading the Max Planck Institute for Evolutionary Anthropology in Leipzig, Germany. And Tomasello has derived an answer:

> The 6 million years that separates human beings from other great apes is a very short time evolutionarily. . . . Our problem is thus one of time. The fact is, there simply has not been enough time for normal processes of biological evolution involving genetic variation and natural selection to have created, one by one, each of the cognitive skills necessary for modern humans to invent and maintain complex tool-use industries and technologies, complex forms of symbolic communication and representation, and complex social organizations and institutions. And the puzzle is only magnified if we take seriously current research in paleo-anthropology suggesting that (a) for all but the last 2 million years the human lineage showed no signs of anything other than typical great ape cognitive skills, and (b) the first dramatic signs of species-unique cognitive skills emerged only in the last one-quarter of a million years with modern Homo sapiens.
>
> There is only one possible solution to this puzzle. That is, there is only one known biological mechanism that could bring about these kinds of changes in behavior and cognition in so short a time—whether that time be thought of as 6 million, 2 million, or one-quarter of a million years. This biological mechanism is social or cultural transmission [in this case a] species-unique mode or modes of cultural transmission. (1999, 2–4)

As with many of these issues, there remains some controversy over these observations, but Tomasello's main argument carries particular importance. It represents a cornerstone of a growing scientific consensus and resolution to the long-standing debate about the relative importance of nature versus nurture (Pinker 2004). The consensus dismisses the controversy over the relative importance of each factor by pointing out that it is the interaction of the two from which collective cultures (and also from which individual personalities) spring. The idea is captured by the term *co-evolution* (Durham 1991; King 1994; Richerson and Boyd 2005).

Why Is Writing So Recent? Given the celebrated creativity, plasticity, and problem-solving capacities of the fully human cerebrum that evolved in its current form a half million years ago, we are drawn to the second of the questions above—why did the capacity to transmit and accumulate human knowledge by some form of writing have to be effectively deferred for 99 percent of those half million years before it was in evidence? Scholars are in general agreement about this aspect of the historical record, but Princeton historian and Near Eastern specialist Michael Cook (2003) has a particularly effective way of putting it. He makes the case that writing basically requires two things—the hardware (some sort of writing instrument and an appropriate medium on which to write) and the software (some sort of coding system to transpose something we hear into something we see). Both should have easily been within the grasp of the well-organized tribes of hunters and gatherers, but there is virtually no surviving evidence of its evolution until about 3000 B.C. in Mesopotamia and Egypt and in about 2000 B.C. in the Indus Valley of India and (probably independently) in China in 1000 B.C. and (certainly independently) in Mesoamerica in 500 B.C. Cook continues: "It was not the inherent difficulty of procuring the hardware or developing the software that stood in the way. Rather, it was the need for an appropriate social structure. Somebody had to have a strong need for this information technology, and a willingness to pay handsomely for it by maintaining a community of otherwise unproductive scribes. Such a need and such a willingness are hallmarks of a complex society. To put it crudely, early writing presupposes a powerful state" (47).

Writing was a full-time profession—the profession of scribe. Most of the kings and generals were not themselves literate. And because the capacity to read and write was a source of power, the scribes were careful indeed not to unnecessarily share the secrets of their profession. The early scripts in Sumer (the wedge-based cuneiform) and the Egyptian hieroglyphics were nonalphabetic and linked to phonetic speech by complex rules and thus extremely labor intensive to learn. When the Egyptian scribes, it is reported, were confronted with a reformed and simplified (more alphabetic) version of hieroglyphic coding system, they rejected it for fear a more easily learned language would challenge their status and power (Logan 1986, 33).

Why the Current Profusion? If we have had the capacity for reading and writing for seven thousand years, why have we witnessed an informational explosion in only the past few centuries? This is a straightforward question and it has an easy answer, well two answers—the industrial mechanization of written communication (printing and data processing) and the growth of mass literacy. In the ancient cultures of Mesopotamia, Egypt, Greece, and Rome and in the independently evolved (and remarkably parallel evolution of) pre-Columbian America tribal cultures, only the elite and frequently a small fraction of the elite were actually literate. A few copies of an important document or hand-transcribed book-length scroll would be more than sufficient for the relative few in society who could make sense of it. The manually produced supply and literacy-constrained demand were in inadvertent but nonetheless harmonious balance.

Much has been made of Gutenberg's "invention" of the printing press. As is frequently the case with dramatic historical events, this one is clouded by a fair amount of academic controversy. Some point out that many of the elements of printing and of mechanical press technologies preceded Gutenberg (from wine and cloth presses) and that his principal contribution drawing on his skill as a metallurgist was in primarily perfecting the quality of the movable metal type (known in Korea since 1403) rather than "inventing" printing (Febvre and Martin 1997). Others claim that the historical significance of Gutenberg's press has also been exaggerated (Briggs and Burke 2009). Of particular resonance for those of us interested in the current technical revolution is the complexity and cultural delicacy of this now famous "invention." The idea of both mechanical printing and moveable type had been around for many centuries. There is controversy surrounding the possibility that these technical developments in China, Korea, and Japan actually did migrate to Europe because the Islamic authorities in the Middle East and Asia were famously opposed to any form of printing that might threaten its incumbent power, defining it as a capital crime (Briggs and Burke 2009). The perception of enhanced public communication as a potential threat to established regimes is a recurring theme in history (Innis 1950; Eisenstein 1979). Perhaps more interesting than the Islamic prohibition was the fact that printing as a form of public communication in Asia was less prohibited than simply unimagined. "The purpose of printing among the Chinese was not the creation of uniform repeatable products for a market and a price system. Print was an alternative to their prayer-wheels and was a visual means of multiplying incantatory spells" (McLuhan 1969, 34). Printing as the capacity for public communication in

Gutenberg's case resulted from the unique historical chemistry of a guild culture, emerging capitalism and technological refinement.

University of Michigan historian Elizabeth Eisenstein built her career on the study of the social impact of the Gutenberg press. She had read McLuhan's popular, anecdotal, and speculative book *The Gutenberg Galaxy* (1964) and became convinced that this nonhistorian was onto something important and was highlighting a development largely missed by traditional historical scholarship. She labeled it the "unacknowledged revolution" and proceeded over her career to establish the cultural importance of Gutenberg's work (Eisenstein 1979). Careful and resourceful in scholarship, she avoids a simply attribution of technical determinism and focuses on technical-cultural interaction. A principal case in point is the centrally important trends in public literacy. She notes that when there was virtually nothing available to read, the capacity to read was not terribly salient to the public at large. The explosive growth of a diversity of books and pamphlets in the vernacular (above and beyond the Bible, of course) from fiction to cookbooks was connected to a parallel historical growth in literacy (Eisenstein 1979). Historians and pundits like McLuhan sometimes succumb to celebrating a dramatic "causal" technical turning point, but of course, as is clearly the case in our era, the process almost always represents a delicate interactive, causal spiraling over time. Further, it is typically an interaction alternatively constrained and/or stimulated by local cultural and political conditions. So Eisenstein is hesitant to assert that printing caused literacy any more than growing literacy caused the invention of the printing press.

A second case in point, even more controversial than the first, is the interaction of printing and the growth of the vibrant public sphere that ultimately nurtured the Reformation, the Renaissance, and the Scientific Revolution. Eisenstein's case is both modest and immodest. On the immodest side she argues with extensive supporting historical detail that the vernacular Bible, now both inexpensive and widely available, was the foundation on which Luther's revolution was to be built. She asserts that work of scientific and religious elites once in close concert became separated by the advent of printing. Theologians and astronomers of the early Middle Ages linked studying "how the heavens go" with "how to go to heaven." But the advent of widely available print communication drove a wedge between the religious and scientific communities and propelling them in different directions (Eisenstein 1979, 696). She concludes, almost apologetically but powerfully, that printing was a pivotal factor in the introduction to what we have come to call modernity, a muscular exemplar of historical scholarship indeed. The argument is straightforward. Progress requires rational/critical thought that in turns requires the capacity for a comparison of alternatives. We must recognize, she asserts, "the novelty of being able to assemble diverse records and reference guides, and of being able to study them without having to transcribe them at the same time. If we want to explain heightened awareness of anomalies or discontent with inherited schemes then it seems especially important to emphasize the wider range of reading matter that was being surveyed at one time by a single pair of eyes" (686).

On the more modest side, again stepping back for a crudely deterministic form of technical history, she posits that with little effort the priests and princes, had they been so inclined, could have harnessed the new technology to promote, celebrate, and protect the established hierarchies and harnessed public energies to ward off heresy rather than the technology of public communication. It could easily have been otherwise. Rather than technological determinism, she suggests, we encountered some good luck, historically speaking.

Eisenstein's (2011) work, still very much in progress, has spawned a small industry of scholarship on printing and the public sphere all useful to those of us trying to make sense of the following stages of industrialization and digitization (McNally 1987; Febvre and Martin 1997; Man 2002; Baron, Lindquist, and Shevlin 2007). What started it all? Eisenstein confesses in the preface to her masterwork that it started early in her career in 1963 in response to a rather apocalyptic presidential address at the American Historical Society that proclaimed that "runaway technology" was severing all bonds with the past and that modern thinking suffered from collective amnesia, a loss of history altogether. It is our old friend information overload. The question defined a challenge she undertook as she set out on her career as an historian—is it inevitable that overload will lead to incoherence (Eisenstein 1979)? The printing press in the context of a culture of entrepreneurial capitalism, an openness to public education, and growing literacy and with only modest efforts at prohibition and censorship in the mid-fifteenth century in central Europe would represent the first wave of industrial mass communication, but clearly only the first of many waves each an exponentiation of its predecessor.

Why Are We Not Utterly Overwhelmed? We come now full circle to the originating question for this chapter. And our principal answer remains the same. We are not overwhelmed because we are extraordinarily good at being inattentive, or perhaps more precisely—selectively attentive. A few pages ago we observed with Charles Jonscher, the hunter-gatherer surveying the environment and selectively attending to a select few of the many thousands of visual and auditory cues on the verdant horizon. Those were the cues he had learned (and his forebears had learned) were useful for survival. The cognitive mechanics of selective attention have remained virtually unchanged from that era. The lessons from our more recent forebears may have evolved dramatically, but further significant change in our environment may require us to rethink these lessons anew. It is the defining question of communication scholarship. *The better we understand both the cognitive mechanics and accumulated cultural norms building up from recent centuries, the better we can counterbalance our ingrained and accumulated propensities to miscommunicate and misunderstand.*

The Dimensions of the Profusion

At the dawn of the digital age in the early 1980s, the pioneering student of media technology Ithiel de Sola Pool based at MIT published a series of studies attempting to quantify the growing flow of information focusing on case studies of the American and Japanese mass media (Pool 1983; Pool et al. 1984; Neuman and Pool 1986). Pool had been working with Japanese and American colleagues over the previous decade in an effort to quantify the increasingly electronic media supply in meaningful terms and subject the analysis to further theoretical study of how these trends might affect levels of information, diversity of information, and possible polarization within the mass population consuming these media. Pool saw himself as expanding the research agenda and the key methodologies for better understanding the dynamics of the information age. Up to the publication of Pool's work scholars had relied primarily on aggregate economic data focusing on broadly aggregated employment patterns to track

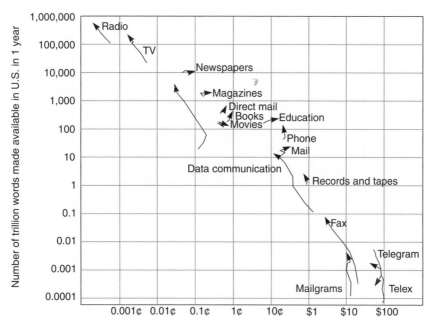

Figure 3.1. Declining Costs (per 1,000 words transmitted) and Increasing Volumes of Communication in United States, 1960–1977
Source: Pool (1983), p. 99.

the transitions from the agricultural to the industrial and in turn to the information age (Machlup 1962; Porat 1977; Bell 1979).

Pool knew it was important to understand how much information was "out there" but equally important to understand how much was actually being consumed by the population at large. The key variables of analysis were the number of words supplied and consumed yearly at a national level and the average price per word in various common media. His findings were dramatic and led to an obvious conundrum (see Figure 3.1). First, he noted that the flow was increasingly electronic. Second, the price per word was falling radically. Third, the supply was growing at an impressive compounded rate of 8.8 percent per annum. Fourth, the consumption was also growing at impressive rate, in this case 3.3 percent per annum, compounded and thus generating a growing disparity between information supplied and information consumed.

A conundrum? Well, there might not be a technical limit on supply, but there are only twenty-four hours in the day—a clear-cut limit on individual

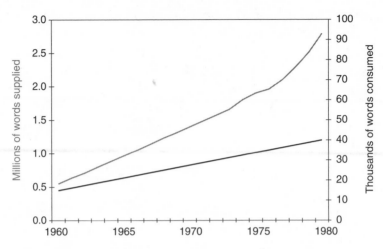

Figure 3.2. Increasing Supply Outpaces Consumption of Communication in United
States, 1960–1980
Source: Neuman and Pool (1986), Figure 5.3, p. 82.

consumption of mass media and social media (see Figure 3.2). Pool and
colleagues speculated about information overload, information diversity,
and the economics necessary to sustain vibrant creative industries in journalism and popular and high culture. So the basic theoretical proposition
of this research tradition was to challenge the generally unquestioned tenet
in the historical analysis of media trends that more is necessarily better.
They propose that some levels of media quantity might make informed
choice impractical and perhaps even frustrating (Miller 1960; Bell 1979;
Blumler 1980; Eppler and Mengis 2004). Further, this research challenged
the notion that new media replace or partially replace older media, a notion generally referred to as relative constancy theory (McCombs and Eyal
1980; Dupagne and Green 1996). Relative constancy theory relies primarily
on expenditure data rather than words or minutes of usage, but the underlying argument is the same—time budgets and financial budgets constrain growth of media supply and media use. Much of this work took place
a decade before the introduction of the Internet-linked PC that would, of
course, dramatically reinforce these trends of expanding media supply and
raise new questions about supply, demand, overload, attentional dynamics,
and economic viability (Gitlin 2002; Hindman 2009).

A research team at the University of Michigan had the opportunity to
pick up where Pool's data collection left off in 1980. We made a few adjust-

ments and minor corrections and focusing on the U.S. case, carried the data collection forward to 2005 with a research report published in 2012 (Neuman, Yong, and Panek 2012). The study is briefly summarized here.

In the years following 2005 we have been witnessing an even more dramatic digital convergence as traditional media flows from recordings, newspapers, books, television, and movies are increasingly being delivered by the web itself, making the distinction between, say, watching broadcast TV and a digitally delivered video program an increasingly subtle one. Accordingly, the trends we have been able to track empirically so far represent just the earliest stages of what we anticipate will play out as a fundamental structural shift in the technology and institutionalization of mass (and interpersonal) communication. Pool's original Japanese and American work focused on the assessment of media volume measured in quadrillions of words per medium per annum at a national level. It is a useful metric for international comparison of trends and infrastructure, but our focus in this analysis is a more human-level metric addressing the dynamics of choice and attention. Quadrillions of anything would be hard to conceptualize. So we present all annual data divided by 365 days and the total number of households in the nation for an assessment of flow of information into the typical home on a twenty-four-hour day measured for the most part in the thousands, in our view, a more interpretable and accessible metric. We also switched from words to minutes as the principal measure. Because we are analyzing print media that are measured in spatial terms—column inches, thousands of words—and broadcast media that are measured in temporal metrics of minutes and hours, this type of analysis requires a common metric. We follow Pool and take the average adult American reading speed of 240 words a minute to approximately equate space and time. The original analyses of Pool and associates made a practical strategic choice and focused just on the flow of words, ignoring the proverbial "elephant in the room" represented by still and motion imagery and graphic representations. We live in a world of increasingly high-resolution graphics and expanded video displays. They warrant close attention and analysis. As a starting point, however, we pick up where Pool left off and set aside the graphic component for the present analysis.

We took Pool's original data and measurement definitions as our starting point, turning to new data sources as necessary and dropping a few media such as telex and telegrams to focus on the historical continuity of the primary mass and interpersonal media. The key to the measurement of supply pivots on what is available at a particular historical interval to a typical household. So we picked a pooled average of the median-size cities for the United States (Charlotte, North Carolina; Indianapolis, Indiana; San Diego, California; Raleigh-Durham, North Carolina) for estimates of the typical number of available over-the-air broadcast television stations that calculate out to four stations in 1960 growing to nine stations by 2005. We did similar calculations for radio, newspapers, and the like. This averaging obscures the fact that the number of channels available to the typical urban household is often much higher than the typical rural one. Such differences, however, become less distinct with the increasing reliance on cable, satellite, and the Internet rather than traditional over-the-air transmission and local printing as the primary medium of content transmission.

This is a period of significant growth for the United States. The United States population grew from 181 million to 296 million individuals, from 52 million to 113 million households and fell from an average of 3.29 to 2.63 persons per household during these forty-six years. So if the number of movie screens in existence stayed constant, the number of screens available per capita (or in our case per household) would have declined reflecting a relative decline in supply. That did not happen, however. The growth in the number movie screens outpaced population growth significantly growing from 12,291 to 38,852 screens (more theaters and more screens per theater) typifying a growth of supply characteristic of almost all media for this period. Other patterns of increasing supply included an increased availability in the household (average number of working TVs from 1 to 2.7, radios—including portable and automotive—from 5 to 8). Also people (perhaps a function of affluence, or at least the perception thereof) were buying more magazines and books, although, importantly, fewer newspapers. Our primary sources of data include industrial trade associations, audience measurement firms, academic studies, and government analyses. A full list of the sources and the formulae for calculating supply and consumption for each medium is detailed in Neuman, Yong, and Panek (2012).

Because there is some significant money involved, the assessment of media supply is pretty carefully monitored and vetted as various commercial media outlets keep an eye on the competition. The matter of consumption is somewhat more difficult as individuals rely on their memories to fill out viewer diaries or recall how many minutes they spent "reading a newspaper yesterday." Our strategy was to record all available measures, assess measurement biases associated with each, and, as appropriate, compute a weighted average. Take, for example, the difficult assessment of number of minutes per day of radio listening. John Robinson's well-known twenty-four-hour recall time budget survey reveals an average total of four minutes a day per individual (Robinson and Godbey 1997). The official Arbitron commercial radio ratings estimate an hour and twenty minutes per day per individual. The difference is significant but easily understood. Robinson asks people to recall what they were doing in his survey as they proceed through the day hour by hour. Most radio listening from a bedside, bathroom, kitchen, and car radio is, in fact, a secondary or tertiary activity and unlikely to be mentioned as the primary recalled activity of the hour in Robinson's methodology. Arbitron uses mechanical devices to assess radio listening and diaries listing favorite radio stations that establish very different grounds for recall. Ball State's recent extensive field/ethnographic study following typical media users from morning to night reveals the Arbitron measures are more accurate (Papper et al. 2005). But since we are focusing on trends rather than static metrics, the parameter calculation is less important. Notably, both Arbitron and Robinson report equivalent and steep declines in radio listening over this period.

The patterns Pool uncovered continue in recent years and in many cases accelerate. The broadcast media of radio and television continue to be the primary sources of information and entertainment for the American public. Although radio listening has decreased somewhat as the result of competition from other audio media such as the Walkman and iPod, the supply has increased as a result of more radio stations, more hours broadcast per day, and more radios in the home and car. The growth of television supply is accounted for in a small degree by a larger number of broadcast stations but primarily by the growth of cable and satellite TV. In 2005, 84 percent of American television viewers used cable or satellite delivery as their primary source for television. In 1960 the figure was only 1 percent cable subscribers. Commercial satellite TV was not yet available. In 1960 the typical number of cable channels available on most systems was eight. By 2005 it had grown to 110 channels. Our data tracked the recently exacerbated pattern and the matter of some concern in journalism—the

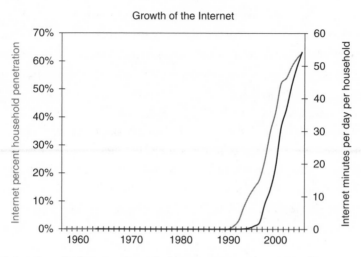

Figure 3.3.   Growth of Internet Household Penetration and Minutes of Use, 1960–2005

steady decline of the daily newspaper (Meyer 2005). In this case we report on the parallel slopes of declining demand and declining supply. Fewer households subscribe or buy newspapers from newsstands. In 1960 there were on average 1.1 newspapers per household. In 2005 the number is .5 per household. And the number of minutes of newspaper reading per day declines from eighteen minutes to seven. The newspaper reading habit is largely restricted to older Americans. Relatively few younger citizens read newspapers at all, so this decline is primarily cohort demographics and is fueling an exodus of youth-seeking advertisers that may intensify the economic decline of the industry (Pew Research Center 2015).

Figure 3.3 illustrates the dramatic and relatively recent growth of the Internet as a home-information and entertainment medium. The left vertical axis records the percent penetration of Internet access (narrowband and broadband combined in this case) and the right vertical axis and the corresponding curve to the right depicts the levels of actual usage. Note that the usage curves in minutes per day reflect the usage for all households including those without Internet access and according zero minutes per day of use. As noted above the daily use of the web in Internet households is closer to one hour and thirty minutes a day. So we find that in just a decade, the Internet has already begun to compete with radio and television in its usage levels. We note, however, that this notion of "competition" will

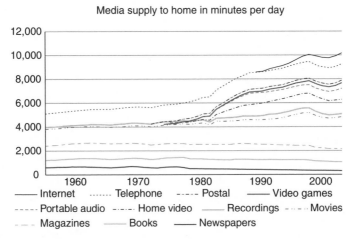

Figure 3.4.   Media Supply to Home in Minutes per Day, 1960–2005

recede, as it will become less clear to the user whether the video they are watching or the music to which they are listening is being piped over traditional media or over the net.

Figure 3.4 provides an overview of the growth of supply of the remaining traditional media and the Internet (excluding radio and television) from 1960 until 2005. The traditional mass media of books, magazines, and movies hold their own against the growing competition in media supply. Recordings hold strong but dip in the past few years, apparently losing out to Internet-based competition (legal and otherwise). The interpersonal media of first-class postal communication and telephonic communication maintain a constant supply. In the case of telecommunication there is a dramatic shift from wireline to mobile communication and an actual decline in wireline household beginning in 2003 and accelerating after 2005. The evolving media of home video, portable audio, video games, and Internet each grow to significant sources of supply beginning in the late 1980s and 1990s. We find that the earlier studies had only begun to scratch the surface of an explosive pattern of growth of supply.

Adding up the contributions of each of the media permits us to address a fundamental issue of the digital revolution—the ratio of media supply to consumer demand (as measured by actual viewing, listening and reading). Our key conclusion is captured Figure 3.5—the ratio of media

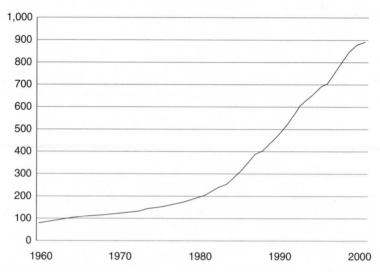

Figure 3.5.   Ratio of Media Supplied to Consumed in Minutes per Day per Household, 1960–2005

supply to media demand over time. Such a curve follows, naturally enough, from the disjuncture of an order of magnitude increase in supply paired with a modest linear growth in consumption. But it is worthwhile to pause briefly to consider the actual metrics we have been at some pains to calculate. Take the ratio of supply to demand in 1960. It is 82 to 1. That represents the number of media minutes available in the typical American household in 1960 divided by number of minutes of actual consumption. It represents the fundamental a metric of choice. And it is a human scale choice. In 1960 there are typically 3.4 television stations available, 8.2 radio stations, 1.1 newspapers, 1.5 recently purchased books, 3.6 magazines, and so on. It is relatively easy for one to know where the country music station, the public broadcasting station, and the rock station are on the radio dial. It is a choice situation that with appropriate chunking, labeling (Miller 1956), habitual behavior, and radio-button-setting (Papper et al. 2005) that can be intuitively managed by the human cognitive system. But if we take the ratio of supply to demand in 2005, we find a very different metric. The ratio is 884 to 1, a little less than a thousand minutes of mediated content available for every minute to be consumed. It could be argued that that is *not* a human-scale cognitive challenge; it is one in which humans will inevitably turn to the increasingly intelligent digital technologies that created the abundance in the first place for help in sorting it out—search engines, TiVo's recommendation systems, collaborative filters (Adomavicius and Tuzhilin 2005). We see this as a historical variant of Beniger's (1986) widely cited "crisis of control" in the nineteenth century. Briefly, Beniger argued that the growth of automated intelligent control systems in transportation and manufacturing were not just a technical artifact but a necessary development as mechanized process speeds and complexity challenged the capacity of individual humans to control them. He cites frequent train crashes in the late nineteenth century resulting from human error as a particularly dramatic exemplar. We may not be confronting equivalent dramaturgy in the realm of media flows, but it represents nonetheless a critical shift in how individuals will negotiate the mediated world. Because it is a gradual process and lacks the obvious urgency of rail accidents, we may underestimate its structural significance.

Media abundance has led from a dynamic of *push* to one characterized here as *pull*. In the traditional one-way broadcast and publishing media, the audience accepts that the newspaper editors determine the headlines that audiences will read and the network executives pick which program is on at 8:00 p.m. Push media. In a world of approximately one thousand choices for any given minute, audience members are less likely to passively wait to see what's on at 8:00. They use the evolving technologies to pull what they want to watch and read.

This logic leads us to look to the search engine and to social media as increasingly coming to define the architecture of media access in the future. Google is widely acknowledged to be an increasingly important factor in connecting potential online customers with online vendors. To date, that has been the prime factor behind the companies celebrated market value of currently more than $100 billion. Facebook is just getting into the game, and as of this writing is valued at a mere $36 billion. What is now coming under appropriate scrutiny is how Google, Facebook, and similar portals to the broadband world will exercise their powers of control in directing attention, cueing fashions in popular culture, and influencing public opinion and commonly held information in the future (Pariser 2011).

As noted above, for practical purposes in this analysis we have treated the Internet as if it were a single medium "competing" with traditional media rather than serving as a digital gateway to all media sources. In recent years Google has reportedly been monitoring approximately 8.5 billion web pages (Gulli and Signorini 2005). The number now is certainly higher. Such a number dwarfs our calculations of words and minutes. As analysis of the "pull media" interface proceeds, we will need new metrics and need to rethink some of our most cherished theoretical tools in the study of media effects. Agenda-setting studies, for example, are typically derived from comparing public opinion with media headlines and broadcast lead stories (McCombs 2004). Two-step flow analysis draws on opinion leadership from personal conversation rather than online mediated recommendation systems (Rogers 1973). Much of media effects theory, upon examination, turns out to be premised on a notion of push media (Katz 2001).

From these findings we conclude that the growth of the web and the digitization of traditional media is not just another step of growth or another technical refinement. Theories of media exposure and media effects need to be reexamined from first principles. We confront not just a few more media channels, but a new media environment and, it appears, a fundamentally new interface between media and audience.

Our analysis which concludes with data for the year 2005 may be among the last that can be appropriately designed around measuring information flow by individual medium of delivery in the Pool tradition and calculating corresponding supply and demand within a single twenty-four-hour period. It is increasingly evident that counting the number of books printed or the number of television broadcasts transmitted on the VHF and UHF spectrum bands is akin to estimating travel trends on the basis of horseshoe manufacturing data. Currently only about 10 percent of American households still rely on over-the-air television broadcasts, with the great majority relying on cable, satellite, and increasingly broadband streaming of on-demand video (Nielsen 2014).

There are hints in the literature that despite the size of the digital cornucopia, what people are attending to now is an even smaller sliver of available information with less diversity of sources and perspectives than at the height of the dominant industrial-age push media of publishing and broadcasting (Hindman 2009). So the fundamental methodological (and theoretical) challenge ahead as Lyman and Varian (2003) foresaw, is not just the measurement of information quantity but a meaningful assessment of information diversity.

The Public Response to the Profusion

The working hypothesis is that the profusion is indeed upon us, but although we may enjoy complaining about too many emails and about "five hundred channels and nothing on," the refined human cognitive capacity for selective attention means that we are not, in fact, overloaded, overwhelmed, and underwater in a flood of incomprehensible information. To explore the diversity of public responses to the evolving information environment, I took advantage of an opportunity to work with Eszter Hargittai and Olivia Curry from Northwestern University to conduct a series of focus group interviews with Americans from across the country. The informality of the interchange among participants, and between participants and the moderator in a focus group setting, helps reveal the nature of people's perceptions and interpretations qualitatively. Rather than responding by selecting from among a limited set of questionnaire item options, the natural language of the discussion permits the identification of ambivalence or ambiguity or, at times, emphatic responses among participants. Focus group research is particularly useful in identifying unanticipated responses to the subject matter at hand and in sharpening hypotheses for more systematic experimental and survey research down the road.

Focus groups, however, are not designed to derive representative samples and project quantitative parameters to larger populations. Although the demographic characteristics of a participant may be indicated in the research report illustratively, the typically smaller focus group samples are not appropriate for assessing differences in attitudes or behaviors by demographic categories. Because of the relatively public character of group participation, focus groups are not ideal for inquiring about socially sensitive or potentially embarrassing domains of human activity. So, for example, a study of the use of pornographic web content or illegal online gambling might be better suited for one-on-one in-depth interviews. But given our interest in strategies for finding news, entertainment, and gossip in public media and online social networks, the focus group technique was particularly promising. The following is drawn from our analysis of these interviews published independently as "Taming the Information Tide: Americans' Thoughts on Information Overload, Polarization and Social Media" in *The Information Society* (Hargittai, Neuman, and Curry 2012).

We conducted seven focus groups with nine to twelve participants each over a period of three days in October 2009, at CBS Television City, a state-of-the-art focus group research facility in Las Vegas. The location was chosen because of its ability to draw together a diverse group of participants from across the country. Individuals with little or no experience online were excluded. A full report on the study is available in Hargittai, Neuman, and Curry (2012).

The core questions we posed to focus group participants were the following:

(a) How do you keep up with what's going on in the world?
(b) How do you feel about the amount of information out there?

We began the focus groups with the moderator asking everyone present to comment on how they keep up with what is going on in the world, and what their strategy is for dealing with the information. We tried to get both practical information such as what resources people use for news consumption, and emotional information such as how people feel about the plethora of choice available to them, out of participants in all groups. We guided the discussions by asking questions based on participants' responses or asking the same question of various individuals. Occasionally, we jump-started a new topic by doing another round-robin in which every participant answered a question. Often the moderator would pose a question to the group at large and wait for any participant to respond, such as "Are you guys much smarter than you used to be because of all this information coming to you?" The sessions lasted fifty-eight minutes on average, yielding transcriptions totaling just under ninety thousand words. We had about equal representation of gender. Ages ranged from people in their twenties to over sixty, with the majority under forty. The group was relatively well educated with almost half possessing a college degree and close to a fifth with a postgraduate degree in contrast to just over a quarter who had some college experience but no degree and just a handful of people with only a high school degree. The participants came from across the United States with almost a third coming from the Northeast, a similar number from the South, just under a quarter from the Midwest, and the rest from the West. Almost half of the participants lived in suburban areas, many others in urban areas, with just over 10 percent from rural areas.

In response to our central questions about satisfaction with "keeping up" and concern about being overwhelmed, only a few of our participants noted any unease with the new media environment or sense of being overwhelmed. The overall tone of the discussions was largely positive and enthusiastic. Instead of feeling burdened by choice, many participants expressed appreciation of the freedom it brought, especially the range of information available online. Respondents of all age ranges used a wide variety of technologies, and many of them owned smartphones and relished the accompanying mobility. The identifiably negative responses to the new media environment took three forms: (1) frustration with the sensationalistic and partisan pronouncements increasingly found on some cable news channels; (2) annoyance at the distracting trivialities associated with social network sites such as Twitter and Facebook; and (3) a general sense that with the diversity of professional and nonprofessional voices online, it is hard to know whom to trust. One participant, a college student explained:

> [I am] not really sure where to turn for the most accurate information. But I guess I have more of a negative association with it because I tend to more shut down and not really know where to turn, so I don't really turn anywhere.

Another student, this one from the urban South, explained her feelings this way: "I do find it overwhelming. I don't like it. When I try to find something or research something, I never know what is the accurate information, like she said," referencing the other student. When the moderator asked her to put this sense of being overwhelmed into emotional terms, she replied: "It's frustrating." In sum only eleven from among seventy-seven participants acknowledged any sense of being overwhelmed by the volume of information available in the new media environment even though we probed this question repeatedly in all groups. Others tended to express nothing less than delight when discussing the ways in which they used media to find information, and many more simply seemed neutral on the subject or had mixed feelings that balanced out in the end. Many participants acknowledged the high volume of information but did not find it problematic. A banker in his twenties from the urban South was unsurprised by the enormous volume of information disseminated in the current media environment. He said, "I think there's media everywhere, and you can't really get away from it." However, instead of finding this overwhelming, he replied: "No, I think it's good. I think it exposes people to different ideas and attitudes." A counselor in his thirties from the urban South said: "There's so much information; it's very helpful at times. You just have to decipher what is a good source and what's not, and I think once you get that down, it's fine." A woman in her thirties from the suburban South who worked as an IT department database manager, said: "I love it. You know, I have the Internet on my phone. I have Internet at the house, at work. We have satellite television. I love being able to access any information whenever I want." Another participant chimes in:

> You can pull [the news] up whenever you're ready, and you can look at it. So if you miss it at the hour block that it comes on, you can always go back, you know, any time, and pull it. So, I mean, for me, I love it. [pause] On the go.

In comparison with TV news, the online news environment seemed to generate almost uniformly positive responses. A physician in his thirties from the urban South thinks we are better informed in the Internet age:

> Nowadays you can jump on the Internet read in German, you know, French or whatever else you want. So you're definitely better informed with an extra sort of different point of view from that side. And also I would say faster informed, you know, something happens in Southeast Asia you find out about [it] right away.

A woman in her forties working in customer service in the suburban South agreed, saying, "There's certainly more tools and information out there available for everyone to have easier access to."

The rationale for participants' relative enthusiasm about the Internet as a resource for news in comparison with traditional media appears to be the issue of personal control. A diversity of sources and a cacophony of video, audio and textual streams online require audience members to "pull" what they want rather than simply sit back and allow the media professionals to decide what is important and "push" the headlines out to passive audience recipients. Pulling involves occasional errors and takes effort and some evolved skill at manipulating the digital environment. All but a few of our participants, it appears, were motivated to invest a bit of effort and get over the skill "hump" to locate and manipulate routinely the information they wanted and needed with some success. It is likely that the nature of our methodology and sampling underrepresent those who are financially or experientially marginalized from the digital domain. As a result, although we are not equipped from this study to estimate the size of the strata still marginalized by limited skills or limited technical access to online resources, it remains an important issue for analysis and public policy (Hargittai and Hsieh 2010).

Findings from our focus group interviews suggest that Americans are getting their news from an increasingly diverse set of sources and are actually quite pleased about it. They complain from time to time about too much fluff and sensationalism, but that may not be a recent development. The Internet is seen as a helpful source of information about current events, while television news, particularly cable news, attracts more criticism because of sensationalism and the constant stream of repetitive stories. Only a scattered few participants expressed a sense of being overwhelmed by the volume of information or the type of media they encountered.

Our participants acknowledge witnessing a fundamental shift in the interface between the media environment and the individual audience member moving from the characteristic "push" of a fixed broadcast schedule and daily news headlines to a "pull" dynamic characterized by an online search, although they don't use that language explicitly. Furthermore, between push and pull they acknowledge an intermediate form of interaction characterized by recommendation engines, collaborative filtering, short messages on current events, and email attachments from the mainstream media that could be characterized as the electronic equivalent of the classic two-step flow.

Where once Americans might have gossiped over the back fence or sitting around the cracker barrel at the general store, online gossip and commentary through social network sites is now all the rage. Twitter and Facebook get mixed reviews with oversharing by some as a subject of particular scorn and humor. Online social networking is relatively new, and, as is often the case in social diffusion, it may be some time before the norms of appropriate use and skills at filtering start to stabilize. Currently those who are technically savvy report setting up their media usage in a way that represents their preferences, while those less savvy simply tune out completely.

On the issues of fragmentation and polarization, the focus groups did not reveal evidence of individuals retreating into a partisan silo or "daily me" of one-sided information. On the contrary, reinforcing recent survey and experimental research (Garrett 2009), our participants indicated an interest in understanding more about how "the other side" felt and the logic of their arguments.

The casual use of the concept "information overload" for consumers of traditional mass media and the increasingly prevalent digital media may be misleading and would benefit from some conceptual clarification. These focus group participants musing in 2009, for the record, expressed near unanimous enthusiasm about the new media environment. Those in rural areas with only dial-up access were looking forward to getting hooked up to broadband and getting more reliable cell phone service. When frustration is mentioned, it typically takes two forms: (1) individuals have not yet perfected skills at mastering the searching and filtering that enables them to find what they want; and (2) they find much content to be sensationalistic and lacking in seriousness. This may be a manifestation not at all new in media behavior represented by the oft-told story—I watched three hours of gossip and fluff on TV last night. It was awful. I plan to watch another three hours tonight.

4

Pondering Polysemy

Normally, whenever we hear anything said we spring spontaneously to an
immediate conclusion, namely, that the speaker is referring to what we
should be referring to were we speaking the words ourselves.

—CHARLES OGDEN AND IVOR RICHARDS (1923)

The notion that we can express to our deaf selves, let alone communication
to any other human beings, blind, deaf, insensate as they are, a complete
truth, fact or sensation is arrogant folly.

—GEORGE STEINER (1968)

Codes are by their very nature full of gaps, inconsistencies and are
subject to constant change.

—WENDY LEEDS-HURWITZ (1993)

MARSHALL McLUHAN has contributed to communication theory in
some interesting ways, frequently provoking us to think freshly about
media and messages. His notion that "the medium is the message" and his
popularization of Canadian economist Harold Innis's observations about
media systems continue to inspire students of the field (Innis 1951, 1952).
But for me, one of his most intriguing observations is an elaborated example in his lesser-known *The Gutenberg Galaxy* (1969). The point he was
trying to make, as I understand it, was that individuals need to be "trained,"

that is, to be familiar with cultural conventions of viewing to view a motion picture properly. I found his comments on "nonliterate" movie viewers less than convincing and his comment on "natives" rather awkward even for the era in which this was published. But I find his example of a failure of communication and polysemy utterly and enduringly fascinating. He writes as follows:

> Why non-literate societies cannot see films *or* photos without much training . . .
>
> Let us turn to a paper by Professor John Wilson of the African Institute of London University. For literate societies it is not easy to grasp why non-literates cannot see in three dimensions or perspective. We assume that this is normal vision and that no training is needed to view photos or films. Wilson's experiences arose from trying to use film in teaching natives to read: The next bit of evidence was very, very interesting. This man—the sanitary inspector—made a moving picture, in very slow time, very slow technique, of what would be required of the ordinary household in a primitive African village in getting rid of standing water—draining pools, picking up all empty tins and putting them away, and so forth. We showed this film to an audience and asked them what they had seen, and they said they had seen a chicken, a fowl, and we didn't know that there was a fowl in it! So we very carefully scanned the frames one by one for this fowl, and, sure enough, for about a second, a fowl went over the corner of the frame. Someone had frightened the fowl and it had taken flight, through the right hand, bottom segment of the frame. This was all that had been seen. The other things he had hoped they would pick up from the film they had not picked up at all, and they had picked up something which we didn't know was in the film until we inspected it minutely. . . . The film was about five minutes long. The chicken appeared for a second in this kind of setting.
>
> Question: Do you literally mean that when you talked with the audience you came to believe that they had not seen anything else but the chicken?
>
> Wilson: We simply asked them: What did you see in this film?
>
> Question: Not what did you think?
>
> Wilson: No, what did you see?
>
> Question: How many people were in the viewing audience of whom you asked this question?
>
> Wilson: 30-odd.
>
> (36–37)

I find this exposition particularly telling for three reasons. First, it is a pretty dramatic example of polysemy, in this case miscommunication, or perhaps more accurately unintended communication. After five minutes of painfully slow demonstrations of removing pools of standing water to prevent mosquito infestation, the audience declares that they have witnessed a film about a chicken, in this case the foot of a chicken observable at the extreme corner of the frame for approximately one second of a three-hundred-second film. The creators of the film who had presumably seen it many times were so puzzled that they had to inspect the film frame by frame to find the chicken. Polysemic communication indeed. Second, it would appear that for this audience of farmers, chickens are what they know and what is relevant to their livelihood and survival. The business about mosquitoes, apparently, had not yet been made relevant. What a wonderful example of resonance with audience expectations and identity. Third, the communicators were surprised, befuddled, and frustrated by their failure to communicate successfully. They describe themselves as more or less dumfounded by the answer they received when they asked their patient and cooperative audience what they had just seen. It is yet another vivid example of the fundamental semantic fallacy that senders strongly presume messages are received as senders intend and as the senders themselves interpret a complex symbolic stream. The Africanists and McLuhan himself imply that the "natives" are in fact interpreting the film "incorrectly" and need to be schooled in the correct way to interpret that film as well as films in general.

Pondering Polysemy

The distinguished anthropologist Clifford Geertz (1973, 2000) has pondered polysemy. He describes such an enterprise as being at the very core of what ethnographers aspire to accomplish. Borrowing terminology from the Oxford philosopher Gilbert Ryle, he characterized this task as "thick description." The term stuck and is now firmly associated with Geertz's eminent repute. Thick description is "sorting out the structures of signification" and "rendering mere occurrences scientifically eloquent" (Geertz 1973, 9, 28). It is a semiotic undertaking. "Believing, with Max Weber, that man is an animal suspended in webs of significance he himself has spun, I take culture to be those webs, and the analysis of it to be therefore not an experimental science in search of law but an interpretive one in search of meaning. It is explication I am after, construing social expressions on their surface enigmatical" (5).

Geertz develops a series of examples of the delicacy of trying to interpret culturally embedded communication. Particularly memorable is the story of Cohen the trader in the remote Moroccan mountains in 1912 where the French Foreign Legion had just begun to try to assert authority. Geertz does not explain his motives; he simply jumps into a reportage from his field notes as the story was conveyed to him in 1968. It is a bit of colonial history, a comedy of misunderstanding bordering on farce as Cohen is robbed by one tribe of Berbers and with friends from another Berber tribe steals sheep in the dead of night for revenge, ultimately settles his differences under the grounds of the traditional 'ar only to have the disbelieving French find Cohen with his sheep and assume he had stolen them. So they take the sheep and throw Cohen in jail as a spy for good measure. For the full details of this engaging situation comedy, see the first chapter of Geertz (1973). It is a tour de force of miscommunication among the Jews (who incidentally speak fluent Berber), the various Berber tribes, and the newly arrived French military—a dramatic multiplicative demonstration of the disjunctures between messages as sent and as received in this case across several cultural divides.

Similarly, the renowned communication scholar Everett Rogers (2003) begins his widely cited and now classic study of communication and innovation with a case study about failed communication and resistance to innovation. It is a story about boiled water in the relatively remote San José de los Molinos, three hundred kilometers south of Lima along the desert coast of southern Peru. The health authorities mounted an extensive and intensive campaign to persuade the villagers simply to boil the water for drinking from their polluted water source as a full-scale sanitation system was viewed as impractical. The campaign involved multiple personal household visits to the two hundred families of the village with careful explanations of how boiling kills the germs that plagues so many of the villagers with disease and infection. After two years and many in-home conversations and demonstrations only eleven of the two hundred households adopted this critically important innovation of health behavior. How is such a failure of communication possible? Rogers explains. It has to do with the symbolic meaning of "cooked water" in local culture. Only persons who are already sick drink cooked water, and that is for ritual rather than health-related reasons. It was simply counterintuitive in this cultural setting for a person not yet sick to behave like a sick person. Look at the health message from the point of view of the local village housewife: "[She] does not understand germ theory. How, she argues, can microbes survive in water that would drown people? Are they fish? If germs are so small that they cannot be seen or felt, how can they hurt a grown person? There are enough real threats in the world to worry about-poverty and hunger—without bothering about tiny animals one cannot see, hear, touch, or smell. [Her] allegiance to traditional village norms is at odds with the boiling of water. A firm believer in the hot-cold superstition, she feels that only the sick must drink boiled water" (Rogers 2003, 4).

Further: "Most poor families saw the health worker as a 'snooper' sent to Los Molinas to pry for dirt and to press already harassed housewives into keeping cleaner homes. Because the lower-status housewives had less free time, they were unlikely to talk with [the health worker] Nelida about water boiling. Their contacts outside the community were limited, and as a result, they saw the technically proficient Nelida with eyes bound by the social horizons and traditional beliefs of Los Molinas. They distrusted this outsider whom they perceived as a social stranger" (5).

One wonders how it was possible for professional fieldworkers and their supervisors over a two-year period not to recognize that their germ-theory approach was not working. But they did not adjust their persuasion strategy; they just kept at it with a resultant and stunning lack of success. Rogers's strategy in opening his book with this detailed case study was to draw his readers' attention to the importance of the social context of communication and the complex attributions of motivation between speakers and listeners.

McLuhan's, Geertz's, and Rogers's stories, of course, speak of relatively exotic cultural contexts of polysemic miscommunication. Surely, the sophisticated persuasive media messages of commercial advertisers are more successful at getting a clear message across and avoiding the countervailing, and misdirecting polysemic diversity of perceptions. Actually not so. It is perhaps the dirty little secret of the industry of professional persuaders. They are not very good at it. Research pointing to such conclusions is painfully uncomfortable so it is dismissed as flawed research or most often simply ignored. Two stunning examples—one from a billion-dollar six-year National Youth Anti-Drug campaign in the early 2000s and another, a meta-analysis of 389 consumer product television commercial campaigns (most about a year in length) from the 1990s also representing about a billion dollars of advertising efforts. The first, the health campaign, had no measurable effect whatever on its target audience and some modest evidence that the campaign may have actually increased drug use. The second, the compendium of TV commercials, demonstrated a statistically significant increase in sales in only 17 percent of the cases. In both cases the research subjects fully acknowledge seeing the persuasive messages, indeed many times over many months. It just was not received and interpreted as salient or even relevant, and in all but a small fraction of the cases not connected to any behavioral response. Let's take a closer look.

From 1998 to 2004 the White House Office of National Drug Control Policy in partnership with the nonprofit Partnership for a Drug-Free America contracted with Ogilvy and Mather and a number of other advertising agencies to oversee a National Youth Anti-Drug Media Campaign (Hornik et al. 2008). Congress allocated nearly a billion dollars for the effort. The announced purpose was to educate and enable America's youths to reject illegal drugs, to prevent youths from initiating use of drugs, especially marijuana and inhalants, and to convince occasional drug users to stop. The campaign represented a wide-ranging social marketing effort focusing on youths aged nine to eighteen years, their parents, and other influential adults. The campaign mobilized a diversity of media channels including local, cable, and network television, radio, Internet, magazines, and movie theaters. It was an immense undertaking. The design called for an average of 2.5 ad exposures per week primarily on TV over the course of the campaign—that is about eight hundred ads for each of twenty-five million young people in the target audience. In the final years of the campaign it increasingly focused on the attempt to prevent marijuana use among teens. The effort including one visually powerful poster that included a picture of a plumber's pipe with the caption indicating that is what "pipe" means to a sixth grader. Beneath that image was another, in this case a drug pipe with the caption—the meaning of "pipe" for a seventh grader. Another ad had a paper sandwich bag and a plastic bag of marijuana with similar captions (Office of National Drug Control Policy

2000). The problem is that the impression the youngsters apparently derived from these marketing messages was that many older kids were involved with drugs, especially the cool ones. The researchers discovered a boomerang effect, a statistically significant highly level of reported marijuana use for those with the highest levels of exposure to the advertising campaign. That particular finding, not surprisingly, got the agency in a lot of hot water. But the predominant finding for the campaign as a whole was simply no correlation at all between exposure and any drug-related behaviors (Hornik et al. 2008). Message received eight hundred times. Message ignored or misinterpreted eight hundred times.

Commercial advertisers have a difficult time assessing the effectiveness of their ads. As often as not when they increase their investment in advertising of a brand, corresponding sales will either stay the same or actually decrease (Ashley, Granger, and Schmalensee 1980). Other factors like price fluctuations, in-store promotions, or the marketing behavior of competitors complicate the analysis. The expensive but highly reliable technique for measuring the effectiveness of television ads is called the split cable methodology known in the industry as Behavior Scan, a commercial service of Information Resources based in Chicago (Lodish et al. 1995). The technique pairs up identical neighborhoods in various locations around the country that independent cable systems and run ads on one and none at all on the other. Brand purchases are measured by automatic UPC statistics at local grocery, pharmacy, and department stores in the corresponding neighborhoods. Most of the comparisons in the review are yearlong advertising campaigns of various intensities. The analysis of all available studies to date in the mid-1990s by Leonard Lodish and his associates at the Wharton School concluded that only in 17 percent of the cases did the commercial campaign result in a sales increase statistically distinguishable from zero (using in most cases relatively large sample sizes). The statistic is stunning. Like the antidrug campaign, these are not one-shot field experiments, but long-term over-time studies of repeated exposure to advertising messages. It is a study of representative, real-world commercials from companies like Campbell Soup, Colgate-Palmolive, Best Foods, Frito-Lay, General Mills, Kraft, and Pepsi—attractive if not overtly sexy youthful role models consuming desirable products with evident satisfaction, plain folks testimonials, kitchen dramas, celebrity testimonials, colorful text overlays reminding viewers of desirable product attributes—the most sophisticated large-scale corporate marketing money can buy. But if this carefully conducted Wharton study is to be credited, it is money utterly wasted six times out of seven—tens of billions of dollars a year. *Successful propagandizing is harder than it looks.* This insight is counterintuitive and in fundamental opposition to the paradigmatic foundations of communication effects research. Propaganda is designed to create a singular worldview. The fundamentally polysemic character of most human communication is powerfully antithetical to such singularity.

I could fill this tome from beginning to end with diverse examples of the vagaries of communication and the fragile connection between sender and receiver. What strikes me as puzzling is why so few of the efforts of communication researchers are devoted to understanding the systemic patterns of miscommunication as well as the more traditional patterns of successful communication and persuasion. I have posited that roots of this defining predisposition of the communication research tradition may lie historically in the fixation with the power of propaganda during and following World War II. But it may be worthwhile to explore a little further why such a predisposition to minimize the potential significance of polysemy and miscommunication has been sustained so strongly for the following decades of scholarship.

Two Theories

In my view there are two candidate theories for explaining the relative inattention to polysemy in communication research. The first is that communication researchers as human observers are subject to the very same cognitive biases and attributional errors they may study in the communication of others. I noted above the fundamental semantic fallacy that communicators routinely assume that a message was interpreted by listeners as the sender intended. Communication researchers are surprisingly unreflexive about how communication theorizing may itself be subject to the evolved cognitive biases of human perception. The noted communication theorist Klaus Krippendorff of the University of Pennsylvania published a brief paper on the topic in 1994 entitled "A Recursive Theory of Communication," but it is only infrequently cited and has not been as widely read as it deserves to be. He begins his piece:

This is an essay in human communication. It contains "communication," mentions and is, hence, about communication, but, what is important here yet often overlooked in other essays, it also is communication to its readers. This exemplifies that no statement, no essay and no theory can say anything about communication without also being communication to someone. Among the scientific discourses, this is an unusual fact—fact in the sense of having been made or realized—and I suggest it is constitutive of communication scholarship that its discourse is included in what it is about and, therefore, cannot escape the self-reference this entails. If I had to formulate a first axiom for communication research I would say that to be acceptable,

Human communication theory must also be about itself.

Although this seems obvious, I understand that many writers on the subject do not recognize this axiom and talk about communication as if their own use of language had nothing to do with communication. I suspect the reason for this omission lies neither in bad intentions nor in an inability to understand this phenomenon, but in the unquestioned commitment to certain ontological assumptions and vocabularies that in effect prevent these scholars from facing themselves in their own constructions. (1994, 78)

I find his observation persuasive and might add that the relatively few citations of the paper in the decades following may indicate that scholars continue to avoid facing themselves in their constructions.

The second theory about the relatively weak role polysemy plays in media effects research may be an even more powerful explanation. Dealing with polysemy is an extremely difficult challenge to quantitative methodology. It makes studying the potential covariance of exposure to communication with patterns of attitudes or attitude change problematic. When we confront a net attitudinal or behavioral change in the intended direction of a persuasive message, we take that to be the effect. But clearly in virtually all real-world and experimental simulation of real-world communication situations a persuasive message has a distribution of positive, null, and contrary responses among different individuals exposed. Researchers tend to study only net effects. The accompanying null and "reverse" effects are treated as noise and random variation. But null and reverse effects probably reflect important systematic patterns of individuals' varying levels interest and sense of a messages' relevance. Another paradox.

Some readers proceeding this far may have been become increasingly uncomfortable with my assertion that communication theory deals only fleetingly with polysemy. For some readers that would be because the humanistic tradition of communication scholarship focusing on cultural studies and textual analysis take polysemy very seriously indeed. It is, in fact, central to theorizing in these traditions. So let us take a brief look at the celebrated gap between the humanistic and social scientific traditions of communication scholarship. What would it take to bridge the gap? What would result if survey researchers and experimentalists took polysemy seriously?

The Disconnect between the Cultural Studies and Media Effects Traditions

The central argument of this chapter is that the phenomenon of polysemy represents one of the central problematiques of human communication and of its study. The challenge of dealing with the sometimes dramatic variances between the interpretation of a complex message as sent and as received and the variance among audience members of the mass media distinguishes the field of communication from sociology, political science, and anthropology, although each of these disciplines from time to time addresses similar structural puzzles. In this section we take special note of the two distinct traditions that have evolved for struggling with the polysemic character of human communication. The first tradition of social scientific communication research as noted in Chapter 2 has been under way now for about seventy years. The second tradition, which is most commonly referred to as cultural studies, began to crystallize in England in the 1960s and 1970s as a critical reaction and counterpoint to ongoing social scientific efforts is thus about forty years old.

The mutual hostility between the two traditions of scholarship remains as strong as ever and is reinforced by differences of perspective on several dimensions of epistemology and methodology, which we explore in turn in the pages ahead (Willis 1980; Rogers 1985; Jensen 1987, 1990, 2011; Schroder 1987; Lull 1988; Wolf 1988; Livingstone 1998). My view is that the competitive hostility between these traditions is both unfortunate and unsurprising. Academic scholarship is awash in competing and more often than not overtly hostile schools of thought. Sociologist Randall Collins

(1998), in fact, has done an exhaustive study of schools of thought in the field of philosophy over recorded history and demonstrates predictable dynamics as competing paradigms wax and wane and even form strategic alliances. A number of scholars have addressed the divide between cultural studies and media effects, and each seems to have a somewhat different emphasis on what the core differences actually are.

One of the first and enduringly influential essays goes back to the Rockefeller Foundation era and a principal actor in that drama, Columbia sociologist Paul Lazarsfeld. From interaction with a number of fellow European intellectual émigrés, particularly Max Horkheimer and Theodor Adorno, Lazarsfeld contributed to Horkheimer's journal *Studies in Philosophy and Social Science* with an essay entitled "Remarks on Administrative and Critical Communication Research" (1941), which introduced that unfortunate terminological binary. Although Lazarsfeld first published the term *administrative,* his humanistic counterpoint at that time, Theodor Adorno (1969), could not recall whether the term was Adorno's and picked up by Lazarsfeld or actually invented by Lazarsfeld. Later work, particularly that of Todd Gitlin (1978), further polarized the divide. That development was particularly ironic because the original Lazarsfeld essay conveyed both a great sensitivity to the historical and political underpinnings of the cultural studies tradition and a strong endorsement of how each tradition could benefit from cross-fertilization. "Only a very catholic conception of the task of research can lead to valuable results. . . . The writer, whose interests and occupational duties are in the field of administrative research, wanted to express his conviction that there is here a type of approach which, if it were included in the general stream of communications research, could contribute much in terms of challenging problems and new concepts useful in the interpretation of known, and in the search for new data" (Lazarsfeld 1941, 16).

His remarks were tempered by a realistic assessment of the numerous impediments to rapprochement "As long as there is so little experience in the actual cooperation of critical and administrative research, it is very difficult to be concrete" (Lazarsfeld 1941, 14). But in the essay he set up a set of parallel and complementary foundational questions for the two traditions. For the effects tradition: Who are the people exposed to the different media? What are their specific preferences? What are the effects of different methods of presentation? One who uses media of communication is in competition with other agencies whose purposes are different, and thus research must also keep track of what is communicated by others. Finally, communications research has to be aware that the effect of radio, print, or the movie, does not end with the purposive use which is made of it by administrative agencies.

For the cultural studies tradition, a critical student who analyzes modern media of communication will look at radio, motion pictures, the press, and will ask the following kinds of questions: How are these media organized and controlled? How, in their institutional setup, is the trend toward centralization, standardization, and promotional pressure expressed? In what form, however disguised, are they threatening human values? He will feel that the main task of research is to uncover the unintentional (for the most part) and often very subtle ways in which these media contribute to living habits and social attitudes that he considers deplorable.

The unfortunate element of the labels Lazarsfeld invented, as I noted above, is that "administrative" conveys the sense of managerial bureaucrats in thoughtless service of and taking direct orders from self-interested government and corporate executives. Such a practice was not characteristic of the Bureau of Radio Research or of Lazarsfeld's independent scholarship, which could be highly critical and historically grounded (for a notable example, see his work with Robert Merton, "Mass Communication, Popular Taste, and Organized Social Action," 1948). That creative academic empirical research would be meaningfully utilized by media professionals was Lazarsfeld's aspiration, but was not to become a reality (with only a few exceptions) in his era or since. As a result in the years since the publication of his thoughtful and hopeful essay, the terminology today is used only as a dismissive disparagement by cultural studies specialists impatient with work of their empirically inclined colleagues.

Some years later James Carey would introduce another pair of concepts to try to capture the distinctions between the two traditions. Drawing on John Dewey's (1925) *Experience and Nature,* he contrasted the concept of communication as transmission versus communication as ritual:

> The transmission view of communication is the commonest in our culture—perhaps in all industrial cultures—and dominates contemporary dictionary entries under the term. It is defined by terms such as "imparting," "sending," "transmitting," or "giving information to others." It is formed from a metaphor of geography or transportation . . .

The ritual view of communication, though a minor thread in our national thought, is by far the older of those views—old enough in fact for dictionaries to list it under "Archaic." In a ritual definition, communication is linked to terms such as "sharing," "participation," "association," "fellowship," and "the possession of a common faith." This definition exploits the ancient identity and common roots of the terms "commonness," "communion," "community," and "communication." A ritual view of communication is directed not toward the extension of messages in space but toward the maintenance of society in time; not the act of imparting information but the representation of shared beliefs. (1989, 15–18)

The transmission concept treats the movement of "information" like the movement of people or goods—a straightforward and concrete process subject to systematic empirical verification. But the ritual concept gets at something deeper and to Carey more important. Although he cites primarily literary rather than sociological sources, his argument is firmly rooted in the extensive literature of the sociology of knowledge (Mannheim 1936; Berger and Luckman 1966). He argues that rather than the traditional notion that the words of communication are symbols each designed to represent some object or process in reality we should see it the other way around. That is, that language is necessary for and prior to perception— "reality is brought into existence, is produced, by communication-by, in short, the construction, apprehension, and utilization of symbolic forms. Reality, while not a mere function of symbolic forms, is produced by terministic systems. . . . We first produce the world by symbolic work and then take up residence in the world we have produced. Alas, there is magic in our self deceptions" (25–30). His purpose in this exercise, he explains, is to remove the taken-for-granted quality of transmissive communication and identify the phenomenon of human communication appropriately as "a far more problematic activity than it ordinarily seems" (25).

In a widely cited analysis, John Durham Peters (1999) picks up where Carey leaves off and expands the dichotomy philosophically and histori-cally using the terms *dialog* (roughly parallel to *ritual*) and *dissemination* (largely equivalent to *transmission*). I won't address Peters's thoughtful and complex analysis here except to note one particular insight he emphasizes that Carey does not. He starts as Carey does with the observation that communication is problematic—"with all its misfires, mismatches, and skewed effects" (6). But then he turns the terminology on its head and argues that the phenomenon of "miscommunication" precedes and makes possible the prospect of communication itself. "The potentials for disruption in long-distance 'communication'—lost letters, wrong numbers, dubious signals from the dead, downed wires, and missed deliveries—have since come to describe the vexations of face-to-face converse as well. Communication as a person-to-person activity became thinkable only in the shadow of mediated communication. Mass communication came first" (6).

One other terminological distinction emphasizes not different understandings of what communication is, but very different purposes in the motivation of scholarship on communication. The distinction is between the aesthetic and interpretive impulse and the scientific analysis of causes and effects (Schroder 1987). Like Rogers (1985) Schroder describes the distinctive approaches as historically associated with European (interpretive) and North American (social scientific) traditions in the study of social and cultural dynamics more broadly defined. Virtually all of these observers note and take as self-evident that the social science tradition is older, more dominant, better funded than the latecomer cultural studies. The enthusiasts among these observers insist cultural studies must resist the dominant discourse and fight for recognition and independence. Schroder and Rogers in particular emphasize the need for and the possibility of rapprochement and cooperation. One thing is clear. Both traditions are strong and healthy today with recognition within the universities, academic publishers, professional associations, and journals. So the resist-the-dominant-discourse argument is much less compelling at this stage. But an important question remains unclear. Are the two perspectives actually and fundamentally incommensurate?

Some may disagree, but I am going to argue at some length in the pages ahead that the answer is no. The two traditions are not, as some have insisted, different ways of knowing or independent epistemologies. Both aspire to understand human communication as it is historically rooted and to influence the real-world social structures and technologies so they are better able to serve agreed-on ends. Both are empirical although styles of empiricism vary. The fundamental difference is really one of only emphasis. I am going to try to make the case that both approaches are partial and incomplete; that each could draw substantial insight from the other. Briefly, my argument is that *the cultural studies tradition focuses on media texts with a deep interest in their polysemic character but only rarely on audience reception. Social science research focuses on audience reception but only rarely on the character of the message itself and then even more rarely on polysemic character of complex messages. Neither pays much attention to institutional creation of mass mediated messages although both acknowledge its importance.* From my point of view it is an elegant and dramatic reprise of the blind men and elephant. I am unable to discern a fundamentally intrinsic reason for this essential divide in the systematic study of human communication. What is fundamental is that *it is tremendously difficult to address the polysemic character of complex messages, the diversity of audience responses and perhaps exponentially more difficult to study both in concert. But clearly that is what is needed for a systematic and coherent discipline of communication scholarship.* A central theme of this book is that polysemy is one of the defining paradigmatic challenges communication research must confront. In my view that means it is not sufficient just to study the central tendency of audience reception. The variance in reception is equally important. The central hypotheses of communication theory must focus on the conditions under which each tends to vary. Let's take a look at the sometimes warring camps and the prospects for a demilitarized zone between them.

Cultural Studies

Cultural studies had no John Marshall and Rockefeller Foundation to nurture its growth, but it certainly had prolific, committed, and charismatic founding scholars. And it had an institutional base in the Centre for Contemporary Cultural Studies at the University of Birmingham in the UK. The center was founded in the mid-1960s by sociologist and literary critic Richard Hoggart who was well known for *The Uses of Literacy* (1957), a critical review of the weakening of close-knit working-class communities in northern England following World War II and their replacement by a centralized and manipulative mass-mediated culture. Perhaps most influential was Stuart Hall, who joined the center early on and became director in 1968. Raymond Williams, professor of drama at Cambridge University and author of the seminal *Culture and Society* (1958), was also active in the founding of the center. All three had working-class or mixed social class upbringings and sensibilities and all, especially Hall and Williams, were active in the politics and publications of the British Left.

The center was always small (a staff of three and perhaps twenty graduate students) and struggling for funding, and it was ultimately disbanded amid predictable controversy in 2002. But its influence on the evolution of cultural studies in England and globally is hard to overemphasize. The earlier work emphasized the Gramscian neo-Marxism of the 1960s under Hall with a focus on the hegemonic power of the media, particularly the conjuncture of popular culture and the reproduction of social power structures in England. Later work moved beyond class-based power relations focusing on media depictions involving race and gender.

I am going to argue there are four foundational elements of the cultural studies perspective that motivate the energies of researchers in this tradition and their critique and ultimate rejection of what they take the be the dominant and misguided social science paradigm of media effects. Reviewing these four will take us a reasonable distance toward understanding the tensions between the perspectives and the prospects of a peace process.

1) A Focus on the Text. Cultural studies, most notably in the early work of Hoggart and Williams, evolved out of literary studies, a tradition with a rich appreciation of and interest in the multiple meanings in text. And, again reflecting the literary tradition, there is a great deal of attention to the interpretations of the analyst and speculations about the author, social and cultural contexts of authorship, and possible authorial intentions, all worthy and important endeavors. There are data under analysis as there is in the social science tradition. The data are the text. But only occasionally are there any data about audience sense making and the prospective multiplicity of senses audiences may make. Williams's *Culture and Society,* for example, reviews the works of forty literary authors spanning two centuries with a provocative neo-Marxist thematic about how conservative thinkers had appropriated the very definition of what we take to be culture. The concept of audience is central, but it is an audience as imagined by Williams. The notion of audience is even more central to Hoggart's *The Uses of Literacy* (1966), but again it is audience as remembered from the early postwar years of Hoggart's working-class youth in Leeds in the industrial West Midlands of England. He moves attention from literary classics to a study of pulp fiction, popular magazines, newspapers, and movies and the theoretical centrality of the imagined audience remains strong but unburdened by actual interaction with contemporary working-class readers and listeners.

Hall's highly influential paradigm articulated in "Encoding/Decoding" (1980), however, upped the ante by moving the question of whether audiences did or did not make sense of media content by "oppositional" rather than "preferred" readings generally intended by media elites to center stage. He emphasized more subtle forms of hegemonic power rather than older Marxist ideas of a deterministic effect of capitalist propaganda leading to widespread false consciousness. Hall felt a naturalistic empirical assessment of how audiences in context negotiated media texts was a worthy undertaking although he was highly critical of typical survey and experimental quantification. Indeed several members of the center community undertook empirical studies, notably David Morley and Charlotte Brundson (1999) in their studies of the BBC's *Nationwide* documentary series. The growing recognition of empirical, primarily ethnographic analyses of polysemic audience responses was signaled when Hall's 1980 compendium of center research from the preceding decade was released with an entire section devoted to "Ethnography at the Centre." But, unfortunately, although highly celebrated, studies that actually move beyond the text to the audience in this tradition were extremely rare. Morley and Brundson were able to study responses to only two of the nearly four thousand Nationwide programs aired and were able to interview only limited nonrepresentative samples of college students and adult education students in small groups. In the compendium of center work in 1980, of the five chapters in the section on ethnography, only two involve actual ethnography and only one of them deals with the media (the other focuses on a scout camp). There were several notable and extraordinarily influential ethnographic studies of popular media at this time (although most were conducted quite independently of the center). Among them was Janice Radway's *Reading the Romance* published in 1984. Radway, a young and ambitious literature scholar in the American Civilization Department (in the tradition of American Studies) at the University of Pennsylvania, befriended and interviewed a bookstore owner named Dot in the pseudonymous midwestern town of Smithton. Dot had many customers who, like herself, were devoted fans of romance fiction. Radway conducted taped free-ranging focus groups and individual interviews with twenty-one of the most devoted romance readers in the community and persuaded forty-two of the bookstores customers to fill out a small survey. In the tradition of ethnography, there was no reason for these informants to be a representative sample of a larger population because the purpose was to explore the meaning of these novels in the lives of these women, not to calculate global parameters for any specific variables or populations.

The book was an instant hit and became core reading in communication studies and women's studies. Radway, however, had intended her book for a quite different audience in literary studies and to contribute to the debate between the formalist New Criticism school and its detractors. The New Criticism movement made a special point that one should study just the text itself and was dramatically dismissive of those who gave attention to the presumed intentions of the author or readings of an audience. Radway explains in an introduction to a later edition of the book that at the time she was writing she was simply unaware of the Birmingham School but then not quite ten years later was delighted to discover it and she had no objection to the fact that her work was "hijacked" (her word) for other disciplinary purposes (Radway 1991).

Another instant hit was Ien Ang's *Watching Dallas* published in 1985. As a doctoral student at the University of Amsterdam, Ang simply put an ad in a Dutch woman's magazine asking for volunteers to write her a note on why they liked or disliked the American prime-time soap opera that was being rerun in Holland (and around the world) and causing a bit of a cultural stir at the time. In response to the small ad, forty-two women described their thoughts and feelings in varying levels of detail, providing the more than sufficient raw material for a provocative and thoughtful dissertation and book on the cross-cultural, commercial, and gender issues raised by this interaction of audience and text. This study, along with the multinational study of *Dallas*'s audiences by Liebes and Katz (1990), became almost instantly canonic in the domain of cultural studies.

Small and casually compiled samples, but richly detailed, freewheeling recorded conversations with audience members are analyzed with a sympathetic ear and a bountiful array of well-grounded theories about historical context, elite manipulation, power relations, gender, and identity. Is there a problem? Actually, yes. The difficulty is that although the literature of cultural studies has grown large, studies like these that involve interaction with actual audiences, although celebrated, are extremely rare, about as rare as examples of social scientists who pay much attention to the texts they use in effects studies. A close reading of a text is a difficult endeavor. The analysis of the diversity of close readings of audience members, it appears, is sufficiently daunting to make its undertaking exotically infrequent.

2) An Emphasis on the Holistic. In their efforts to accumulate a series of lawlike and generalizable statements about the nature of human communication, social scientists reduce complex and contextualized interactions to their basic elements and decontextualize them. This sort of theorizing is by its very nature reductionist. The reaction of one or several individuals to a specific communication in a specific context is paradigmatically beside the point. Social scientists seek generalizations about human reactions to a class of communication, historically, as we have seen, the success or failure of propaganda to persuade. This perfectly reasonable undertaking, however, drives humanists to distraction. In their view such an approach encompasses the following:

- It is reductionist.
- It treats the richness and subtlety of human communication in a mechanical way, much like the notions of force, mass, and velocity in Newtonian physics.
- It is based on a naive pretense that the values and cultural and political commitments of the social scientists do not color their choice of research topics and research methods.
- It decontextualizes.
- It recodes complex phenomena into variables and their quantified relationships.
- And, of course, it pays almost no attention to the nature of the texts involved.

Yes. Pretty much guilty on all counts. Most members of the cultural studies community have little or no objection to the scientific study of public health dynamics or economics where quantitative assessments and generalizations about the relationships of, say, diet and disease or supply and demand do not violate the fundamental character of the phenomenon at hand. A recorded death is an unambiguous datum in the study of public health. In economics, a dollar is a dollar. The view of the community of humanistic communication scholars, simply enough, is that the traditional social scientific methods are not well matched for the study of the phenomenon of human communication. I examine whether with some effort they could be better matched.

3) A Critical Perspective. The normative valence of the cultural studies tradition is deep and strong. It has strong roots in Marxism, neo-Marxism, and a critique of capitalism, commercialism, colonialism, neocolonialism, neoliberalism, and the inequities and stereotypes associated with gender, gender preference, and race. Following Foucault, it is frequently asserted that all cultural activity and communication is inherently infused with the political. Accordingly, the analysis of elite and popular culture is an unavoidable political act. In turn, the social scientific pretense of value-neutrality is seen as naive and or some cases disingenuous. As noted above, Hoggart, Hall, and Williams had working-class backgrounds (although Hall's father eventually worked his way up through the ranks of the United Fruit Company in Jamaica), and they used the experiences and identities of their youth frequently and prominently in their critique of the role of commercial mass culture in reproducing inequities.

Larry Grossberg captures the spirit of political engagement in the introduction to his *Cultural Studies in the Future Tense* (2010), entitled "We All Want to Change the World." He reviews and rejects a series of definitions of cultural studies in the introduction and then presents his own as follows: "Cultural studies . . . is concerned with describing and intervening in the ways cultural practices are produced within, inserted into, and operate in the everyday life of human beings and social formations, so as to reproduce, struggle against, and perhaps transform the existing structures of power. . . . It investigates how people are empowered and disempowered by the particular structures and forces that organize their everyday lives in contradictory ways. . . . Cultural studies is concerned with the construction of the contexts of life as matrices of power, understanding that discursive practices are inextricably involved in the organization of relations of power" (8). A little later on he notes that within cultural studies and especially at the seminal Centre at Birmingham there is an insistence on "a fundamental refusal of the demand, so powerfully enforced in the academy, that one bracket one's passions, one's biographical sympathies, and one's political commitments, in the name of a (spurious) intellectual (read scientific) objectivity. Cultural studies knew . . . that without such investments in the world, in our lives, and in the lives of others, there is no desire, need for, or possibility of knowledge. Knowledge always depends on a visceral relevance" (18).

Social scientists have a different mantra for dealing with the difficult complexities of how the personal values of the human investigator may impinge on their conduct of research. The search for the confirmation of a hypothesis in social sciences has to be independent of what the investigator would prefer to find, otherwise what would be the point of the exercise. It is perfectly reasonable for an investigator (with appropriate public acknowledgment) to select research topics based on their values and personal interests. But the goal is to be open to disconfirmation. Disconfirmation is a valuable as confirmation and as often as not will send the researcher creatively back to the drawing board to figure out what dynamics are at work in the data at hand. Grossberg is skeptical of the typical working social scientist's ability to remove the effects of their values from the way they design and interpret their studies thus his reference in the quotation above to "spurious scientific objectivity." He may have a point about how some media effects researchers conduct their business, but that does not necessarily undermine the importance of the underlying scientific principle.

4) A Historically Grounded Perspective. I noted at some length about how the early experimental and survey work on propaganda was deeply rooted in the historical circumstances of World War II and the Cold War that followed. With the exception of Harold Lasswell and John Marshall himself, the leaders of the Rockefeller initiative saw the historical grounding of their work as obvious but as incidental. They were scientists. The war may have drawn their attention to possible pathologies of mass communication and public opinion, but they were seeking to understand the fundamental dynamics of human psychology and collective behavior—an understanding that would hopefully transcend the immediate historical environment and provide useful insights for generations to follow. They would not hesitate to acknowledge that broadcasting technologies represented something new and important, but the question at hand would be seen as the interaction of a new technology with the enduring psychological properties of individual and collective behavior.

Such a perspective struck the humanists as misguided and naive. In tandem with the notion of a critical and politically engaged scholarship, the notion of historically grounded intellectual work fundamentally defines the humanistic traditions of the cultural studies enterprise. With strong roots in the Marxist conception of historical stages, Raymond Williams's seminal *Culture and Society* and *The Long Revolution* were cultural histories (Williams used the phrase *social history*). Likewise Hoggart's central theme focused on postwar changes in working-class culture and mass media. And Stuart Hall, drawing on Marx, the Frankfurt School, and particularly Gramsci, was explicitly historical (and, of course, political) in his influential role as center director and productive scholar. Colonial history and conceptions of modernism and postmodernism continue to central organizing themes in this literature. So more than just a potentially dry academic historicism, as in Grossberg's memorable "we all want to change the world" it is a politically engaged historical turn.

It becomes possible then both to understand why the two traditions talk past each other so frequently and why they also can be seen to represent the iconic blind men stubbornly grasping on to their treasured extremity of the very same elephant. By emphasizing the complexity of the text in social and historical context, cultural analysts can speculate about its meaning and its effects provocatively but in so doing are unable to subject their tests to traditional tests of verification. In turn, the social scientist's proclivity to look only at a few decontextualized and abstracted attributes of communication at a time allows for verification but at the loss of historical and cultural context and the polysemic richness of text under study. Clearly researchers should aspire to understanding communication in context with full capacity for empirical verification. It is frustratingly difficult, which explains why the gap is so seldom bridged. But it is not impossible, and in light of the dramatically changing media environment, it might be an ideal time for a rapprochement.

Stuart Hall, Meet Albert Bandura

To better understand the great divide in communication scholarship and to further explore prospects for convergence, or at least partial convergence that in my view would benefit both traditions immeasurably, let us review the intellectual trajectories of the fields over the past half century.

My thesis is that both the cultural studies and media effects traditions as they developed independently of each other in midcentury went to considerable lengths to break free of their equally independent forebears—cultural studies from literary studies, which systematically ignored popular culture, and media effects from a radical behaviorism that systematically ignored human cognitive and symbolic behavior. The intellectual birthing of a new tradition, of course, involves a youthful rebellion against the parental tradition. The case I attempt to make is that this process of rebellion and transition led to a disjuncture between the models of cultural studies and media effects scholarship, but possibly only a temporary one as the trajectories of both may be, or at least should be, moving toward each other. The key to this prospective convergence, I argue, is the acknowledgment of the fundamentally polysemic character of communication. To make the case I'll contrast the histories of two iconic figures, one in each of these traditions—cultural studies pioneer Stuart Hall from the acclaimed Birmingham School and equally acclaimed experimental psychologist from Stanford, Albert Bandura.

Stuart Hall was born in Kingston, Jamaica, in 1932 to middle-class parents (Rojek 2003; Proctor 2004). His father worked his way up to become chief accountant at the American-owned United Fruit Company, a major employer and political player on the island. Jamaica's racial makeup then as now was about 90 percent black. But there is an elaborately differentiated system of racial hues in Jamaican culture, and by that set of traditions Hall was not black but rather a "brown man." Such racial distinctions would not be made when Hall reached England in 1951, when, according to local norms at Merton College at Oxford, he, as a man of African heritage, was simply black. The colonial traditions were very strong in Jamaica during his youth, and Hall attended schools modeled on the British tradition right down to the blazers, caps, and cricket athletics designed, as Hall put it, to create a black Englishman. He studied modern literature and poetry and the works of Karl Marx and won a Rhodes scholarship to Oxford, where he initially sought to pursue a PhD dissertation focusing on the relationship between Europe and America in the novels of Henry James, drawing on the theories noted literary critic F. R. Leavis and a bit of Marx. So like the others now credited with founding British cultural studies, his starting point was literary criticism. But as he became more active in leftist circles in London in the late 1950s he left Oxford, became a supply teacher in South

London, and ultimately the first editor of the newly created *New Left Review* from 1960 to 1962. Tiring of the doctrinal squabbles among the many factions of the Left, he retreated to teaching, at this point lecturing on popular culture at the University of London, where he caught the eye of English professor Richard Hoggart, who invited him to join the newly established Birmingham University Centre for Contemporary Cultural Studies in 1964 as a research fellow. The center was one of a kind. There was no tradition of popular cultural studies in the UK, so many at Birmingham and elsewhere were critical of the upstart institution undertaking the initiation of what was to become an entirely new field. There was no university support for the center; modest financial support was recruited from various sources, including the sympathetic publisher of Penguin Books. Hall succeeded Hoggart as director of the center four years later and drew the intellectual and political engagement of the center further to the left during its "golden age" for the next decade, after which Hall left to join the Open University in London. The center continued with some ups and downs under new leadership after Hall left in 1979 and was closed down in a controversial move by the university in 2002. At the height of the center's famous decade, Hall wrote a short essay on the fundamentally polysemic character of communication *Encoding/Decoding* (1973), although not published until 1980, it became a highly influential and foundational model for theory and methodology in cultural studies. Interestingly the original paper, a "Stenciled Occasional Paper" in the center's self-publishing tradition, was a critical reaction to the behavioristic work at a sister research center working in the media effects tradition at Leicester and notably its well-known director James Halloran. When the "Encoding/Decoding" paper was published as a book chapter, much of that material was edited out and the original essay did not become fully available until 2007 when the center's complete papers were published in book form (Gray et al. 2007). We take a close look at that paper in contrast with a parallel foundational paper by Bandura.

Albert Bandura was born in 1925 and raised in rural poverty by immigrant parents of Ukrainian and Polish decent in western Canada near Edmonton (Evans 1989; Pajares 2004). Although his parents had limited educational experience themselves, they instilled a respect for education. Young Albert set out to study biology at the University of British Columbia but discovered psychology as something of a fluke in searching for a class that better matched his commuting schedule. It was intellectual love at first sight, and when Bandura graduated in 1949 he asked his adviser where the "stone tablets of psychology were kept" so he could go to graduate school there and was given the answer—the University of Iowa.

Departments of psychology at midcentury were still dominated by somewhat doctrinaire versions of Freudian theory and strict behaviorism, which, drawing on Ivan Pavlov, Clark Hull, John Watson, and B. F. Skinner, dismissed cognition and communication as distractions—the orthodoxy dictated that experimental psychologists should simply observe stimuli and behavior and the resultant operant conditioning. This is important because although Bandura was an enthusiastic student of experimental empiricism, he increasingly moved away from behaviorism of his graduate school days to participate in the cognitive revolution in psychological thinking of the next half century. He joined the department of psychology at Stanford in 1953 and still continues his work there today as an emeritus professor. It is a fascinating intellectual trajectory, actually. His early work on aggression and his development of Social Learning Theory reflected roots in behaviorism and modeling of behavior especially by children of role models in the media and day-to-day experience. This work included the classic Bobo doll experiment that demonstrated that children observing adults hitting the inflatable doll were more likely to behave aggressively in modeled behavior with the doll and with other children. Bandura became increasingly sensitive to the criticism that "social learning" was thoughtless mimicry and changed his labeling from "social learning" to "social cognitive" theory and progressively came to embrace the somewhat awkward term *agentic behavior* to emphasize the importance of human cognition, choice, and agency.

Written more toward the conclusion of his career, the essay "Social Cognitive Theory of Mass Communication," published in 2009, represents Bandura's counterpoint to Hall's "Encoding/Decoding" as a prescriptive methodological overview and master theory of communication process in this case breaking free of psychological behaviorism rather than literary criticism. Both Hall and Bandura from a radically distant starting point, focus on the fundamentally polysemic character of communication and the need to move away from a definition of media depictions as of singular meaning and deterministic effect.

Both essays incidentally address the impact of violence in the media on audiences. Both reflect an openness to the empirical assessment of the diversity of audience interpretation although Hall makes clear his skepticism of the prospect of simple experiments succeeding in such an enterprise. Table 4.1 demonstrates the striking similarities among these two foundational documents written at different times, from different traditions and for different audiences yet highly convergent in defining an agenda for research.

Each scholar started with a difficult battle with their parenting tradition—Hall with literary studies that had no patience for the serious study of popular culture, Bandura with behavioristic psychology that had no patience for the serious study of what individuals think (or say they think) as opposed to their observable behavior. Both scholars won those battles and become celebrated as founders of new and vibrant academic traditions—cultural studies and media effects research. Both men viewed their scholarship as an extension of their fundamental moral values—Hall focusing on the power inequities of modern industrial systems and Bandura on media violence and on the conditions under which individuals turn to violence and morally disengage. Hall was more overtly "political" and "critical" in his style of scholarship, but both men spent comfortable careers as dutiful employees of large academic institutions as each felt he was able to pursue his primary moral concerns. But in the end, both turned out not to have fully broken free from their parenting discipline. Hall's tradition was literary studies, which studied the text and only the text. Bandura's was behaviorism, which studied the behavioral response but typically took the text as a given. Both were deeply concerned about the depiction of violence in the popular media. One can only wonder what might have evolved if by chance they found themselves in adjoining academic offices, rather than an ocean and a continent apart. The Jamaican and Canadian found new academic homes for their careers in Birmingham and Palo Alto. They might have also found intellectual common ground had their trajectories converged. But, as luck would have it, their paths never crossed.

What If Effects Research Took Polysemy Seriously?

It was a hot and humid summer night in Ann Arbor, Michigan. A particularly talented graduate student in political science at the University of Michigan had a hunch. He decided to take a box of surveys home for the evening—the original mimeographed questionnaires with the various textual answers

Table 4.1   Comparing Hall's and Bandura's Theories of Communication Process

	Hall (1980) "Encoding/ Decoding"	Bandura (2001) "Social Cognitive Theory"
Fundamentally polysemic character of communication	"Before this message can have an 'effect' . . . it must first be perceived as meaningful discourse and meaningfully decoded . . . one of the most significant political moments . . . is the point when events which are normally signified and decoded in a negotiated way begin to be given an oppositional reading."	"An extraordinary capacity for symbolization provides humans with a powerful tool for comprehending their environment . . . cognitive factors partly determine which environmental events will be observed [and] what meaning will be conferred on them."
Social context of audience interpretation	"Effects . . . are themselves framed by structures of understanding as well as social and economic structures which shape its realization at the reception end."	"Human self-development, adaptation and change are imbedded in social systems. Therefore personal agency operates within a broad network of sociostructural influences."
Critique of simplistic behavioristic media effects	"The use of the semiotic paradigm promises to dispel the lingering behaviourism that has dogged mass media research for so long."	"Human behavior has often been explained in terms of unidirectional causation . . . [however] people are self-developing, proactive, self-regulating and self-reflecting not just reactive organisms."
Hegemonic power of the media	"The hegemonic interpretation of . . . politics are given by political elites: the particular choice of presentational occasions and formats, the selection of personnel, the choice of images, the 'staging' of debates."	"Televised representations of social realities reflect ideological bents in their portrayal of human nature, social relations and the norms and structure of society."
Case study of violence in the media	"Representations of violence on the TV screen are not violence but messages about violence . . . with its clear-cut, good/bad Manichean moral universe, its clear social and moral designation of villain and hero."	"In televised representations of human discord, physical aggression is a preferred solution, is acceptable, is usually successful and socially sanctioned by superheroes triumphing over evil by violent means."

hastily scribbled in by interviewers during the old-style, in-home, face-to-face interviews. He wanted a closer look. The year was 1958. Once survey answers were coded numerically on IBM cards, nobody ever went back to the "raw" questionnaires (except in the rare case of an evident coding error in order to correct the IBM card). The student was Philip Converse, and he was about to make a discovery that would define his career-long reputation as a scholar and soon win him the rare honor of coauthorship with his professors on a major project, a tenure-line position in the department there, and, incidentally, start a major debate in behavioral political science that would rage for many decades.

He took the questionnaires to the basement (it was cooler down there) and pored over them late into the evening. His suspicion was at first a vague notion that there was something wrong with the way he and his colleagues were interpreting voters' sentiments in the 1952 and 1956 presidential campaign election surveys. As he proceeded to review the actual language of the transcribed remarks of potential voters concerning what they liked and disliked about Eisenhower and Stevenson and their respective parties, he was struck by an anomaly. Politicians and journalists (and, of course, political scientists) routinely utilize the all-purpose political yardstick of the liberal–conservative continuum to judge issues, parties, and candidates. What drew Converse's attention was the relative rarity of such a vocabulary in the day-to-day language of voters answering questions in the middle of a heated presidential campaign. Voters heard the terms *liberal* and *conservative* mentioned perhaps even daily in the media, but such abstract concepts did not seem to resonate.

So Converse persuaded his colleagues to undertake an unusual enterprise—asking respondents if they recognized the terms, and if so, what the terms *liberal* and *conservative* meant to them. It was unusual because it is the nature of the survey research enterprise to simply assign a singular meaning to each of the multiple choices among which the respondents are to select. Otherwise it becomes impossibly complicated to interpret the statistical results. It is understood that people "misinterpret," but it is common practice to assume that the misinterpretation is random noise, so "measurement scales" or "indexes" are built from strings of questions with alternative wordings averaged together to minimize noise. What is interesting about the notion of misinterpretation is the conception of a singular "correct interpretation" and various, random, and irrelevant corresponding misinterpretations. But Converse, following his hunch, was hot on the trail of a stunning finding.

Ultimately when he analyzed the verbatim transcriptions of what the terms *liberal* and *conservative* meant to average voters, he concluded that only about one out of every six respondents could give a complete and accurate answer and fully half the sample had no idea what the terms meant (Campbell et al. 1960; Converse 1964). Many in between those two groups equated conservatism simply with environmental conservation and "saving things" and liberal with being a spendthrift. What Converse had insisted on was an empirical examination of the polysemic character of one of the most basic concepts of the political sphere, and the results were stunning and triggered both a debate and a serious rigorous research projects to better understand the political sophistication and working political vocabulary of the mass electorate (Nie, Verba, and Petrocik 1976; Neuman 1986; Converse 2000). Ironically most of the debate about political sophistication was based on different interpretations of correlations among issue opinions and between issue opinions and reported voting behavior rather than returning as Converse did to an in-depth analysis of the everyday vocabulary of citizens discussing politics.

The behavioral social science paradigm relies heavily on surveys and experiments with reported attitudes and behaviors identified by predefined multiple-choice options. On rare occasions researchers include questions permitting "open-ended" or natural language responses that are transcribed verbatim, but routinely (as in the case of the election study Converse was working on) a coding process converts the natural language to a categorical system reflecting the distinctions that interest the researchers. Open-ended questions are generally avoided because they are messy and expensive and frustrating because so frequently respondents respond in a mind-set or vocabulary that is at odds with the researchers' expectations. Ironically there is no intrinsic reason that surveys and experiments could not address questions of polysemy. Such matters have simply traditionally been at the margin of scholarly attention and debate, in effect outside the received "normal science" paradigm of media effects experiments and surveys as Thomas Kuhn (1962) might describe it.

We turn now to a major thesis of this chapter. In behavioral research polysemy has traditionally been treated as an awkward inconvenience, but there is no logical or empirical requirement that the phenomenon be marginalized in this way. In fact, as a central element of the complexities of human interpersonal and mass communication, it should and could be central to the research enterprise. What would happen if behaviorists took polysemy seriously? Well, it is actually rather straightforward. *Polysemy would be treated as a central analytic variable.* Sometimes the media message is received and interpreted in virtually the same sense it was sent and intended and sometimes at dramatic variance. What are the conditions that are associated with the former or the latter? In communicative exchange some utterances are straightforward and perhaps simple or concrete, and the meaning intended will be received without ambiguity or distortion among others in that language community. In the context of an evening meal among English speakers, for example, the utterance "please pass the salt" is unlikely to provoke confusion or misunderstanding. (It should be noted that writing under the nom de plume Murdock Pencil, Stanford communication graduate student Michael Pacanowsky published a hilarious send-up of overly literal behavioral research in the *Journal of Communication* in 1976 concluding: "Conclusive evidence on the effects of the utterance 'please pass the salt' are sadly lacking" [Pencil 1976, 31]. It should be required reading for all graduate students in the field.) But as we have been reviewing at some length in this chapter, most of what constitutes the content of the public sphere is complex, contested, multivalent, and prone to diverse and systematic variations of interpretation— indeed the very essence of what constitutes an empirical variable of potential interest.

The normal science paradigm of effects research, as noted in Chapter 2, calls for the analyst to posit an "effect" associated with a particular type of media content and accordingly to test whether increased exposure to the content (measured in various ways) is associated with increased evidence of the posited attitude or behavior (also measured in various ways). Within such a paradigm the potential broad diversity of interpretations, reactions, and behavioral responses is simply beside the point. It brings to mind the oft-told tale of the scientist who wanted to assess the distribution of the sizes of fish in the sea and sailed the oceans systematically trolling using a net with a two-inch grid concluding confidently that his data revealed that there are no fish in the sea under two inches in size.

It turns out there are some scattered efforts in the communication and related literatures that have in various ways utilized polysemy as a theoretical variable. But the paucity of systematic work in this area is striking. I'll identify four diverse examples, each with a distinct theoretical focus and method. Because of their different foci and methodologies it is not surprising that a casual observer would overlook the commonality that each is systematically assessing the distribution of variation in the meaning of particular symbolic object—a single word, narrative, public event, or text. Most of these assessments are primarily descriptive because coherent and systematic analytic techniques remain in their infancy.

Tamar Liebes and Elihu Katz were struck by the global popularity of the American prime-time soap opera *Dallas,* so they sat down with a series of discussion groups (typically three couples and the interviewer) who had just watched an episode of the program. In Israel *Dallas* is broadcast in English with Hebrew and Arabic subtitles. "Our subjects are persons of some secondary schooling drawn from four ethnic communities in Israel—Arabs, newly-arrived Russian Jews, Moroccan Jews, and kibbutz members—and non-ethnic Americans in Los Angeles" (Liebes and Katz 1990, 114). This was not, of course, a systematic or representative sample of any of these ethnic populations and the research does not attempt to project precise quantitative distributions of belief. The study design engages an informal loosely structured group interview that invites the participants to retell and discuss the story they have just encountered. They described their research focus: "We have been studying the ways in which members of different ethnic groups decode the worldwide hit program Dallas. . . . What we do wish to do here is to distinguish, first of all, among different types of understanding. Then we wish to show that these types of understanding are related to different types of involvement. Finally, we will argue that programs like Dallas invite these multiple levels of understanding and involvement, offering a wide variety of different projects and games to different types of viewers" (114).

They proceeded to demonstrate that although the narrative that each group was watching was virtually identical the "meaning" and resonance of the narrative was distinctly different among the various cultural groups. The Russians, for example, tended to draw on metalinguistic questions comparing the story structure with Tolstoy (negatively, of course). To conclude that the rich are unhappy is exactly what the (well-to-do) producers of the program want you to believe, the Russians asserted. The Arab-Israelis, however, were much more literal and concrete in their viewing, following the ups and downs of the characters as important life lessons in morality.

In the second example Klaus Bruhn Jensen and a group of international collaborators (which included this author) studied one day's news (May 11, 1993) around the world. Their technique had two components, a systematic content analysis of the news broadcast for May 11 and the days leading up to that date and loosely structured interviews with news audience in seven countries around the world—Belarus, Denmark, India, Israel, Italy, the United States, and Mexico. "While the purpose of this core element of the study was to explore how different national audiences may interpret and apply news within their specific cultural settings, the study also examined the respondents' assessment of the quality of the different available news media and their ranking of the most important events in the world at the time of the study. In addition, a content analysis was conducted of the national news programs in each country during the week leading up to 11 May. Each form of evidence provides a perspective on the meaning of citizenship in an age of global communications as citizenship is reenacted on a daily basis in the interaction between media and audiences" (Jensen 1998, 16).

Interviewers in each country would ask respondents to retell some of the main news stories of the day in their own words. Jensen and his associates use the notion of "superthemes" to demonstrate how "viewers may arrive at interpretive themes which are quite different from the journalistic themes that the media might expect audiences to reproduce—'the viewer's story' may appear to be incompatible with 'the journalists' story'" (Jensen 1998, 19). In this case there were overlapping international news stories but much of the local news the respondents watched was regional. The interviewers probed: "Please tell me about the one story in the news we just watched that was most important to you. To you, what was the main point or event of that story? What did you think about when you were watching that particular story? How important is the story to this country?" The study included a quantitative content analysis of the respective country news broadcasts, but the main conclusion from the comparative study was one of striking similarity of superthemes. The specific players varied from regional newscast to newscast (in Italy the Pope was a major topic, in Belarus it was the recent liberation from Soviet domination) but stories of the dynamics of power, problems of corruption, and hope for improvement were strikingly common superthemes. Jensen concludes by quoting Gripsrud (1992, 196): "If the world looks incomprehensibly chaotic, it is only on the surface. Underneath, it's the same old story."

The third example draws on sociologist William Gamson (1992), who has devoted his career to studying social movements and political communication. When is it, he asks, that an individual watching a newscast feels that the problem at hand is one of those issues he or she can do something about? It is question of "issue framing," an analytic concept for which he has become famous, with the individual perhaps accepting or rejecting various official journalistic frames. When is a public problem subject to a "remedial action" frame or just one of those things you learn to accept? This is, of course, akin to Stuart Hall's conception of an oppositional reading. But as a sociologist, Gamson's impulse is to systematically analyze the media content and interview the audience members at length and then compare the two. He picked four ongoing issues in American public life at the time—affirmative action, nuclear power, troubled industry (recent plant closings), and the Arab-Israeli conflict—as a focal point. He systematically analyzed news coverage of these issues in television network news, national news magazine accounts, syndicated editorial cartoons, and syndicated opinion columns and in parallel conducted thirty-seven focus groups (he calls them peer group conversations) around Boston among 188 respondents to discuss the same four issues. His goal was to transcribe and analyze the process of people from similar backgrounds constructing and negotiating shared meaning, using their natural vocabulary. His peer groups were all working-class adults—cooks and kitchen workers, bus drivers, medical and lab technicians, nurses, firefighters, and auto service workers. Half the groups were African American, half were white, only a few reflected mixed ethnicities. Like Liebes and Katz and Jensen, he emphasized that audiences were active interpreters who would frequently interweave what they were hearing in the local newscast with their personal life experiences, political beliefs, and bits of popular wisdom about human nature. Rather than use terms like the *effect* of a particular journalistic news frame, he spoke of the presence (or frequently the absence) of a "resonance" of the news frame with audience sensibilities. "Not all symbols are equally potent. Some metaphors soar, others fall flat; some visual images linger in the mind, others are quickly forgotten. Some frames have a natural advantage because their ideas and language resonate with a broader political culture" (1992, 135).

Two themes about the polysemic character of public issues dominate his study. The first is the observation that many of the common frames with which audiences interpret the news are completely absent in the official journalistic news frames themselves. For example, "in spite of its absence in media discourse," he reports, "43 percent of the conversations on troubled industry spontaneously referred to collective action by workers" (1992, 67). In another example, there is apparently a widely shared belief that the boring nature of work in nuclear power plants leads to chronic alcoholism among the workers, including drunkenness on the job with resultant potential safety risks, again an element not present in the media coverage of the issue (154).

Fourth and finally, in my own work with colleagues Marion Just and Ann Crigler, we studied five issues in the news of the day, contrasting the dominant journalistic issue frames with those spontaneously invoked by citizens in semistructured in-depth interviews (Neuman, Just, and Crigler 1992). Our results dramatically challenged the confident agenda-setting and issue-framing schools of effects research by demonstrating numerous exceptions to the rule—"Extensive news coverage of apartheid in South Africa, for example, which was especially prominent on television, was not mirrored in public concern. At the same time, the public's deep concern about drug abuse and related crime problems was only weakly reflected in media coverage" (111). Further we found repeated examples of how journalists would frame public issues in terms of conflicts among various prominent political entities (akin to the horse race style of election coverage), which was only rarely reflected in public framing. And we found that the public frequently used a moral frame, life lessons of good and evil, a frame almost always eschewed by the journalists as inappropriate and unprofessional.

All of these studies have several common elements theoretically and methodologically. On the theory side, the researchers avoid narrow notions of predefined "effects" and opt for a more nuanced notion of active audience, negotiated meaning, and the presence and absence of a resonance with the persuasive or informative media message and the audience response. Further each systematically explores the diversity of responses—

that is the fundamental polysemic character of human communication, in this case mediated politics, news and popular culture. The methods employed engage a parallel analysis of media content and audience interpretations as assessed by a variety of semistructured interviews or transcribed group discussions. This approach does not yet have a distinct and commonly accepted label. I tried to popularize the term *parallel content analysis* and promoted this methodological approach in a review chapter (Neuman 1989), but the term and the related logic did not catch on and the publication was cited only infrequently. That chapter, incidentally, presages many of the arguments here and traces the history of a number of scholars championing an approach of this sort going back to the work of Max Weber (1910), Harold Lasswell (1963), George Gerbner (1969), Morris Janowitz (1976), Steven Chaffee (1975), and James Beniger (1978). It is, in my view, a very distinguished pantheon of social science scholars each of whom has published extensively and influentially. It did not occur to me then, although it should have, that every one of those scholars who had done extensive empirical work, work that made each of them very well known, had not in fact followed through themselves and conducted the research they had championed but relied on more traditional survey and historical methods. If it is a promising approach to study public communication as they profess, it is clearly also frustratingly difficult, expensive, and time-consuming.

There are some more straightforward and less daunting techniques of empirical inquiry that have attracted some limited attention in the study of communication. Again, the effort is to engage some form of quantitatively accessible natural language on the part of the respondent rather than assess differences in predefined self-report survey items querying behavior or attitudes. The broadest term for these techniques is protocol analysis popularized by Anders Ericsson and Herbert A. Simon in studies published in 1984 and 1993. They review the variety of techniques and interpreting natural language and semistructured interviews primarily in the domain of psychology (see also Mishler 1986). Related methodologies include the think-aloud technique (Eveland and Dunwoody 2000) and thought-listing protocols (Cacioppo and Petty 1981; Iyengar 1987; Cacioppo, von Hipple, and Ernst 1997; Price, Tewksbury, and Powers 1997; Valkenburg, Semetko, and De Vreese 1999). But attention to such approaches is sporadic.

We conclude this review by returning to the central argument. The phenomenon of polysemy has not escaped the attention of the media effects tradition, but because it is so frustratingly difficult to address empirically, it has been addressed only fleetingly, incompletely, and largely discursively. The strong commonalities and convergent findings among pioneering studies are only vaguely acknowledged. And perhaps most important, this direction of research has not yet been incorporated into the paradigmatic models that dominate the effects tradition.

What If Cultural Studies Took Polysemy Seriously?

At first glance, although it mimics the preceding section heading, the heading for this section of the chapter may strike some readers as puzzling. The media effects tradition rather famously sidesteps the messiness of polysemic communication and would benefit from confronting it more seriously, but the cultural studies tradition, if anything, would appear to be obsessive about debating the alternative meanings of text. Cultural studies students, it is understood, are prone to argue late into the night over what a text really means and (my personal favorite) whether the meaning of a text resides in the text or in the reader (or perhaps in the author's intent). Is the argument here that cultural studies has not taken polysemy seriously? Indeed, my argument is precisely that.

The proposition in a nutshell is as follows. Polysemy implies attention to the enduring importance of multiple meanings of a text as they evolve among multiple audience members and audience groups. To assert that a text really means this or should mean that is to propose monosemy. For cultural studies to take polysemy seriously it would need to address the structure of variation in meaning. In my view, that is what the call to arms of Stuart Hall's celebrated essay "Encoding/Decoding" (1980) necessitates. Hall identifies four ideal-typical modes of culturally situated meanings of popular narratives: (1) a *preferred* or *dominant* code reflecting the interests of the extant elite; (2) a *professional* code of the intermediary journalists, publishers, and broadcasters; (3) a *negotiated* code of some audience members, what Hall describes as "a mixture of adaptive and oppositional elements"; and (4) a fully integrated *oppositional* code. My reading of Hall is that this is a call to colleagues (many or most, of course, within the Marxist tradition) that the work of cultural studies is not simply to critique the dominant code but to try to understand the conditions under which the dominant code is accepted pretty much at face value and when it is opposed in various ways by various groups. Further, when does the professional code tend to reinforce or to seriously question the dominant code? Such questions are fundamentally empirical in nature—what are the historical and structural conditions of acceptance and resistance? It is rarely a dichotomous either-or of acceptance or resistance, but almost always a culturally rich, polysemic distribution of views and interpretations. And the structural patterns may vary systematically across issue space, moving from matters concerning economics, gender, race, sexual identity, religious identity, and so on.

So to take Hall seriously, my argument proceeds, platoons of cultural studies students should be moving out from documenting once again the perceived inauthenticity of televised cultural depictions they have dutifully recorded and analyzed and move into the field to assess the contours of real-world polysemy. I return to a theme, noted above. Cultural studies celebrates ethnographic exploration, but actually undertakes the enterprise only on extremely rare occasions. David Morley and Charlotte Brunsdon, as young colleagues of Hall at Birmingham, set out explicitly to explore the Encoding/Decoding model with audience groups watching tapes of an episode of the BBC's *Nationwide* documentary series (Morley and Brunsdon 1978; Morley 1980). Their work has become a model for the ethnography of group viewing and transcribed loosely structured discussions. Morley returned to the field for his 1986 study of family television viewing, but his publications since have been primarily republication and expanded commentary on the original field studies and Brunsdon has returned to her roots as a film scholar and feminist critic. Likewise Ien Ang's (1985) study of *Dallas* viewers and Janice Radway's (1991) study of romance readers have stimulated abundant praise but virtually no fellow travelers. Like Morley, both Ang and Radway have published frequently about reception analysis as an analytic practice referring to it approvingly as, for example, the "new audience research" (Ang 1996, 98) or the "ethnographic turn in audience studies" (Ang 1996, 138) or "an altered research practice (Radway 1988, 362) but unfortunately without conducting any. The work of Morley, Brunsdon, Ang, and Radway (along with Liebes and Katz) are inevitably cited, universally praised, and extensively commented upon in the cultural studies literatures but almost never imitated. Why such an anomaly?

Well, part of the answer parallels the circumstance in the media effects tradition and represents a stark if not terribly exciting reality. Field research is time-consuming and expensive. The tradition of textual analysis and exegesis is long and distinguished, so the typical inquiry in the cultural studies tradition simply follows that inertial direction. That may be simplest and most powerful explanation of the anomaly, but I believe there are several other suspects that may merit some attention.

First, there is an enduring if vague suspicion about the slippery slope of "data collection" that may seduce cultural analysts into an awkward and inexcusable scientism. Even the word *data* itself is capable of inspiring a chilling of the spine. Ien Ang, for example, while championing a convergence of empirical and critical research, cannot withhold the extended expression of some of her concerns and second thoughts: "A troubling aspect about the idea of (and desire for) convergence, then, is that it tends to be conceptualized as an exclusively 'scientific' enterprise. Echoing the tenets of positivism, its aim seems to be the gradual accumulation of scientifically confirmed 'findings.' . . . In other words, this scientific project implicitly claims in principle (if not in practice) to be able to produce total knowledge, to reveal the full and objective 'truth' about 'the audience.' The audience here is imagined as, and turned into, an object with researchable attributes and features (be it described in terms of preferences, uses, effects, decodings, interpretive strategies, or whatever) that can be definitively known" (1996, 43–44). Larry Grossberg reiterates and extends her concerns: "We should not assume that scientism is no longer a part of the problem. On the contrary, the situation is more contradictory than we usually admit, for science still seems to hold sway not only in the university but also in a variety of public arenas. And, increasingly, many cultural intellectuals who should be suspicious of the continuing power (and reductionisms) of science have hitched their wagons to what appear on the surface to be more sympathetic paradigms . . . because they use language that sounds similar to our own . . . and of course, without sharing their research grants" (2010, 46–47).

There remains in this literature a sense that the relationship between traditional cultural and empirically oriented scholarship is unavoidably zero-sum, and that to even acknowledge the latter somehow diminishes the other and what appears to be an odd comfort in sustaining the misleading "dominant" versus "critical" binary.

Second, there is a discomfort in treating audience members as objects of study. This is a response similar to the first, but it has its roots not in a competition among schools of thought but rather a genuine discomfort in the unavoidable imposition of participant observation and invasion of others' private existence. Consider, for example, Thomas Lindlof's remonstrations about the human being as research subject: "In quantitative audience research, the human subject is formulated as a specimen of whatever audience construct has already been decided is relevant. . . . Qualitative researchers can take justifiable pride in their attempts to 'decolonize' the human subject—that is, to accord personal being status to those whom they study and to engage them as partners in the enterprise" (1991, 33–34).

The traditions of ethnography in anthropology and participant observation in sociology are so well established that young scholars gamely approach fieldwork as a potentially fear-inducing task they simply must jump into. Cultural studies, however, has not developed such a tradition and offers numerous alternative models of textual analysis with no field or reception component to complicate the undertaking.

And third, perhaps most interesting, there is a deeply felt resistance to reception analysis because it often fails to provide supporting evidence for the canonical notions of strong hegemonic effects of media. This awkward problem, of course, has numerous parallels in the not-so-minimal faux debates among researchers in the effects tradition. The cultural critique of reception analysis as insufficiently illustrative of hegemonic control is amply illustrated by several papers of Glasgow University Media Group director Greg Philo, who takes Morley to task for neglecting to note that audience "conceptual structures include 'knowledges' about what typically occurs and assumptions about the rationality and legitimacy of action which may already have been subject to prior exposure to media messages. There is little room in the encoding/decoding model to investigate such a possibility" (Philo 2008, 541). Philo continues: "The main problem which I have with the encoding/decoding model [and the reception analysis it engendered] is the impact which it had on the subsequent development of media and cultural studies. The view which many took from it was that audiences could resist messages, safe in the conceptual boxes of their class and culture, and renegotiating an endlessly plyable language. This led eventually to the serious neglect of issues of media power" (541).

Ironically, the research Philo cites to support his argument of media power is primarily from the effects tradition. Philo is by no means alone. There was a robust debate in the literature over whether the new work on the diversity of audience readings, "the new revisionism," according to James Curran (1990), was actually a step back and a reinvention of the pluralist–empiricist wheel. John Corner (1991) developed a rather lengthy list of the blind spots of this new tradition, and David Morley (1992) mounted a spirited defense. The debate strikes me as particularly puzzling, however. It is reminiscent of the irresolvable sectarian debates in literary study over whether meaning resides in the text, the author, or the reader. The question of how a particular media depiction at a particular historical moment generates a mixture of persuasive and oppositional responses is (1) the most fundamental and central question at hand, (2) fully amendable to empirical inquiry by diverse methods, and (3) simply not appropriately subject to predetermination by doctrine.

Toward a Theory of Polysemy

Our exploration of these diverse literatures, it would appear, has led to a singular conclusion and strategy for moving forward. Let us briefly retrace our steps. The human capacity for miscommunication is incredibly vast, indeed, a worthy competitor in size and diversity to the widely celebrated human capacity for communication itself. Noam Chomsky (1972) was so mystified by the powers of young children to acquire language he came to posit an ingrained "deep structure" in human cognition to explain such a capacity evidently unparalleled in any other species. But part of that fundamental human capacity is sensitivity to the intent of the "other" in a communicative dyad and a rich sense of the evolved social structures that color and contextualize communication. So we might quickly conclude that such a term as *miscommunication* may ill serve further inquiry as it seems to imply a singular correct meaning and an inchoate and generally unimportant collection of miscommunications or essentially errors. Humans miscommunicate in systematic and important patterns. It would be folly to label them errors. Humans have distinct, socially, and culturally situated propensities to interpret and perceive at varying distances from the original intent of authors. So we turn from the sterile binary of communication–miscommunication and posit the phenomenon of communicative polysemy as a core analytic question for scholarship. So rather than a singular effect (in the effects tradition, typically assessed as a statistically significant increase in "agreement" in response to a persuasive message) or singular correct reading (in the literary studies and cultural studies traditions derived from textual analysis), I posit rather than the study of meaning as the study of the distribution of meaning—polysemy.

We encounter an effects tradition that is impatient with the polysemic character of text and a cultural studies tradition uncomfortable with the need to engage audiences as well as texts. An attentive reader will have already noted that the answer to the question *what if effects research took polysemy seriously?* and the question *what if cultural studies took polysemy seriously?* is essentially the same—it is the practice generally labeled as reception analysis. For a book already brimming with paradoxes, I add yet another. The effects tradition simply does not use the term nor acknowledge the phenomenon as a coherent research methodology, although it is pleased to cite and sometimes celebrate the occasional study within the tradition that undertakes the practice. We noted studies by Liebes and Katz, Jensen and Gamson, among others as examples. The cultural studies tradition recognizes the term *reception analysis* (although its use sometimes dusts up a bit of controversy) and celebrates its promise, but only very rarely its actual practice. We noted studies by Morley, Ang, and Radway, among a few others as examples.

It is difficult to integrate these diverse perspectives, but we may be in a position to take some first steps.

- Treat polysemy as a central analytic variable in the study of human communication.
- Treat the social, cultural, economic, and political structures that influence the distribution of polysemy as central analytic variables in the study of human communication.
- Treat the polysemic text and the polysemic response as equally important constitutive elements of human communication.

Theodor Adorno, Meet Paul Lazarsfeld

Two legendary figures in the history of communication study came together at the Princeton Radio Research Project in New York City in 1938, both recent émigrés from the looming conflict and holocaust in Europe (Lazarsfeld 1941; Adorno 1969; Rogers 1994; Scannell 2007; Cavin 2008). Lazarsfeld, of course, the consummate empiricist and the progenitor of the original critical-administrative binary had arrived in the United States first and was entirely in his element working on building new research tools to assess the radio audience. He had heard about Adorno's reputation as a scholar and an expert on music and was enthusiastic about putting him to work on the musical component of the audience research. Adorno was a bit overwhelmed by his new environs but at the outset gamely attempted to bridge his critical and aesthetic sensibilities rooted in the Frankfurt School tradition with the idea of empirically grounded audience research. It would prove to be a bridge too far and Adorno escaped to join other Frankfurt School émigrés in California. The debates and exchanges between these two highly committed and larger-than-life intellectuals became the source of many anecdotes and probably by now much elaborated legend about the incommensurate character of the effects and the cultural studies traditions. They tried to make it work and abandoned the effort. I have a theory about why.

It was because each saw himself as a central player in a new and exciting approach to the study of human communication that needed to establish both its independence and its bona fides. To blunt the sharp edges of each of these new approaches, they each calculated, perhaps correctly, would potentially weaken their capacity to become established. It was too early for rapprochement.

We have reviewed the parallel efforts in the succeeding generations of scholars by comparing the work of Stuart Hall and Albert Bandura. As circumstances would have it, although they were aware of each other's work, these two scholars reported to me that they never met and through the late decades of the twentieth century it would still not yet be an appropriate time if they had met for them to collaborate despite overlapping interests.

As we enter digital revolution and a series of dramatic institutional changes in the structure of communication institutions and the very definition of mass communication, it is time to give rapprochement another opportunity. Who knows where it might lead?

5

Predisposed to Polarization

The individual has always had to struggle to keep from being
overwhelmed by the tribe.

—ATTRIBUTED TO FRIEDRICH NIETZSCHE

It is much too simple to say that any system of communications is desirable
if and because it allows individuals to see and hear what they choose.
Unanticipated, unchosen exposures, and shared experiences, are important
too.

—CASS SUNSTEIN (2001)

It is probably no exaggeration to suggest that derogation of outgroups is one
of the most fundamental and universal features of all societies and cultures.

—DAVID J. SCHNEIDER (2004)

HERE IS YET ANOTHER PUZZLE about the human condition to con-
template. The word *communication* is based on the concept of commonness, shared understanding, and community (Carey 1989; Peters
1999). Furthermore, communication is often posited as the preferable
alternative to conflict and conflict correspondingly founded in a "failure
to communicate." How is it then that conflict and polarization are so central to the enterprise of communication scholarship?

One way to respond is to argue that there is nothing really paradoxical about this. Since communication is the reduction of uncertainly and the correction of misunderstanding, conflict and its reduction would naturally be theoretically central for this field of inquiry as the counterpoint to communication. Such a response, however, is hardly satisfactory. For one thing, the notion of a communication–conflict binary is rather crude and simplistic. Communication frequently enhances the probability of conflict. Recall the grounding of the newly created field of communication research in concern about propaganda's ill effects in rousing ethnic prejudice and bloodlust during and following World War II. For another, the communication–conflict binary sidesteps the central question of polysemy that occupied our attention in Chapter 4. *It is because culturally and historically embedded human communication is so richly polysemic, that it is frequently the case, for example, that a speaker's protestations of benevolent intent strongly confirm the worst fears of just the opposite among those listening?*

It is often observed that the history of human existence is a history of human conflict (Durbin and Bowlby 1939; Walker 2001). Lest there be some doubt—historian and archeologist Lawrence H. Keeley (1996), who specializes in such matters, concludes that approximately 90–95 percent of known societies throughout history engaged in at least occasional warfare and many fought constantly. He notes that among the indigenous peoples of the Americas, only 13 percent did not engage in wars with their neighbors at least once per year. It does not seem to be getting any better. Summerfield's (1997) analysis counts 160 wars and armed conflicts since 1945. True, many are minor border skirmishes, but in the 136 recent wars and civil wars analyzed in Wikipedia, the average total of military and civilian casualties per war is two million individuals (ranging from 60 million casualties in World War II to 907 in the Falklands Conflict in 1982 and 62 in Slovenia's Ten Day War of Independence in 1991). Key to all of this is the collective social identity of the tribe and nation (Isaacs 1975; Anderson 1983; Schlesinger 1991).

The Central Concept of Social Identification

Perhaps a more fruitful way to approach the issue of the highly variable and complex role of interpersonal and mass communication in alternatively exacerbating or mitigating human conflict is to turn to the central underlying basis of both polysemy and variable attentiveness to communication in our evolved cognitive systems—*the cognitive dynamics of identity and identification.*

The task of the communication researcher is theory building and theory testing, that is, the thoughtful and systematic accumulation of knowledge about human communication. A central and foundational question is—was the act of communication successful? Was the stream of symbols attended to and interpreted as intended by the sender? Communication science aspires to a set of well-tested and agreed-on insights about the generalizable conditions under which communication is or is not successful. Importantly, when do individuals choose to pay attention in the increasingly complex profusion of words and images? I propose that a key generalizable analytic concept in answering such question is social identification. Social identification is central to the dynamics of how a narrative engages an audience member. It is central to how individuals pay attention to several stories in the news out of the thousands that drift by without notice. It is central to the mechanism by which nationalist or racialized propaganda influences its audience. Social identification with the in group is thus the natural and necessary counterpoint to prejudice and aversion toward the out group.

Social Identification and Evolution. It is widely observed in the evolutionary tradition that humankind represents a particularly social animal, attentive to the moods and motives of others, and quick to make attributions (Barkow, Cosmides, and Tooby 1992; Hauser 1996; Deacon 1997). Cooperative and collaborative skills are seen as being of particular importance for survival for a species that is not particularly strong, quick or in possession of unique innate physical properties for self-defense or aggression. Food gathering and particularly collaborative hunting benefited from sharing information and the coordination of effort. So if acquiescence to evolved social norms and strong tribal identity increased survival odds, over hundreds of generations such properties would be strengthened and reinforced in the genome according to the conventional logic of this tradition (Reynolds, Falger, and Vine 1987). There is a bit of a catch-22 with this strand of evolutionary theorizing, however. It is hard to refute. Without independent reliable data on survival over long periods of time for those more and those less "socially" or "tribally" oriented, we simply posit the cause and resultant effect. Furthermore, such notions represent a troublingly simplistic and deterministic model of the human condition. That said, however, the proposition that the evolved human cognitive system is prone to social identification and polarization can be and should be subject to rigorous empirical examination whatever its historical roots. The extent to which social groups behave particularly tribally is likely to vary with conditions and that variation, if we can capture it reliably, represents a potential mother lode for the scientific study of human communication. If some communication practices, institutions and norms reinforce and others demonstrably mitigate polarization across boundaries of social identity, such findings are both normatively and scientifically valuable.

Communication and Power. The seminal thinkers in the social scientific tradition of communication research (notably Lasswell, Lazarsfeld, and Merton) and the cultural studies tradition (notably Williams, Hoggart, and Hall) were all deeply concerned about the relationship of the central institutions of mass communication and the maintenance (and, at times, the challenge to) established political, economic, and cultural elites. From Marx's rudimentary notion of false consciousness through Gramsci's more sophisticated dynamics of hegemony, Goebel's theories of propaganda, and Lazarsfeld and Merton's study of Kate Smith's charismatic promotion of war bonds, there was sustained attention to the psychology of nationalism and social class identification (Monroe, Hankin, and Van Vechten 2000). In more recent decades, the relationship between power and the politics of racial and gender identification have served to expand the topical breadth of this research tradition. Understandably, the sympathies of the researchers for the marginalized groups and a corresponding critical stance toward the apparently successful self-protective mediated machinations of various elites led to a rather asymmetric appraisal of how the dynamics of social identification work in complex industrial societies— social identification among minority groups is praiseworthy, among majority groups it is more often characterized as an unthinking bias. Simply being critical of elite behavior is a rather modest aspiration for communication scholarship—better to theorize its successes and failures. Sometimes elites are stunningly successful at getting their way in the public sphere. For example, one might point to the vilification of inheritance taxes as "death taxes" as having successfully convinced many working-class

Americans to oppose inheritance taxes because when they too become rich they may wish to pass on the benefits of their labor and luck to their offspring (Bartels 2005). Sometimes self-serving elite mythmaking does not succeed. One thinks of King Louis XVI's dramatically unsuccessful efforts to defend the centuries-old tradition of the French monarchy during the French Revolution and the many somewhat less dramatic cases of protest, unrest, and occasionally revolution against the established order. Comparisons across cultures and centuries are difficult—but, again, understanding the variation in the capacity of elites to reproduce and sustain inequality is a key element in the evolving puzzle. *Hegemony is not a constant; it is a variable.*

Furthermore, most of the work in communication particularly on the psychology and cultural dynamics of identification and prejudice is focused on a particular demographic, especially on ethnicity, gender, and nationality. Identification with the in group is positively portrayed. The corresponding aversion or suspicion toward those who by the definition of the in group are accordingly not included is negatively portrayed. This asymmetry is, as I note, understandable, but needs to be problematized and incorporated into the theory building. It may be possible to root for the Red Sox and not despise the Yankees. It may even be desirable on some level. But it may also be relatively rare. Being patriotic is valorized. Being xenophobic is demonized. These evolved and fundamentally tribal cognitive proclivities need to be realistically understood. Tribalism in the era of globalization moves front and center in the list of practical concerns that motivates research on the structure of communication systems (Castells 1997). In that spirit, let us take a look at how recent work in the psychology of identification and aversion may inform communication research. Furthermore, few who work on, for example, racial depictions in the media generalize about the dynamics of identification and prejudice beyond their specialized case study of racial attitudes so the accumulation of findings and a more generalizable theory of social identity may be impeded.

Let me be clear. This is not an effort to champion some imagined benefits of human prejudice. Quite the contrary, it is a critique of a deeply ingrained tradition of research that simply demonstrates the existence of stereotypical prejudgment in human behavior and then condemns it and with corresponding invidiousness condemns those social strata that exhibit these traits most strongly, usually the less well educated. Perhaps condemnation should simply be left to the artists and editorialists. If this work is to be useful, it needs to focus on how the structure of communication, the social practices, institutions, and norms contribute to variation in identification and prejudgment—alternatively strengthening or diminishing its effects.

The Psychology of Social Identification

There is a seminal study in the psychology of identification that still inspires attention and evokes thought. In 1954, Muzafer and Carolyn Sherif designed a field study to explore the dynamics of identification and conflict in social groups. They organized a summer camp in Robbers Cave State Park in Oklahoma for research purposes and recruited twenty-two twelve-year-old boys of similar backgrounds. They were picked up separately in two buses; each bus carried half the campers. Neither group knew of the other's existence. The boys were assigned to two camping areas at sufficient distance that each group remained unaware of the other for the first stage of the experiment to permit in-group bonding and group identification, which happened spontaneously in the first few days, including hierarchies within each group. One group decided to name themselves "The Rattlers" and the other "The Eagles." At this point the researchers brought the groups into a common area of the campground for a series of sports competitions and contests between the two newly formed social groups. The Rattlers' response to the competition was to proclaim confidence that they would emerge the victors and spend a day in preparation talking about the forthcoming events and proposed to put a "Keep Off" sign on the playing field. They decided to put their Rattlers flag on the field instead and threatened revenge if anybody touched their flag. When it was arranged for the two groups to eat in the same mess hall there was spontaneous name-calling and singing of mutually disparaging songs. The flags of each social groups became increasingly salient symbols as each group raided the other's cabins and threatened to burn the other's newly created symbolic identity. The competitions were intense, the children almost came to blows, and there was much name-calling and demonstrable holding of noses in the presence of the other group. It is safe to say that the Sherifs, although they carefully documented this nearly instantaneous bonding and out-group derogation, became concerned that their exercise turned out to be a little too successful, so they confiscated pocketknives and other potential weapons and quickly moved to a series of cooperative activities to get the groups working together and did so with some success (Sherif et al. 1961). A great deal of research on social identity and out-group derogation has been undertaken since the Sherifs, but the melodrama of their summer exercise still resonates and motivates this research effort (for overviews of this research tradition, see Tajfel and Turner 1986; Dovidio, Glick, and Rudman 2005; Trepte 2006; Kinder and Kam 2009).

It is amusing to note that when the Sherifs and their staff realized how quickly and powerfully they had succeeded in creating team social identity and energetic competition between the teams they became concerned about incipient violence and proceeded to commandeer potential weapons and tone down the competition. The cooperative activities did succeed in keeping the competition healthy and in appropriate bounds. *The impulse toward polarization is a given, but structure of social interaction matters.* Accordingly in the pages ahead we review the constituent elements of the predisposition to polarization and in the chapters ahead the structural elements that facilitate its constructive power and limit its destructive capacity. The focus here is on the transition between traditional one-way broadcasting/ publishing industries and the cacophonous multidirectional new media environment, so we will focus on five psychological dynamics of particular relevance:

1. Humans seek familiarity.
2. Humans seek identity reinforcement.
3. Humans rely heavily on categorical heuristics.
4. Human categorical heuristics tend to be invidious.
5. Humans seek communication for intrinsic enjoyment.

1) Humans Seek Familiarity. Humans are creatures of habit with an enduring preference for the familiar in their immediate environment. It is true, of course, that such dynamics are punctuated with the occasional impulse for novelty and curiosity, but the familiar routine is comforting and dominant. The seeking of the familiar is perhaps the most prominent dynamic of audience behavior and media economics—movie stars earn millions more than unknown actors precisely because they are already "known" to potential audiences (Rosen 1981). A successful movie is likely to spawn a sequel (Waterman 2005). Book publishers bank on recognizable authors producing modest variations on the familiar plot lines of their chosen genre (Compaine and Gomery 2000). Television is built on the episodic program with familiar characters returning each week to respond in familiar ways to modestly novel circumstances (Goodhardt, Ehrenberg, and Collins 1980; Napoli 2003).

Psychologists do not fully understand the neurophysiology of this cognitive proclivity, but it is clearly not a trait limited to humans. It is evidenced in mammals and even birds and insects (Harrison 1977). The classic exemplar of the human appetite for the familiar is a series of studies conducted by Robert Zajonc (1968) at the University of Michigan in the 1960s. The terminology that caught on in the literature is the *mere exposure effect*. In the classic study Zajonc exposed his subjects (none of whom had any knowledge of Chinese) to a diverse set of Chinese ideograms. Then he exposed his subjects to a second set of ideograms, including some already seen and some novel and asked his subjects to rate each on how much they liked the characters they were viewing aesthetically. Those characters previously viewed were consistently more frequently highly rated, thus the notion of "mere exposure" as a basis of preference. In Zajonc's view, such a dynamic is fundamental and understandably linked to evolutionary survival.

Researchers originally posited that the mere exposure effect may be in evidence only under limited circumstances, but subsequent studies revealed it to be quite robust and consistent (Harrison 1977). Interestingly, the studies demonstrated that the familiarity–liking linkage is evident instantaneously in fractions of a second even before the more complex cognitive mechanisms of recognition, memory and appraisal kick in (Yonelinas 2002). Further research may reveal that the interaction takes the form of an inverted U with preference increasing as familiarity increases and then reversing and declining when phenomena of satiation and boredom are engaged in response to overexposure and intense repetition (Harrison 1977). Tolerance for repetition may still be quite high as evidenced in research on musical preference (Bradley 1971).

Key to our purposes here is the finding that this perfectly natural and functional cognitive proclivity appears to be linked to social identification, stereotyping, and out-group hostility (Tajfel and Billig 1974; Smith et al. 2006; Förster 2009; Housley et al. 2010). The linkages are complex but fundamental. For example, Tajfel and Billig tested two groups of subjects: one group was made familiar with the social and physical setting of the experiment in a situation closely resembling the actual experiment, and came back for a second session in which the actual experiment was conducted; the second group came only for the actual experimental session. The results demonstrate that the "familiar" group engaged in more out-group discrimination than the "unfamiliar" one. The initial hypothesis was that discrimination against the out group could be understood to some degree as resulting from the subjects' general uncertainty in the situation and could attempt to reduce this uncertainty and provide a familiar meaning to the situation by drawing on habitual and stereotypical intergroup categorization in the experimental situation. But the experimenters found just the reverse to be the case. Rather than uncertainty reduction, the stereotyping and categorization increased with repeated exposure. Tajfel and Billig posit that this negative spiral results from the fact that subjects feel more confident and extreme in their in-group–out-group judgments over time.

These dynamics are likely to become more salient in the modern media environment as abundant choice, socially networked exposure patterns, and more narrowly targeted informational and entertainment content interact with what people know and how they feel about their social environment. There is speculation about increased social cocooning with familiar and reduced interaction with those outside one's immediate circle (Campbell and Kwak 2010) and decreased exposure to those with whom one is likely to disagree (Sunstein 2001). All of this is reinforced in the American case by the particularly strong partisan mood in Congress and among party activists, which is taken by some to be a possible trend rather than just another cycle (Bafumi and Shapiro 2009; see also DiMaggio, Evans, and Bryson 1996; Fiorina, Abrams, and Pope 2010).

2) Humans Seek Identity Reinforcement. In the episodic radio narratives of Garrison Keillor, we are reminded from week to week that in Lake Wobegon—"all the children are above average." The humor is evident enough, but the underlying psychology is serious and important (Hoorens 1993; Ehrlinger and Dunning 2003). People aspire to positive self-concepts and the wishful thinking associated with the maintenance of self-esteem leads to systematic bias in self-evaluation and in evaluation of one's primary

identification groups. This dynamic may be independent of preference for the familiar, but it represents a cognitive predisposition that functions to complement and reinforce familiarity seeking. The most influential theoretical developments in this area has been associated with the work of social psychologist Henri Tajfel and his associates and is identified by the acronym SIT for Social Identity Theory (Tajfel 1982; Tajfel and Turner 1986). The basic idea is that the individuals' self-concept is intimately linked to the primary social groups of which they perceive themselves to be members. These social groups, in turn, are perceived to be superior, while other or competitive groups (usually termed *out groups*) are inferior and/or malicious. While some researchers have noted that in real-world situations of scarcity and competition, objective group interests represent an important source of group conflict and intergroup derogation, further research revealed that subjective group identity alone was sufficient to generate conflict and prejudice (Sherif et al. 1961; Campbell 1965; Kinder and Kam 2009). Tajfel and his associates in Bristol, England, replicated the Robbers Cave exercise in the lab with arbitrary and whimsically based groupings of local schoolboys who were also quick to adopt group identity and out-group hostility in what became known as the minimal group paradigm. They organized groups based on those who overestimated versus underestimated in counting a pattern of dots, those who preferred Klee to Kandinsky, and on the basis of a ritualistic flipping of a coin (Tajfel et al. 1971). The minimal group finding has since been replicated with numerous settings and groups internationally (Billig and Tajfel 1973; Locksley, Ortiz, and Hepburn 1980). That said, there is no inevitable mechanical linkage between the strength of in-group identity and out-group bias, so further research seeks to understand the conditions that may foster the positive aspects of social identity without the negative aspects of out-group hostility (Brewer 1999).

Scholars have had more success at the macro level and in cultural analysis in probing these dynamics. Among the notable and frequently cited historical analyses is Benedict Anderson's *Imagined Communities* (1983), which argues persuasively that the market motivations of print capitalism to maximize sales in printing in the vernacular led, perhaps inadvertently, to shared ideas and perspectives and a surprisingly strong sense of nationalism and national community in the evolving nation-states. Likewise Manuel Castells's influential *The Power of Identity* (1997) takes the argument further into the era of networked globalization. Perhaps some evolving linkages between the macro and micro perspectives on identity dynamics will be fruitful.

3) Humans Rely Heavily on Categorical Heuristics. Some might say that the phrases like *categorical heuristics* represent a nice way of referring to a stereotype, bias, or prejudice. They have a point. The phrase is chosen because it is a way of emphasizing that such cognitive dynamics are natural and functional as well as subject to perilous misapplication in our collective coexistence. So sternly lecturing humanity about the evils of prejudicial thinking is not likely to be very effective. Trying to better understand how these cognitive dynamics work and what structures of communication and social organization might facilitate the functional and sidestep the dysfunctional aspects about generalizing about social categories might be a better bet.

Perhaps the most influential thinker to characterize categorical heuristics as inevitable is the American journalist and political philosopher Walter Lippmann. His pathbreaking book *Public Opinion*, published first in 1922 is still in print and widely read today. In the beginning of the book he introduces a frequently quoted variant of Plato's shadows in the cave—in Lippmann's vocabulary it is the distinction between the complex real world outside and the much simplified "pictures in our heads." "The real environment is altogether too big, too complex, and too fleeting for direct acquaintance. We are not equipped to deal with so much subtlety, so much variety, so many permutations and combinations. And although we have to act in that environment, we have to reconstruct it on a simpler model before we can manage it" (16). Lippmann did not use the word *heuristic*; he used the word *stereotype* and in fact introduced the word to the social science community by means of his book. In Lippmann's conception stereotype was a relatively neutral concept referring to necessarily simplified conceptions of complex realities rather than ethnic prejudice. As a journalist, Lippmann's style was discursive and engaging. He would cite an occasional psychological experiment but relied primarily on commonsense examples to highlight how expectations affect perception. Perhaps the classic exemplar captured in a brief vignette: "Anyone who has stood at the end of a railroad platform waiting for a friend, will recall what queer people he mistook for him" (1922, 76).

Table 5.1   Trait Stereotypes. *Source:* Extracted from Katz and Braly (1933), Table 1, pp. 284–285. Reprinted courtesy of the American Psychological Association.

Traits Checked Rank Order	No.	Percent
Germans		
Scientifically minded	78	78
Industrious	65	65
Stolid	44	44
Italians		
Artistic	53	53
Impulsive	44	44
Passionate	37	37
Negroes		
Superstitious	84	84
Lazy	75	75
Happy-go-lucky	38	38
Jews		
Shrewd	79	79
Mercenary	49	49
Industrious	48	48
Americans		
Industrious	48	48.5
Intelligent	47	47.5
Materialistic	33	33.3

But like the concept of propaganda, given the zeitgeist and world events of the 1930s and 1940s, the concept of the stereotype became increasingly associated with the distortions of exaggerated notions of "national character" and ethnocentric bias as it began to take hold as a focus of systematic research inquiry. The first efforts were simple efforts to demonstrate that stereotypes exist. Daniel Katz and Kenneth Braly, for example, in 1933, queried the opinions of one hundred Princeton undergraduates asking them to pick the five most characteristic adjectives to describe ten national, religious, and ethnic groups. Interestingly reflective of the times the authors refer to these religious and nationalities as "races" and they helpfully report both the number out of one hundred students who gave each answer and the percent, as well (see Table 5.1). The results are crudely fascinating and provoke a reader to wonder where such generalizations come from, how they might change over time, and whether students today would even feel comfortable about sharing such views. Katz would later characterize these stereotypes as basically public fictions, describing them as fallacious and absurd, as well as contradictory (Katz and Schanck 1938).

Perhaps one of the most influential social scientific studies of all time represented the next stage in research on stereotypes, Adorno et al.'s *The Authoritarian Personality* published in 1950. This team working at the University of California, Berkeley, just after the Second World War started the movement in research from describing the stereotypes to trying to understand the psychological dynamics of those prone to prejudice. Their efforts, drawing on a tactfully downplayed Marx and Freud, characterized the roots of ethnic prejudice as a collective personality disorder of rigid adherence to authority and a traditionally defined collective identity.

The timely popularity of the book and its easily replicated survey scale of propensity to authoritarianism, the F-(for Fascism) Scale, was no doubt rooted in the historical shadow of the holocaust and the related question of whether Germans were particularly authoritarian and whether under certain conditions similar popular movements could arise elsewhere perhaps in the context of the evolving Cold War. The core of the theory of authoritarianism they proposed is clearly evident in the scale items particularly such traits as conventionalism, submission to authority, aggression, superstition, and toughness. For authoritarians, obedience and respect for authority are the most important virtues children should learn.

A subset of F-Scale items are listed below:

- Every person should have complete faith in some supernatural power whose decisions he or she obeys without question.
- What the youth needs most is strict discipline, rugged determination, and the will to work and fight for family and country.
- An insult to our honor should always be punished.
- Most of our social problems would be solved if we could somehow get rid of the immoral, crooked, and feebleminded people.
- People can be divided into two distinct classes: the weak and the strong.
- Most people don't realize how much our lives are controlled by plots hatched in secret places.

- The true American way of life is disappearing so fast that force may be necessary to preserve it.

For a while the character of the F-Scale and what it was actually measuring became the subject of academic controversy but without clear resolution the matter faded into obscurity although related issues continue to pop up in the literature from time to time (Christie and Jahoda 1954; Kirsht and Dillehay 1967; Kreml 1977; Jost 2009). In retrospect it may appear that Adorno and his colleagues had succumbed to something of a stereotype of those who stereotype in their characterization of authoritarianism as a personality disorder.

The mainstream of psychological research in this area following Adorno and colleagues, however, was to take a new direction with the publication of Gordon Allport's *The Nature of Prejudice* (1954). Rather than a distorted public fiction or personality aberration, Allport emphasized the roots of stereotyping in the normal cognitive function of categorization and abstraction. He called it the "normality of prejudgment": "The human mind must think with the aid of categories (the term is equivalent here to generalizations). Once formed, categories are the basis for normal prejudgment. We cannot possibly avoid this process. Orderly living depends upon it. . . . There is a curious inertia in our thinking. We like to solve problems easily. We can do so best if we can fit them rapidly into a satisfactory category and use this category as a means of prejudging the solution" (20).

Clearly, categories can be more or less accurate and more or less rational. Allport identified the tendency for ethnic and religious groups to "stick with their own kind" as natural, culturally comfortable, and habitual.

4) Human Categorical Heuristics Tend to Be Invidious. The seminal work by Allport and Tajfel notes that the strength of in-group identity was not necessarily related to the strength of out-group hostility and in fact may not even be related to awareness of a corresponding out group. Allport reports:

> In one unpublished study a large number of adults were interviewed. They were asked to name all the groups they could think of to which they belonged. There resulted for each adult a long list of memberships. Family came first in frequency and intensity of mention. Then

followed the specification of geographical region, occupational groups, social (club and friendship) groups, religious, ethnic, and ideological memberships.

When the list was complete the subjects were asked to name "any groups which you feel appear in direct contrast to, or as a threat to, one of the groups you are identified with." In response to this direct invitation only 21 percent of the subjects responded by mentioning outgroups. Seventy-nine percent were unable to name any. Those who did identify out-groups named chiefly ethnic, religious, and ideological groups. (1954, 48)

A key issue for our analysis here is the role of media depictions in these dynamics over time and the prospect for "counter programming" against such heuristics. I address that in the following chapters. So far in this chapter we have been working our way through the literature on social identity in search of a coherent model of socially oriented individual-level information processing. Our interest in such a model is premised on the prospect that it would help us sort out the conditions under which increased communication increases conflict and polarization and those in which communication substitutes for or diminishes physical conflict. Because all of this takes place in a world of overwhelming information abundance, we need to better understand the attentional dynamics. Individuals do not open a magazine, turn on the TV, or open their browser with the expressed intent of confirming their social identity or steering clear of disconfirming facts. Outside of institutional settings of classrooms and training sessions, the key motivational dynamic for engaging the world in print and electronics is clear.

5) Humans Seek Communication for Intrinsic Enjoyment. Communication theorists spend a lot of time thinking about communication effects. The cultural studies tradition, for example, ponders the subtleties of hegemonic influence. The social science tradition maintains a very long list of (mostly) negative media effects from political malaise and loss of social capital to smoking, unhealthy eating, and antisocial behavior, as well as gender and ethnic stereotyping. In all but a few of these dynamics under study the effects would appear to be largely inadvertent and unintended "collateral damage" resulting from communication behavior undertaken for some other reason by both communicators and audience members. Again, a paradoxical collection of circumstances. *As far as executives in the media industries are concerned, what people do with media content is largely beside the point.* These are commercial enterprises and once the movie ticket has been purchased or the television program and associated adverting has been viewed, that is pretty much the end of it. *Further, why they watch is not relevant, only that they watch.* As a result, in commercial audience research there is a great deal of attention to audience interest in and intent to acquire various types of entertainment and information content, but probing the motivations any further than that is not relevant to the bottom line.

This may strike some readers as counterintuitive, so I elaborate briefly. Of course media executives in the motion picture, broadcasting, music, television, and publishing industries consider themselves to be shrewd judges of the aesthetics of their arts and the tastes of their audiences. But the unrelenting and easily fatal grasp of the marketplace bottom line on their professional standing and conditions of employment preclude all but the occasional rogue executive from substituting aesthetic for market judgment in production decisions. It is unerringly evident in the language and deeply rooted epistemology of the respective trade presses of each traditional mass medium. The *New York Times* reviewer seeks to determine for its reader if the movie is worth seeing. The *Variety* reviewer seeks to determine for its readers the ratio between a movie's known production costs and as yet unknown box office return on investment. As in any such cultural industry there is a surface patina discussion of edification, emotional engagement, education, and the provocation of thought among audience members. But all the industry veterans fully understand the distinction between that polite language of cultural aesthetics and the real-world language of the bottom line.

As a concrete example, one might review the several professional consulting firms who work with local television stations news operations to improve their performance. Douglas Drew Media, Frank Magid Associates, and Adams Broadcasting Consulting are exemplars. They are as likely to focus on the color coordination of the newsroom set and the casting of the news anchors as anything dealing with news content. And the meaning of performance is clear and unambiguous—it is the competitive standing of the dinner hour and 11:00 p.m. news programming relative to the other competition in the local television marketplace. It simply would not occur to either the station or the consultant to explore what audiences were learning from newscasts as a component of performance (Neuman 1976).

As a result of this industrial culture, and a curious gap in the academic literature, we know surprisingly little about what motivates individuals to sit down with a book, turn on a radio station in a car, watch TV, or go to a movie. Researchers have certainly made abundant attempts to simply ask audience members about their motivations, but the responses are strikingly vague. Most audience members do not give such issues much thought. They have found they have enjoyed the novels of a particular author and seek her out, or have a favorite episodic TV show, and much of their decision making and behavior is highly habitual. They can identify that they like a particular text or genre, but it is very difficult to articulate why (Zillmann and Vorderer 2000).

To further illustrate this curious gap in the research literature we might briefly return to the classic texts on the "functions" of the media developed at the time structural functional analytics were in high fashion. The most famous characterization of the functions of the media is attributed to Harold Lasswell and is still frequently cited. Writing in 1948 he penned an essay titled simply "The Structure and Function of Communications in Society" that has been reprinted many times since. This is the same essay in which the iconic formula "who says what through what channel to whom with what effect" was introduced. Lasswell goes on: "Any process can be examined in two frames of reference, namely, structure and function; and our analysis of communication will deal with the specializations that carry on certain functions, of which the following may be clearly distinguished: (1) the surveillance of the environment; (2) the correlation of the parts of society in responding to the environment; (3) the transmission of the social heritage from one generation to the next" (38).

Clearly Lasswell is thinking of function at the aggregate level. It is difficult to imagine the individual sitting down to survey the environment or receive a transmission of social heritage. It was not until years later that sociologist Charles R. Wright (1960) would add a fourth function—entertainment—to acknowledge the nature of individual motivations. This curious tension between the notion of collective benefits and the social/cultural level and potentially hedonic motivations at the individual level will represent a recurring theme in the analysis.

We do have an important theoretical starting point for understanding the dynamics of intrinsic enjoyment of media behavior and it comes from our understanding of the comfort humans find in the familiar. The entertainment narrative and the news story have highly predictable structural properties. Motion picture Westerns, murder mystery novels, television situation comedies, and news stories about, say, business corruption have characteristic story lines that vary only at the margins. The textual and literary analysis of these generic forms is well developed, but the linkage between narrative structure and audience motivations and behaviors is still an incomplete link in a potential theoretical chain.

Polarization and Mass Communication

This chapter has been concerned primarily with one dimension of the dynamics of human psychology—why is it that we so quickly cleave to conflict? In common parlance, a term identifying this propensity might be *tribal* as in humans are a tribal sort. In the absence of an actual tribe we will create one out of whatever perceived invidious distinctions we might be able to cobble together and distinguish our identity by contrast with and often conflict with the "other" or frequently, multiple others. We listed four deeply ingrained contributing psychological mechanisms around which we have organized this chapter and a fifth factor that we expect will be interacting with the other four in interesting ways in the new media environment. The first four factors are the proclivities toward (1) the familiar, (2) identity reinforcement, (3) simplifying categorical heuristics, and (4) invidious distinctions. The fifth factor is the curiously seldom-studied and undertheorized aspect of media behavior—the intrinsic enjoyment that drives media use. The importance of the fifth factor is that in a world of greatly expanded choice of media content and the potential fragmentation of public attention, the reinforcement of the tribal is a distinct prospect. So in this final section of the chapter I raise the question of what social structural mechanisms might be explored to compensate for the predisposition to polarization. This question occupies our attention for the remainder of the book.

One particularly useful technique for avoiding conflict with a neighbor from when actual familial tribes were the principal basis of human organization was simply to move away from "the other." This was particularly appropriate when most human groups were nomadic. It got more difficult when domesticated animals and farming replaced hunting and gathering. It got more difficult still when dense urban populations and city-states and nation-states with armies got upset with their neighbors. But one could still choose to ignore one's neighbors—to shun and eschew.

Alas, trying to ignore neighboring tribes is increasingly impractical in a globalized world system. Events in even the remotest corner of the Sulaiman Mountains on the porous border between Pakistan and Afghanistan are instantly videoed back by satellite to viewers in North America if judged to be relevant to American interests. Even the meaning of "neighbor" requires reexamination in the global village.

Anderson's Imagined Communities. The final pages of this chapter attempt to negotiate the transition of level of analysis from the psychological to the social/cultural/political. The guiding theme is that trying to change our evolved psychological proclivities such as they are is not likely to meet with much success. A more promising strategy is probably trying to change and design institutionalized reward systems to realistically account for those deeply ingrained proclivities. A useful starting point for such an enterprise is an unusually polymathic and widely cited historical analysis by Benedict Anderson entitled *Imagined Communities*. In a discursive review of five centuries of world history, Anderson ponders the question of why nationalism and a profound public sense of national identity flourished so strongly in Europe, North America for the seventeenth through the nineteenth centuries and thereafter in postcolonial South America and Asia. While both drawing on the Marxist tradition and distancing himself somewhat from it, Anderson develops a notion of a new and powerful historical force he labels "print capitalism."

> The logic of capitalism thus meant that once the elite Latin market was saturated, the potentially huge markets represented by the monoglot masses would beckon. . . . Liberalism and the Enlightenment clearly had a powerful impact, above all in providing an arsenal of ideological criticisms of imperial and anciens regimes. What I am proposing is that neither economic interest, Liberalism, nor Enlightenment could, or did, create in themselves the kind, or shape, of imagined community to be defended from these regimes' depredations; to put it another way, none provided the framework of a new consciousness—the scarcely-seen periphery of

its vision—as opposed to centre-field objects of its admiration or disgust. In accomplishing this specific task, pilgrim creole functionaries and provincial creole printmen played the decisive historic role. (1983, 38, 65)

The first facilitating historical development was the book, especially the book in the vernacular. Anderson notes that twenty million books were produced in Europe within the first fifty years of the (re)invention of the modern printing press. Until then the church easily won its battles against heresy because it had better lines of communication than its challengers. But when Luther nailed his thesis to the chapel door in Wittenberg in 1517, copies in German were printed up and available in every part of the German countryside within two weeks. Soon after, Luther became the very first best-selling author in modern history, dominating a full third of the German language books published in the next decade (Anderson 1983, 39). What makes Anderson's thesis so interesting for our purposes was the unplanned, undirected, and inadvertent character of these historical forces. Conspiring nationalist ideologists did not plot to print in the vernacular. It was an inadvertent product of the market-expanding strictures of print capitalism. As a singular Roman religious identity throughout Europe was being replaced by divergent and regional protestant schisms, and by secularism and the Enlightenment, new national identities resonated with what might otherwise have been a vacuum the sense of tribe among the mass populations. It may be that the princes of the German city-states gradually began to recognize that religious independence helped the cause of political independence, but Luther's 95 Thesis was a religious document; the politics came later. As Calhoun points out, drawing on Anderson's thesis, once the idea of imagining political communities as nations was developed, it was "modular and could be transplanted into a wide range of otherwise disparate settings" (1993, 216).

Now, five centuries later, religious and national identities continue to be highly salient if taken for granted, but the shrinking globe puts these identities in constant and intense contact and competition. This tension has been acknowledged and highlighted since the publication of Samuel Huntington's (1993, 1996) celebrated clash-of-civilizations hypothesis. Huntington identified a series of cultural fault lines, most notably the tensions between Christian Europe and the Islamic Middle East and North

Africa. The corresponding question of our era is whether modern "broadcast capitalism" instinctively seeking to maximize its audience inadvertently and perhaps even impulsively reinforces the cultural fault lines. This is evident as North American and European media reinforce audience expectations about Arabs ganging up on Israel, and Arab media do just the reverse. If government-controlled and independent commercial media end up doing precisely the same thing in giving audiences what they expect and in some sense what they "want" and spiraling into an increasingly polarized set of conflicting worldviews based on evolved social, national, and religious identities, the digital public sphere is reinforcing rather than compensating for the human proclivity toward polarization. In the pages ahead, I attempt to outline a strategic approach to this issue. For our purposes in this chapter, I briefly review some compensatory approaches that might be suggested but that are unlikely to prove successful because of the deeply ingrained strength of the psychological mechanisms we have just reviewed.

Compensating for Polarization?

The tension between culturally derived norms of appropriate behavior and the impulses of the individuals within the culture is as old as the existence of the very first tribes. A sin, of course, is a violation of social norms, and how often can we find a tribe member who has not sinned? Virtually all religious and civil legal traditions have elaborate rules for defining officially prohibited behaviors and complex mixtures of approbation, punishment, ritual atonement, and divine forgiveness for dealing with them. Whether these systems of authority are actually successful in guiding human behavior down the path of righteousness is an interesting question but a frustratingly difficult one to answer. It is routinely asserted that without these elaborate culturally reinforced systems of detection and enforcement, life would be a Hobbesian chaos of brutish violence and competition for survival. Even if this celebrated social contract establishes a minimal level of social order, if the personal motivations are strong enough or perhaps in turn the collectively determined potential economic rewards are high enough, criminal behavior, organized or otherwise, will be in evidence. So it is sometimes difficult to separate out the dysfunctional spasms of enforced official moral authority such as the early American witch trials, the prohibition experiment, the inquisitions of the twelfth through the sixteenth centuries, and even the Holocaust from normal and necessary enforcement of law and order. We can conclude, therefore, that the effort to counterbalance the human proclivity to distinguish in-groups from others will represent an extremely difficult challenge. We can at this stage review three compensatory strategies that, given the power of these deeply ingrained psychological dynamics, are not likely to be very successful. And then three that research indicates show real promise.

Avoidance. Perhaps this strategy represents the first lesson of Conflict Management 101. To minimize conflict between Billy and Eddie in third grade, as most classroom teachers have learned, put their desks at opposite ends of the room to minimize contact and conflict. When a serious issue arises, either Billy or Eddie may have to be moved to another appropriate classroom. Of course, in nomadic and tribal times, tribes found survival might best be served by going their separate ways. We tentatively conclude, however, that global flows of communication and transportation and a truly globalized economic system make the old avoidance strategy impractical and unrealistic. Nowadays, Eddie can provoke Billy across the classroom with a wireless text message or, for that matter, from another classroom or even in another school.

Simple Prohibition. As the discussion above suggests, outlawing prejudice may sound promising in the abstract, but like many edicts on moral behavior that run contrary to human impulse, successful enforcement is likely to prove to be problematic. By simple prohibition I mean strictures that represent the rule of law but may not have widespread social and cultural normative support. One classic example, of course, is the American experiment in prohibition from 1920 to 1933 (Schrad 2010). Given the seriousness of the Holocaust and related tragedies of World War II, it is understandable that French and German authorities prohibit trade in Nazi artifacts, but it is not clear that these enforcement efforts are more symbolic than consequential (Virzi 2001). The sale of Nazi memorabilia was restricted on the Yahoo! site run in France in accordance with the French Criminal Code, but French citizens could still review items for sale on the U.S.-run Yahoo! site and American courts ruled that French law could not be enforced on the U.S. website (United States Court of Appeals for the Ninth Circuit 2004).

Censorship. Like prohibition, simple censorship of prejudicial speech is an unpromising remedy for such a complex phenomenon. In the American context the courts' strong commitment to First Amendment principles has ruled restrictions on hateful speech as unconstitutional with the exception of defamation, and the technical legal definitions of "fighting words" and "incitement to riot" (Walker 1994). Interestingly, it turns out the American reluctance to restrict hate speech finds few parallels in most other nations and in international fora such as the United Nations (Schauer 2005). But there is little evidence that these diverse efforts at hate speech censorship are successful (Callamard 2005; Hare and Weinstein 2011). Again, in a world of instantaneous global information flow, it is not clear how such attempts at censorship could be successfully sustained. So if avoidance, simple prohibition, and censorship show little promise, what options remain?

The Cooperative Norms Imperative. The psychological literature does provide some encouraging words for those who would like to explore compensatory institutional responses to the human impulse toward polarization and conflict. I review three here and proceed with this thematic in the remaining chapters. The first draws on one conclusion of the literature that although simple top-down prohibitions are unpromising, culturally reinforced norms can make a difference. The key is how the norms are articulated and how well they resonate with the relevant population. The successful exemplars here often can be seen as structuring the character of conflict, channeling it or moderating it, rather than prohibiting it. As differences among individuals and groups arise, norms of playing by the rules and respecting rather than derogating one's "opponent" arise again and again in the foundational literature of sociology (Simmel 1950; Dahrendorf 1959; Lipset 1985; Blalock 1989), the subsequent more practically oriented literature on conflict management in the study of organizations (Rahim 2010), and the dynamics within and between nation-states (Kriesberg 2007). More recently these same themes have been taken up in the civil society literature focusing on the strength of intermediate institutions in addition to those of the state and of the market that reinforce norms of seeking alternatives to conflict (Calhoun 1994; Edwards 2009).

The Fairness Imperative. Complementing the notion of normatively reinforcing the principle of playing by the rules is the shared sense among individuals that "the other side" is indeed playing fair. The perception of fairness, particularly the perception of its absence, turns out to be a fundamental driver of political perceptions and political participation. Strong emotional responses to unfairness, unequal application of the rules, or unequal reward have been consistently and sometimes dramatically demonstrated in children and even in animals (Lind and Tyler 1988; Association for Psychological Science 2008). Collective behavior research has studied the dynamics of when growing frustration with perceived procedural or political fairness reaches a tipping point and public resignation becomes open rebellion (Gamson 1968, 1975; Gurr 1970; Oberschall 1973; McCarthy and Zald 1977; Coleman 1990; Tilly 2004). A parallel line of research has explored cooperation and conflict in organizational settings and legal proceedings focusing on the concept of procedural justice (Lind and Tyler 1988).

Beyond the acknowledgment of an extensive literature on procedural fairness, escalation, and negotiation of potential conflict, our central question here is how the study of communicative dynamics can add value to this domain of scholarship. My proposition is that, although it is not yet fully developed, there is indeed an important role to be played. The key conceptually is the distinction between "fairness" in some objective sense and the perception of fairness. A critical public response to objective unfairness is reasonable enough and in the long run obviously important. The special case of particular interest, however, is the false or exaggerated sense of unfairness because of distorted communication and misperception. Perhaps one notable case example might be the inflammatory rhetoric of conservative talk radio typified by Rush Limbaugh, Sean Hannity, Michael Savage, and others. Representing the conservative point of view, these flamboyant personalities speak only rarely about positive conservative values but find it in the interest of their ratings to talk derisively at great length about the despicable behavior and tortured logic of liberals. The key to their popularity, it would appear, is the somehow satisfying sense of outrage shared with their conservative listeners in observing how liberals are self-serving and prone to violate common norms by cheating, lying, and misrepresenting the facts (Jamieson and Cappella 2008; Neuman, Marcus, and MacKuen 2012).

The Familiarity Imperative. Recalling the Robbers Cave episode, the organizers of the field experiment became so concerned about the competitive energy spontaneously evident between the Rattlers and the Eagles they confiscated pocketknives and organized a series of cooperative activities between the two groups to quiet things down a bit. The idea, of course, is that the better the two groups came to know each other, the less misinformed prejudice and, in turn, the lower prospect of serious conflict. This turns out to be a staple of psychological literature on prejudice and conflict management. Since Allport (1954) it has generally been referred to as the contact hypothesis and has been extensively studied.

Allport (1954) listed what he viewed as optimal conditions of contact for the greatest prospect of reduced prejudicial attitudes particularly that the groups be of equal status, pursue common goals, and have cultural and normative support for the context of contact. A half century of research and a meta-analysis of more than five hundred diverse studies reveals consistent and statistically strong evidence supporting the proposition that increased knowledge about and familiarity with out groups reduces prejudice and aversion (Pettigrew and Tropp 2006). There are several conditions that enhance the contact effect. It is important that the intergroup contact be under conditions of equal or near-equal status. Cooperative activity toward common goals contributes further. General cultural support for the interaction between groups is helpful. And, finally, when groups display behavior contrary to presumed stereotypes, it enhances the contact effect. Although Allport in his original formulation had elaborated these conditions and further research has sustained their relevance, numerous studies reveal at least some reduced prejudice from contact even in the absence of these conditions (Pettigrew and Tropp 2006; Hewstone and Swart 2011). Researchers in this field have speculated that part of the contact process may result from the "mere exposure" familiarity-liking heuristic. Interestingly that has opened up the prospect that virtual contact, mediated contact, and even "imagined" contact can manifest positive effects (Hewstone and Swart 2011; Dovidio, Eller, and Hewstone 2011).

Polarization and Pluralism

This chapter has focused on the psychology of invidiousness. Deep in the human soul is the impulse to judge quickly and to absolve slowly if at all. Deeper still perhaps is the impulse to define our self-worth by means of social membership. It is the psychology of this latter dynamic that I believe offers the most promise. Under some conditions, communication can minimize conflict. A better understanding of those structural conditions is the challenge at hand. Our definitions of *in group* and *out group* are largely socially defined. Our institutions of mass communication and the evolving social media represent an intrinsic part of that process in the digital age. The productive institutional management of differences of opinion, belief, and behavior is associated with the notion of pluralism. The term was ascendant in the late twentieth century in political science and political sociology and associated with the work of Lipset (1985), Dahl (1982), and Lindblom (1977). Productive theorizing and the actual successful implementation of enduring pluralist structures require attention to both the psychological and institutional levels of analysis. The term and associated research and theorizing have not drawn much attention within the field of communication. I believe they should.

6

The Politics of Pluralism

Every human power seeks to enlarge its prerogatives. He who has acquired
power will almost always endeavor to consolidate it and to extend it,
to multiply the ramparts which defend his position.

—ROBERT MICHELS (1962)

Deliberating groups typically suffer from four problems. They amplify the
errors of their members. They do not elicit the information that their
members have. They are subject to cascade effects, producing a situation
in which the blind lead the blind. Finally, they show a tendency to group
polarization, by which groups go to extremes.

—CASS SUNSTEIN (2006)

To argue about the media today is almost inevitably to argue about politics.

—PAUL STARR (2008)

WE TURN NOW to the fourth of the four paradoxes, "the politics of pluralism." The contents of this chapter reflect a somewhat broader focus
and steps back a bit from the individual-level psychology of attention/
interpretation and move to the social level of institutions, collective practices,
and cultural norms. The central organizing concepts here include the
notion of hegemonic media institutions, agenda setting, the public sphere,
the marketplace of ideas, and the delicate challenge of sustaining political

pluralism. The premise is straightforward enough. Maintaining social order in a small, homogenous, and univocal tribal or communal setting is relatively easy, as Wilbur Schramm (1948) has emphasized: "The typical community has until recently been small and relatively homogeneous. From generation to generation its members could expect to live under approximately the same pattern of culture and values. Its members dwelt, worked, worshipped, and played with approximately the same group of persons in the same place. A person could easily comprehend his whole community, and members of the community needed no more than face to face communication to reach agreement and understanding. Now that is no longer possible" (2). Sustaining an open and vibrant pluralism in a diverse industrialized nation-state immersed in a global network of communication and interaction is a difficult challenge. If there are some lessons learned from scholarship in this domain, public policy could potentially put them to good use.

In the preceding pages I have attempted to document what is known about the character of the evolved human cognitive system and how the perception of communicated messages is systematically distorted in largely predictable ways. I turn now to how those patterns are manifested at the collective level and, in turn, how social and political institutions that structure communication flows might be designed to counterbalance known patterns of distortion and polarization and protect an institutional openness to free speech, even unpopular speech.

Six Structural Theories of Communication Effects

At the social, cultural, or institutional level of analysis, one moves up from the individual psychology of a message effect to the prospect that the institutional structure of communication has enduring and important effects on our collective being. Our working list of six structural theories of communication effects is, of course, incomplete—the dynamics of "indexing" (Bennett 1990) or "mediatization" (Mazzoleni and Schultz 1999), among others, for example, might fruitfully have been added—but our working list will serve to illustrate some of the key dynamics at work at the levels above human psychology. In some cases we might come to understand these group-level phenomena as a logical parallel or natural extension from a single individual's cognitive system to a notion of collective cognition (see Table 6.1). One of the six patterns, the notion of "attention space," which examines the constraints on the diversity of the public agenda, the limit to the number of currently debated public issues, could be seen as a natural extension of the attentional limitations of the human brain. But, as we will shortly see, the attentional and dynamic mechanisms at work at the social level have a different character.

This notion of "social facts"—collective phenomena that cannot be reduced to or simply explained from individual level mechanisms—is one of the central concepts of sociology and associated with one of sociology's founding theorists Emile Durkheim (1964) and has been echoed more recently in the work of leading communication scholars (McLeod and Blumler 1987). Each of these social patterns has particular significance for understanding the challenges to maintaining an open structure of communication. Each, in effect, represents a predictable pattern that has the potential to institutionally constrain or distort the open flow of ideas, a balance between majority and minority views, and the capacity to challenge old orthodoxies with new ideas in response to changing historical conditions. The first three, McCombs and Shaw's widely cited agenda-setting hypothesis, Michel's famous "Iron Law," and the Matthew Effect, concern a naturally occurring and strongly conservative bias in cultural dynamics by which established powers try to organize things so they stay established powers. The remaining three, the notion of attention space, the issue-attention cycle, and the spiral of silence, deal with macro-level attentional dynamics. They also represent an interesting conservatizing structural bias that limits the breadth and agility of the public attentiveness.

Later in this chapter I review the concept of a "marketplace of ideas" in an effort to convince the reader of its promise as a normatively grounded and fundamental principle of communication policy that may help counteract these various constraining structural mechanisms (Napoli 2001). The marketplace, of course, is the core concept of neoclassical economics—the celebrated open market as the organizational principle for structuring the exchange of concrete entities such as goods and services. This marketplace for goods and services has a notable mechanism for self-correction—the open entry of competitors when an offered product or service becomes overpriced, shoddy, or lacking in innovation. The

Table 6.1   Six Hypothesized Dynamic Mechanisms of Macro-Level Communication Structure

Pattern	Seminal Theorists	Basic Mechanism
Media Agenda Setting	Maxwell McCombs Donald Shaw David Weaver Shanto Iyengar Donald Kinder	Although media coverage of public issues may not always persuade audience members, the close association between amounts of media coverage and public perceptions of issue salience indicate a dominant media agenda-setting function.
The Iron Law of Oligarchy	Robert Michels Nathan Rosenberg Joel Mokyr	Established elites develop ideological and organizational mechanisms to protect their incumbent status and constrain critical communication and challenge.
The Matthew Effect	Robert Merton Vilfredo Pareto Udny Yule Brian Arthur Albert-Laszlo Barabasi Robert Metcalfe	A set of mechanisms of cumulative advantage, rich get richer, famous get more famous, positive feedback, first mover advantage, preferential attachment, and network effects.
Attention Space—the Law of Small Numbers	Randall Collins Maxwell McCombs Everett Rogers	Limited size of public agenda, mechanism by which new issues supplant, reframe old ones, a natural limit to number of "schools of thought."
The Issue Attention Cycle	Anthony Downs James Stimson Benjamin Page Robert Shapiro	Dynamic model of how issues or issue frames coalesce, peak, and decline, a limited attention span at the macro level.
The Spiral of Silence	Elisabeth Noelle-Neumann Carroll Glynn Vincent Price Michael Slater	Additional dynamic of public opinion whereby public perceptions of dominant or politically correct views reduce the willingness of those with minority views to speak out.

marketplace of ideas would certainly benefit as well from such a self-corrective mechanism, but, as we shall see shortly, practices, institutions, and norms all too often evolve that thwart the open entry of new ideas that might compete with the old. Why would this be so, and why especially in the domain of speech and ideas? The answer is that as human beings follow their natural impulse to create hierarchies, and accordingly some individuals find themselves in positions of power at which point they often come to devote remarkable energies to protecting that institutionalized power from possible intellectual or political challenge. This is perhaps not entirely distinct from the traditional economic marketplace of goods and services as vendors try to undercut and preempt competition through a variety of legal, marginally legal, and illegal means (Williamson 1975, 1985).

Surely this phenomenon of institutional self-preservation manifested itself early in human history as a tribal member who through the exhibition of skill and bravery found himself (most likely a male tribal member) functioning as some form of tribal chief and then decided to proclaim that his chiefdom is divinely inspired and that his offspring will inherit the title and power. It represents a remarkable simple and relatively well-understood mechanism that plagues the openness and free competition in social, cultural and, of course, political organization—those who succeed in a competitive domain use their newly obtained resources and demonstrated talents to distort and prevent any further competition. To criticize the established power structure and its inhabitants becomes defined as a violation of tribal norms reflecting a lack of respect, sedition, rebellion, and perhaps sacrilege. What would have been the fair-minded invisible hand of market competition becomes the fist of the powerful. I first introduce each of the six working theories starting with their intellectual and historical provenance and then attempt to put these ideas to work in analysis of the fragile dynamics of the public sphere in modern industrial democracies with special attention to the potential impact of the new digital technologies.

Media Agenda Setting. We reviewed the evolution of this theoretical tradition and the accumulated findings at some length in Chapter 2 and will not replicate that discussion here. I concluded that because the agenda-setting notion drew attention to issue salience rather than simply attitude change, it quickly drew attention—by one count more than four hundred published studies in the three decades since the seminal paper (McCombs 2004)—as a more sophisticated way of designating potential media effects and offered promise of transcending the particularly troublesome "minimal effects" findings. Further, agenda-setting analyses started to incorporate even more sophisticated notions of issue framing (influencing the salience of particular issue attributes) and issue priming (making an issue or issue attribute more accessible to memory). These extensions were designated as "second-level" agenda-setting effects (McCombs 2004). The weakness (not a weakness of the theory itself, of course, just a recognition of the relatively early stage of theoretical development) was that the sizes of measured agenda-setting effects varied dramatically, sometimes non-existent, sometimes very strong, and the tradition was still in the process of developing an agreed-on auxiliary theory identifying under what conditions and for what kinds of issues effects were most in evidence. Additionally there was the problem that real-world events and trends (the so-called real-world cues, according to Funkhouser 1973; Erbring, Goldenberg, and Miller 1980; Behr and Iyengar 1985; Bartels 1993) might explain both the journalistic agenda and independently public perceptions of issue importance that would make the media–public opinion correlation effectively spurious.

Robert Michels's Iron Law of Oligarchy. Michels was a prized student of sociologist Max Weber and with the publication of *Political Parties: A Sociological Study of Oligarchical Tendencies of Modern Democracy* in 1911, I would have to suspect, he must have made his patron proud. Michels was active in the swirling socialist politics of his day in England, France, Germany, Switzerland, and Italy, and his book was based on a case study of the German Social Democratic Party. The key finding of his analysis, as summarized by Seymour Martin Lipset was the following:

> Political parties, trade unions, and all other large organizations tend to develop a bureaucratic structure, that is, a system of rational (predictable) organization, hierarchically organized. The sheer problem of administration necessitates bureaucracy. As Michels stated: "It is the inevitable product of the very principle of organization. . . ."
>
> Every party organization which has attained to a considerable degree of complication demands that there should be a certain number of persons who devote all their activities to the work of the party." But the price

of increased bureaucracy is the concentration of power at the top and the lessening of influence by rank-and-file members. The leaders possess many resources which give them an almost insurmountable advantage over members who try to change policies. Among their assets can be counted: (1) superior knowledge, e.g., they are privy to much information which can be used to secure assent for their program; (2) control over the formal means of communication with the membership, e.g., they dominate the organization press; as full-time salaried officials, they may travel from place to place presenting their case at the organization's expense, and their position enables them to command an audience; and (3) skill in the art of politics, e.g., they are far more adept than nonprofessionals in making speeches, writing articles, and organizing group activities. (1970, 413)

The choice of a socialist political party was particularly apt because the mission that motivated the party's creation was the empowerment of the working people and certainly not the creation of yet another self-interested power elite. Lipset's engagement with Michels's work was ongoing, and it inspired Lipset's similar case study of oligarchic tendencies in labor unions (Lipset, Trow, and Coleman 1956). But the ongoing focus on socialist parties and labor unions may have obscured a fundamental point in Michels's argument. The mechanism is not restricted to democratically oriented organizations (although that heightens the irony). Michels argued that this was an inherent tendency of all organizations. Once created, the leaders of organizations displace the original goals that led to the creation of the collectivity with self-perpetuation of the organization and their power within it. In fact, the English translation of Michels's title may involve a bit of mistranslation. The last word in the German is not *Democracy* but rather *Gruppenlebens*, that is, group-life, collectivities, organizations—reflecting his development of a more general lawlike phenomenon that extends beyond formal politics. So what has become associated in the literature as a commentary on the political Left is really a broader argument about social organization in general—the displacement of initial collective goals, whatever they might be, with a fundamentally conservatizing impulse for organizational self-preservation. In fact, Michels begins his book with a discussion of one of the more extreme examples of self-perpetuating personal power as noted above—the ancestral monarchy that is characterized by the royal families not as their own convenient invention but the divinely mandated will of God. The emphasis on unions and political parties is understandable given that the Marxist ideal of a people's republic was so dramatically corrupted by the self-serving brutality of the Soviet and Chinese communist parties. That corruption is arguably one of the most prominent and unfortunate historical facts of the twentieth century. But here I want to emphasize the broad generality of Michels's observation. The lesson for the purposes of studying the cultural and institutional structure of communication is straightforward enough—maintaining a truly open marketplace of ideas will be frustrated by those who perceive themselves as benefiting from the current state of affairs and will mount considerable energies against any voices inclined to suggest possible change.

The Matthew Effect. Another widely recognized social mechanism, aptly labeled by sociologist Robert Merton after the biblical proverb in the book of Matthew: "For whosoever hath, to him shall be given, and he shall have more abundance: but whosoever hath not, from him shall be taken away even that he hath" (13:12, King James translation). In the biblical version the mechanism appears to be divine intervention. For Merton and other theorists working on variations of the phenomenon of accumulated advantage, it is instead entirely human in origin and effect. For our purposes here it is an important extension of Michels's point that the established elites generally have more resources to maintain their position than those who would challenge them. Merton (1968) was working on patterns of scientific citation and noted that more eminent scientists will often get more credit than a lesser known or younger researcher for equivalent or collaborative achievements as a phenomenon of reputational cumulative advantage. This is a widely acknowledged phenomenon in the arts, literature, and, of course, the economics of the popular arts, most notably among "bankable" movie stars (Elberse 2006). This lawlike pattern crops up again and again in parallel literatures usually with a fresh label and somewhat different empirical context. In mathematics it is called a Yule Process after statistical pioneer Udny Yule (1925), who worked on mathematical representations of the mechanism. In linguistics it is referred to as Zipf's Law drawing on George Zipf's (1949) statistical study of word use frequencies. In computer science, telecommunications, and economics it has several manifestations, including Metcalfe's Law, Network Externalities, Preferential Attachment, and Positive Feedback Lock-In (Arthur 1994; Economides 1996; Barabasi 1999; Shapiro and Varian 1999). In the business world the same basic pattern is often referred to as a first mover advantage, bandwagon effect, or information cascade (Sherif 1936; Lieberman and Montgomery 1988; Bikhchandani, Hirshleifer, and Welch 1992). All of these are mechanisms of accumulating advantage. In networking, once a network grows large incorporating most individuals or units, any new competitive network, starting small by definition, is less attractive as a means of interconnection. In computer systems, analysts point to Microsoft's near-universal presence in business word and number processing, which poses as unenviable challenge to any potential non-interoperable competitor (Eisenach and Lenard 1999). Such processes are inherently conservatizing, reinforcing the status quo against any prospective alternative. Social inequalities, hierarchies, and stratifications are central to the field of sociology as are their political equivalent in political science. I will not dwell on how these mechanisms play out in each domain (see Newman 2005) but acknowledge the structural parallels here as we embark on the study of its manifestation in the organization of communication and the flows of ideas in a society.

Collins's Attention Space and the Law of Small Numbers. Randall Collins is a prolific sociologist at the University of Pennsylvania with unusually broad-ranging interests. Some years ago he became fascinated with patterns of intellectual debate and decided to study it systematically. The work took decades and ultimately resulted in the eleven-hundred-page tome entitled *The Sociology of Philosophies: A Global Theory of Intellectual Change*, published in 1998. It is an unprecedented and unusual work on many dimensions. The historical and geographic breadth is stunning. He traces in detail the networks of influence and debate among philosophers from ancient Greece, China, Japan, India, the medieval Islamic and Jewish world, medieval Christendom, and modern Europe. But he was less interested in the philosophic content itself than the patterned network structure of multigenerational intellectual influence. Philosophy like many intellectual endeavors, develops "schools of thought," formal or informal collectivities of like-minded individuals. Schools are highly self-aware, conscious of the defining borders between schools, and most often in conflict with each other (in this case intellectual conflict). His dramatic conclusion in reviewing this multi-millennial arch of intellectual work is that there is a consistent pattern of a distinctly limited number of schools a phenomenon he calls the "Law of Small Numbers"—from three to six schools at any one time. Further, he identifies a dynamic mechanism that maintains this law's properties. Simply put, dominant schools tend to divide into factions, while smaller and weaker schools ally and combine to create a stronger competitive capacity to attract attention. It is the ultimate "citation analysis" tracking the multigenerational influences of, by Collins's count, the world's 136 major, 366 secondary, and 2,152 minor philosophers. The key finding of relevance for communication structure in general is the notion of an ongoing competition for attention among members of the field (as in Bourdieu's concept of field) or thought community. The collective dynamics work much like the cognitive dynamics at the level of the individual—at least in the sense that the collective capacity for attention is also limited. In Collins's vocabulary there is within each field an *attention space,* and from his analysis the space proves to be defined by minimum of three schools to a sustainable maximum of six. So the law of small numbers and the concept of attention space echo at the collective/macro level the limited attention of the individual citizen at the micro/psychological level. (This echo is evident in varying degrees in all six mechanisms.) Both levels are of critical importance to understanding the dynamic structure of communication systems and the public sphere. Of course, Collins built his theory analyzing a rarified sample of philosophical dialogues; it remains an open question whether the limits of attention in the public sphere of a modern nation will reflect the same constraints and the same mechanisms that generate the law of small numbers evident in media agendas and public opinion. For the moment I leave that as an empirical question to be addressed.

Downs's Issue-Attention Cycle. Anthony Downs, a noted economist and public policy analyst, was pondering the cycles of policy debate with particular interest in matters environmental and put together a somewhat cynical theory of five stages of public attention and inattention that has broad and significant relevance to delicate dynamics of the public sphere. Downs explains that the cycle

> is rooted both in the nature of certain domestic problems and in the way major communications media interact with the public. The cycle itself has five stages, which may vary in duration depending upon the particular issue involved, but which almost always occur in the following sequence:

1. The pre-problem stage. This prevails when some highly undesirable social condition exists but has not yet captured much public attention, even though some experts or interest groups may already be
alarmed by it. . . .

2. Alarmed discovery and euphoric enthusiasm. As a result of some
dramatic series of events (like the ghetto riots in 1965 to 1967), or for
other reasons, the public suddenly becomes both aware of and alarmed
about the evils of a particular problem. This alarmed discovery is in-
variably accompanied by euphoric enthusiasm about society's ability
to "solve this problem" or "do something effective within a relatively
short time. . . ."

3. Realizing the cost of significant progress. The third stage consists
of a gradually spreading realization that the cost of "solving" the problem
is very high indeed. Really doing so would not only take a great deal of
money but would also require major sacrifices by large groups in the
population. . . .

4. Gradual decline of intense public interest. The previous stage becomes almost imperceptibly transformed into the fourth stage: a gradual
decline in the intensity of public interest in the problem. As more and
more people realize how difficult, and how costly to themselves, a solution to the problem would be, three reactions set in. Some people just
get discouraged. Others feel positively threatened by thinking about
the problem; so they suppress such thoughts. Still others become bored
by the issue. . . .

5. The post-problem stage. In the final stage, an issue that has been replaced at the center of public concern moves into a prolonged limbo—a
twilight realm of lesser attention or spasmodic recurrences of interest.
(1972, 39–40)

To my ear, this sounds very much like a logical extension of the law of
small numbers and the related notion of an attention space. I address the
question of whether the digital world of bloggers, social networkers, and
infinite online "shelf space" for the public display of ideas will ultimately
enlarge the capacities of these constraining agendas. It is far from clear that
the answer is a simple yes. I cite Collins and Downs here rather than
the much more prominent notion in communication research of the
"agenda-setting function" of the press seminally presented by Maxwell
McCombs and Donald Shaw (1972) because the former emphasize the
limited capacity of the public agenda and the latter take that limitation as
an unacknowledged given in their study of potential media effects.

Noelle-Neumann's Spiral of Silence. German public opinion researcher Elisabeth Noelle-Neumann introduced a concept of public opinion dynamics in the 1970s that has become an accepted staple of communication research (Noelle-Neumann 1973, 1984, 1993). In my view it is an empirically grounded variant of Jürgen Habermas's (1990) celebrated notion of an "ideal speech situation," as both concepts emphasize the sensitivity of individuals to speak out when they fear their true opinion may be in the minority and that they will be become embarrassed or socially isolated were they to state their views. In Noelle-Neumann's formulation, she posits a negative spiral of a minority position in public debate as the perception of an unpopular minority status increases in turn as fewer and fewer feel it is safe to speak out. Using over-time public opinion data on voting intentions in Germany of that era, she demonstrated a remarkable ability of voters to sense a changing climate of opinion and bandwagon effects as most notably those with the least confidence in their political views most fear isolation or embarrassment and are inclined to "run with the pack" (1984). Having seen the phenomenon in her own polling work, she turned to the historical literature and found that the idea resonated strongly with the work of Locke, Hume, Rousseau, and especially de Tocqueville:

> I was encouraged when I found a precise description of the dynamics of the spiral of silence in Alexis de Tocqueville's history of the French revolution, published in 1856. Tocqueville recounts the decline of the French church in the middle of the eighteenth century and the manner in which contempt for religion became a general and reigning passion among the French. A major factor, he tells us, was the silence of the French church: "Those who retained their belief in the doctrines of the Church became afraid of being alone in their allegiance and, dreading isolation more than error, professed to share the sentiments of the majority, So what was in reality the opinion of only a part . . . of the nation came to be regarded as the will of all and for this reason seemed irresistible, even to those who had given it this false appearance." (1984, 7)

More recent commentaries on Noelle-Neumann's work and subsequent empirical tests by others reveal that the perception of how much one's own view differs from those in one's community is in fact correlated with a decreased willingness to speak out. But the correlations are quite weak, and the strength of correlation varies across cultural settings, with types of issues, and with the specificity of the perceived "public" holding differing views (Scheufele and Moy 2000). Morally oriented public issues and the definition of the public as a smaller and more proximate social group appear to strengthen spiral of silence effects. Such findings do not necessarily lessen the importance of the concept but rather clarify the conditions under which it is most evident. My reference to spiral effects here is intended to emphasize the macro level of public opinion and climate-of-opinion phenomena. These dynamics are further complicated by the prospect that public perceptions of majority opinions may sometimes be in error, a phenomenon dubbed "pluralistic ignorance" by Katz and Allport (1931) in early work on the social psychology of identity formation and prejudice. So for any issue at some point in time there are actually two "distributions of opinion," the actual opinion distribution and the distribution as perceived by the public and possibly reflected in media coverage as events and perceptions (and misperceptions) of events interact in dynamic ways. The two distributions may or may not be closely related. The term *silent majority* in American political rhetoric and analysis acknowledges that on some issues the two distributions may be at variance (Rosenberg, Verba, and Converse 1970).

We have now reviewed six related macro-level phenomena characterizing the dynamics of the public sphere—media agenda setting, the Iron Law of Oligarchy, the Matthew Effect (and related dynamics of accumulated advantage), collective attention space, the issue-attention cycle, and the spiral of silence. It could be argued that each is a contributing element to the same underlying paradox of pluralism and each helps explain why the much celebrated democratic and pluralistic open marketplace of ideas is hard to sustain and that institutional mechanisms may be needed to protect an open flow of ideas.

Communication Structure and Social Structure: Big Data in Historical Context

It may strike some readers as rather obvious that powerful and wealthy elites are motivated to structure social and communicative relations so they stay powerful and wealthy. Such matters have not escaped the attention of communication scholars. But here, there is a curious development. There is not one literature on this topic but rather two literatures, each almost entirely unaware of the other. The social science tradition of media effects, of course, with its roots in the study of the propagandistic power of statist regimes addresses this issue. And the cultural studies tradition does the same in its study of the hegemonic power of capitalist regimes. But both traditions can be seen as confronting virtually the same underlying social dynamics—systematic distortions in the socially structured flow of ideas. Both posit an overwhelming power to the incumbent elite and a limited capacity to resist on the part of mass audience. And both literatures in turn ignore a third stream of scholarship that addresses the routine policy battles, contested framing of issues and events, the circulation of elites—the field of public opinion and electoral politics. This scholarly balkanization is understandable enough although undoubtedly an impediment to meaningful intellectual progress. Is it possible that with a modest reframing of the analytic question of how to optimally protect against the self-protective impulses of political and media institutions that a confluence of intellectual energies and research would result? Let us review such a prospect.

Hegemony (and Propaganda). Antonio Gramsci as an active communist organizer in Italy in the early twentieth century did not have much time for pondering history. But when Mussolini threw him in jail in 1926, Gramsci found many long hours to think about why it was that the object of his appeals, the Italian working class, seemed to be more attracted to bourgeois or fascist fantasies than a "true consciousness" of their working-class status. His resultant analysis of the "cultural hegemony" of capitalist elites, published posthumously as his *Prison Notebooks* (1933) written from 1926 to 1937, continues to be influential a century later and has become required reading in many domains of cultural studies. It is influential because he initiated a movement away from an incompletely developed and somewhat deterministic notion of superstructure and false consciousness in classic Marxist texts to a notion of a layered, partially contested, multidimensional, historically evolving "common sense" of the mass audiences that may be influenced by but not necessarily controlled by the media institutions (Hall 1986). Such a concept focuses on the possible reinforcement of taken-for-granted social inequities and power relations. It draws attention to subtleties of word usage, issue framing, and causal attribution. It identifies a rich and promising field of research.

Research on "Media Bias." One way to respond to Gramsci's concept is to revert to the original Marxist binary of true consciousness and false consciousness. Following that model the researchers observing the world believe that they have a correct understanding; they proceed to analyze media content and determine that it is at some variance from their view and label that difference to be a "media bias" caused by the self-interests or ideologies or deficient professional practices of media institutions. It is seductively easy to cherry-pick telling examples of media bias, and the condemnation of unfair bias has a resonating urgency with many readers. Accordingly critics on the left (for example, Glasgow University Media Group 1976, 1980; Gitlin 1979; Herman and Chomsky 1988; Entman 1989, 2004; Alterman 2004; Chomsky 2004; Bennett 2011) find evidence of conservative bias generally attributed to the conservative views of media executives and investors. In turn critics on the right (for example, Efron 1971; Theberge 1981; Lichter, Rothman, and Lichter 1986; Kincaid, Aronoff, and Irvine 2007; Groseclose 2011) find a liberal bias usually attributed to left-leaning reporters and academics. The phenomenon of disliking media coverage of contested issues is so prominent that it has acquired a name—"hostile media effect"—whatever your political perspective, you are convinced the institutionalized bias is in the other direction (Vallone, Ross, and Lepper 1985). Critiques of media bias of this sort in the tradition of political advocacy are important and most welcome. Indeed, the fact that such diverse political advocates find compelling evidence of bias might reasonably be taken as evidence of a healthy mix of political views in the marketplace of ideas. My only concern is that such oratory may tend to displace a more serious and systematic examination of Gramsci's thesis and the parallel conceptions in the work of Lippmann and Lasswell on propaganda. A systematic examination would not address just the presumably powerful effects of these biases and distortion but the conditions under which they were and were not successful—the historic cycles of cynicism and enthusiasm, engagement, and resentment, and at times resignation among the mass population in response to events as lived and as depicted in the media.

At this point, we can move the argument forward a bit and proffer a potential solution to the problems of inference and causal direction in the traditional paradigm of hegemonic media effects—big data in historical context. I refer to this approach to inquiry as Communication Trend Analysis. It is not a very catchy terminology, but it is, in my view, a pretty catchy concept and well timed to put the big data created by the digital media environment to very good use (with appropriate protections of personal privacy). The traditional model of public opinion proposes that the establishment media set the public agenda of what issues and events are appropriate to discuss and, in addition, how the issues are generally framed.

Media Agenda Effects

The original formulation of the agenda-setting hypothesis in the media effects tradition had just two elements—the media agenda and the public agenda. The former was measured by a content analysis of news coverage of various issues and the latter by public opinion measures of issue importance. This was the case in the seminal McCombs and Shaw paper of 1972 and in most follow-on studies in the tradition. The high correlation between the two was taken as evidence of a media effect because the reverse direction (public interest influencing media coverage) was taken to be implausible (McCombs and Shaw 1972) a proposition that found some support follow-up research (Weaver et al. 1981; Behr and Iyengar 1985; McCombs 2004). It soon became clear that to answer the question of causal direction adequately, three elements would be necessary: (1) some form of veridical real-world data related to the significance of issues or events, (2) media coverage, and (3) public opinion preferably with meaningful variation in all three over time.

This general direction of research was the media effects equivalent to the hegemonic control hypothesis of the cultural studies tradition and the cultivation effect associated with Gerbner and colleagues and resonated among many researchers as a spirited demonstration of not-so-minimal effects. But it became increasingly clear that a variety of patterns may actually be evident: (1) neither the media nor public may respond to real-world cues, (2) the public may respond to real-world cues although the media do not, (3) the media may emphasize an issue independently of real-world cues and the public may or (4) may not respond, (5) the media may downplay an issue despite dramatic real-world cues and the public may or (6) may not respond. This formulation is a much richer one because instead of evidence or absence of evidence of a "media effect," it proposes to inquire

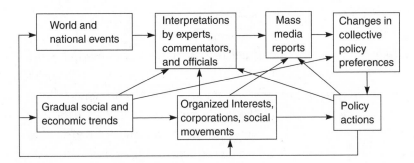

Figure 6.1.   Causes in Changes of Public Opinion
Source: Page and Shapiro (1992), Figure 8.3, p. 354. Reprinted by permission of Chicago
University Press.

under what conditions the media attempt to set the agenda, and, if so, why,
and at times why not, and under what conditions do they succeed.

A number of researchers have tackled an analysis of concurrent communication trends of this sort but because of the tremendous measurement complexities, the full analytic model is only infrequently employed
(Land and Spilerman 1975; Aborn 1984; Fan 1988; Neuman 1989). The systematic collection of reliable public opinion data has not yet reached the
hundred-year mark, so we are constrained historically for systematically
collected data on public opinion and belief. But archived economic and
institutional data and media coverage of concurrent issues and events invites attention, and historians have proved to be remarkably inventive in
finding useful indicators of public sentiment, including but frequently
moving beyond electoral data (Tilly 1970). Of particular interest are the
American and French Revolutions and the American Civil War period
(Davidson 1941; Wilson 1962; Burnham 1970; Tilly 1998, 2002, 2004).
Famously the historical analyses from the eighteenth, nineteenth, and early
twentieth centuries of de Tocqueville (1955, 1961) and Lippmann (1922)
continue to inspire ongoing modern research. Ben Page and Robert Shapiro
have given these issues considerable thought, and their 1991 masterwork
The Rational Public contains extensive analyses along these lines. They are
more interested in the politics of changing opinion rather than the media's
role or media bias, but their underlying model captures the critical need
to independently assess the direct and indirect (mediated) effects of
changing conditions (as illustrated in Figure 6.1). Similar models with virtually the same elements and causal pathways have arisen elsewhere in the

literature, for example, Baum and Potter (2008) and Shoemaker and Vos (2009). In examination of a particular trend we might find that the media and elites influence the public salience and framing of events independently of actual changes in social and economic conditions—clear evidence of hegemonic effects. Or it may be evident that public sentiments follow their own dynamic independent of or perhaps ahead of changes in the media—representing independence or oppositional readings of real-world cues. The challenge becomes comparing and contrasting analyses across diverse issues and political contexts to build an accumulative empirical basis for understanding how distortions in the flow of information arise and, of course, when well-resourced attempts to distort, suppress, or exaggerate are actually unsuccessful.

An Iron Law of Media Hegemony?

Is there a grand conspiracy among the cigar-smoking media barons and the political elites in the modern industrial democracies to suppress speech potentially critical of capitalism and social and economic inequities? Despite recent revelations of the behind-the-scenes maneuvers of prominent media mogul Rupert Murdoch (McKnight 2012) and similar recent corporate media scandals (Halberstam 1979), even the most vehement critics are not inclined to characterize the power structure as a grand conspiracy (Gitlin 1980; Herman and Chomsky 1988; McChesney 1999; Bennett 2011). *The hegemonic pattern, such as it is, is incomplete, self-contradictory, contested, and, interestingly, not particularly secret.* Herman and Chomsky, for example, bristle at the conspiracy-theory label: "We do not use any kind of 'conspiracy' hypothesis to explain mass-media performance. In fact, our treatment is much closer to a 'free market' analysis, with the results largely an outcome of the workings of market forces. Most biased choices in the media arise from the preselection of right-thinking people, internalized preconceptions, and the adaptation of personnel to the constraints of ownership, organization, market, and: political power. Censorship is largely self-censorship, by reporters and commentators who adjust to the realities of source and media organizational requirements, and by . . . the constraints imposed by proprietary and other market and governmental centers of power" (1988, xii).

This pattern is, in fact, much more interesting than a grand conspiracy. Public opinion on issues of politics, economics, inequality, and social questions is complex and surprisingly dynamic (Stimson 1991; McCombs and

Zhu 1995). Sometimes public perceptions seem to move in phase with the inevitable business cycles, punctuated by economic and banking scandals such as the Savings and Loan debacle of the 1980s or the more recent mortgage crisis, while other widely held perceptions about the distribution of power and opportunity remain unchanged (Mayer 1992; Owens 2012). Sometimes perceptions about economics align with social class status while others do not (Lipset 1960; Gilens 1999; Frank 2004; Bartels 2008). Some efforts to defuse public concerns over economic inequities are stunningly successful; others fail rather dramatically (Lippmann 1925; Ginsberg 1986; Zaller 1991; Lewis 2001; Bartels 2008).

Stanford political scientist Shanto Iyengar has provided us with a particularly helpful analytic tool for exploring these dynamics. In a study of the psychology of blame in the domain of public affairs, Iyengar (1991) asks who the public holds responsible when things go wrong and how might patterns of media coverage influence this dynamic. The key distinction is between news coverage he terms *thematic* that emphasizes broader historical and structural conditions versus *episodic* coverage, which takes the form of a case study or event-oriented report and depicts public issues in terms of concrete instances and individuals. He finds an overabundance of episodic coverage and observes experimentally, for example, that those who are exposed to an episodic news story on poverty are more inclined to blame the poor themselves for their circumstances rather than social conditions. The factors in the traditions of journalism that lead to this style of reporting are subtle and complex and may simply be derived from the impulse to tell a good story with concrete exemplars rather than represent a strategic apologia for the power elite (Schudson 1982; Neuman, Just, and Crigler 1992).

Another important concept in this tradition of scholarship is the notion that the poor do not object to inequality per se, just that they have ended up at the wrong end of the unequal income distribution curve. One of the most powerful demonstrations of this counterintuitive but important finding comes from Frantz Fanon's (1963) study of Algerians under French colonial rule in the 1950s. Fanon's work as a psychiatrist led him to marvel at how often oppressed colonials fantasized not of a just and equitable political regime, but rather one in which they could exert the unquestioned power they observed in the French administrators and military. The less dramatic version of this may be the widely cited distaste of middle-class Americans to estate taxes on the very wealthy, as noted above, based in part on the common aspiration among many that one day they will be very wealthy, too (Bartels 2008).

It is also frequently acknowledged that resonances in working-class culture with conservative social issues such as aversion to gun control and religious opposition to abortion and to the acceptance of homosexuality complicate what might otherwise be a more dramatic alignment of public opinion along lines of economic position, although this proposition remains contested (Frank 2004; Bartels 2008).

So the evidence available thus far reveals neither an iron law nor a grand and surreptitious conspiracy. The tradition of cultural studies uses the term *hegemony* to characterize these patterns and relies primarily on textual analysis to repeatedly illustrate its existence. The social science tradition undertakes pretty much the same task using the vocabulary of agenda-setting rather than hegemony and illustrates the existence of media power by means of correlational analysis of issue prominence in the media with issue salience in public opinion data. Both traditions appear to treat the phenomenon as an inevitable constant or law of some kind rather than as a variable of particular interest. Under what conditions and for what kinds of issues does hegemonic agenda setting succeed, succeed partially, simply fail, or actually boomerang? Large corporate motion picture studios and television networks have no reluctance to promote narratives that characterize business leaders and financiers as greedy, dishonest, and vain, especially if such stories prove to be profitable. The fictional Gordon Gekko's "Greed Is Good" speech, for example, has become iconic. Murdoch's own newspapers covered his corporate phone hacking scandal in painful detail. The Occupy Wall Street movement, including its thematic emphasis on economic inequities, the political power of corporations, and lack of criminal charges in the mortgage crisis and related banking scandals, has been fully covered by the traditional and online news media although typically of protest coverage, the potential of violence or illegal behavior on the part of protesters seems to attract the most attention (Alessi 2011; Santo 2011). As one observer put it: "At the moment, the protesters who've been mocked on CNN and Fox News, accused of class warfare by Mitt Romney, and handled delicately by the White House, have . . . decent favorable ratings. Americans are divided on the protestors themselves. Thirty-three percent (33 percent) have a favorable opinion, 27 percent hold an unfavorable view, and a plurality of 40 percent have no opinion one way or the

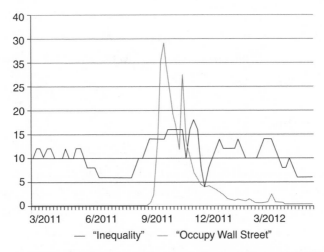

Figure 6.2.   Google Search Frequencies for "Inequality" and "Occupy Wall Street"

other. . . . One way of looking at this is that the Occupy Wall Street movement is more than twice as popular as Congress" (Weigel 2011).

The slogan "We are the 99 percent" appears to have achieved some traction in the public sphere, but the public attention span is short and it remains unclear whether the visceral symbolism of park protesters in Manhattan translates in public pressure for changes in policies and practice.

Figure 6.2 represents a rough approximation of an answer to whether the episodic narrative of the occupation of Zuccotti Park in lower Manhattan near Wall Street translated into an increased public concern with economic inequities in the United States. The figure plots the Google search term frequencies for the terms "Occupy Wall Street" and "Inequality" from the spring of 2011 to mid-2012, utilizing Google's standardized search volume index. The apparent answer is that it did not, but a more complete answer would require a more nuanced inquiry into how the public interprets the motives and character of the protesters and whether the thematic concerns about inequality were or were not obscured by the episodic specifics of the park occupation and the response of law enforcement authorities. In all likelihood some observers saw the protestors as unkempt latter-day hippies with a lot of time on their hands while others saw them as politically energized youth bent on reforming our economic system, and such perceptual lenses are in all likelihood built up over a lifetime rather than a few weeks of news viewing. Again, the pattern is one of distinctive resonances in how public events are interpreted rather than a mechanical cause and effect. Further, the pattern appears to be one of self-reinforcing spirals over time—a phenomenon our research methods are just beginning (and struggling) to address.

Mediated Matthew Effects

The blogosphere does indeed offer the prospect energized grassroots entrepreneurialism—let a thousand Phyllis Schlaflys and also a thousand Gloria Steinems bloom—what a most interesting garden. The phenomenon of the Matthew Effect, however, posits a set of mechanisms of cumulative advantage frequently summarized by the phrase "the rich get richer" that work in the opposite direction. The somewhat more technical language for analyzing such phenomena includes the terms "positive feedback," "first mover advantage," "preferential attachment," and "network effects." The literature is extensive (Merton 1968; Adamic et al. 2001; Barabasi 2002; Watts 2004; Newman 2005; Napoli 2008). Our particular concern in the domain of the public sphere is how the famous become more famous and because of the limited breadth of public attention make it particularly hard for new voices and new ideas to be heard. The basic phenomenon is widely recognized and well understood—the very fact of being on the best-seller list or the box office champion draws self-reinforcing further attention.

But now a new vocabulary has been popularized to suggest how the dynamics of digital profusion might actually engender a significant and largely welcome counterforce to the already-famous-get-all-the-attention dynamic. The seminal thinker here is *Wired* magazine's then executive editor Chris Anderson and the terminology he has popularized focuses on the notion of "the long tail" (Anderson 2004, 2006). Anderson acknowledges that the self-reinforcing psychology of best sellers is still in force but, using the bookstore as the central example, unlike the local brick-and-mortar store that can stock a few thousand unique titles, web-based national giants like Amazon can stock hundreds of thousands of unique titles and now boasts of having sold more than seven million unique titles to date (Rosenthal 2011). I address the dynamics of the long tail in some detail in Chapter 7, but for now it is important to acknowledge that a well-stocked bookstore, surely a good thing of course, does not lead inevitably to an agile and dynamic public agenda and a fully open and equitably competitive marketplace of ideas. Bloggers may well write persuasively and informatively about critically important injustices, but does there remain a threshold of collective public attention for an issue to be taken seriously in the world of politics and public policy and in the mainstream large-audience commercial media? Such inquiries, of course, resolve ultimately to the questions—how large is the attention space, and does the digital environment enlarge it meaningfully?

Mediated Attention Space

Earlier in this chapter I introduced the idea of attention space by reference to sociologist Randall Collins's (1998) study of the history of schools of philosophy. His analysis, it will be recalled, revealed a law of small numbers—in the recorded history of philosophy over the millennia, he concluded, there were no fewer than three and no more than six recognized schools of significant philosophic thought. It was his way of demonstrating within this particular recorded field of inquiry that attention space is limited. And he posited a dynamic sociological mechanism by which the law of small numbers is generally maintained—big schools split up and smaller and less prominent schools combine. In the earlier review of human cognitive psychology we encountered a similar law of small numbers—George Miller's (1956) celebrated magic number seven—the experimentally determined number of considerations or distinctions most of us can manage at one time in our heads. And Miller posited a strikingly similar mechanism of what people do when the number of factors to be considered is much larger—we "chunk"; that is, we group a set of similar factors together as one so the overall number of factors is more manageable.

Is there evidence of a corresponding law of small numbers at work in the public sphere? Does collective attention work like individual attention? Do newly arriving public issues push older ones into obscurity? The short answer is yes, indeed—there is a great deal of evidence to that effect. In studies of the public agenda in the traditional media and by a variety of measures of public opinion analysts have discovered parameters quite similar to Miller and Collins ranging from three to seven (Hilgartner and Bosk 1988; Neuman 1990; Zhu 1992; Brosius and Kepplinger 1995; McCombs and Zhu 1995). The language varies but the underlying idea is the same. Hilgartner and Bosk (1988) refer to what they call the "carrying capacity" of the public arena. In an over-time analysis that demonstrates there is a critical minimum of media attention before public opinion responds, I used the terminology of a "threshold of public attention" to issues in the news (Neuman 1990). Zhu (1992) refers to a "zero-sum theory of agenda-setting," Brosius and Kepplinger (1995) use the terminology of "issue competition" and a model of "equal displacement" as new issues "kill off" older ones. McCombs and Zhu (1995) refer to the "carrying capacity," "diversity," and "volatility" of the public agenda. And most of these scholars tip their hat to the seminal work of Anthony Downs, some citing his classic observation "American public opinion rarely remains sharply focused upon any one domestic issue for very long even if it involves a continuing problem of crucial importance to society" (1972, 38).

But now a new wrinkle—*how will the shift to the informational abundance of digital technologies and the evolution from push to pull media–audience dynamics influence, if at all, the size and character of the public agenda?* It turns out to be an extraordinarily difficult question to address. Our empirical measures and theoretical conceptions of how the dynamics might work are not yet very sophisticated. I'll mention a few of the difficulties that need to be addressed.

First of all, the idea of a public agenda is a tricky one (Wlezien 2005). If an interviewer asks you what the most important issues facing the country are (the classic inquiry used for years by Gallup and the American National Election Study) and waits patiently for you to respond, you are likely to scan your memory banks for a few issues in the news you can recall—that is what other people rather than what you personally might rate as most important. Then after ticking off a few you stop because it is hard to remember and you have politely fulfilled the requirement of the interpersonal exchange by providing a few answers. Perhaps a more deeply probing question would be: "What are a few important issues facing this country which you believe are not getting the attention they deserve?" However, I do not believe such a question has been asked systematically. Further, the finite constraints of a single web page or newspaper front page or broadcast newscast make a finite news agenda a mechanical truism. The delicate but centrally important question of whether the public agenda is responsive to the diversity of issues and framing of issues and deeply felt concerns in the thinking of the citizenry has not yet been fully addressed by the limited tools of empirical scholarship. It could be said that it was the disjuncture between the public political agenda of his day in Italy and his own personal political concerns that motivated Antonio Gramsci's original musings on the phenomenon he in his frustration labeled "hegemonic control."

Second, there is the difficult question of what have been called issue publics—a notion originally popularized by Philip Converse (1964) to acknowledge that even if citizens typically do not study all public issues in-depth, there may be some groups of issue specialists. Examples include African Americans following civil rights (Iyengar 1990), Jews attending to the Middle East (Krosnick and Telhami 1995), and veterans keeping an eye on military-related issues (Jennings and Markus 1977). The logic here is that if a concerned group is keeping an eye out, they will function successfully as a monitor and alert the broader public when trouble looms (Neuman 1986; Schudson 1998; Zaller 2003). Accordingly, research techniques would need to track not only broader public attentiveness to issues but attentiveness among more specialized groups, as well. Zaller (2003) uses the metaphor of a burglar alarm, and the broader challenge becomes to understand under what conditions when, say, a specialized group in the blogosphere rings the alarm over an issue or event, the public at large does or does not notice.

Media Effects and Issue-Attention Cycles

As we discuss the limited breadth of public attention, we are drawn inevitably back to the insights of economist Anthony Downs (1972) and his case study of the issue-attention cycle. He was writing, of course, at the height of the domination of the traditional print and broadcast media, and his informal expectations suggested that these issue cycles would most likely be measured in some modest number of years rather than days or months. But in the era of viral videos and twenty-four-hour news cycles, we may want to consider measures as well in minutes and hours. The case study Downs picked, the issue of ecology, although it is not likely that Downs anticipated it, turned out to be an unusual one. Writing just a decade after Rachel Carson's seminal *Silent Spring* (1962) had begun to stimulate what would become a social movement by raising public attention to the particular danger of pesticides, Downs defined the ecology issue as "cleaning up our air and water and of preserving and restoring open spaces" (1972, 43). He concludes his analysis by noting, "There is good reason, then, to believe that the bundle of issues called 'improving the environment' will also suffer the gradual loss of public attention characteristic of the later stages of the 'issue-attention cycle'" (50). How dramatically wrong Downs turned out to be in this particular case does not diminish the importance of his insight into various forms of issue "wear out," but it points out the importance of more sophisticated models of issue evolution as underlying and enduring concerns take on new valences as events arise and policies develop (Carmines and Stimson 1989; Page and Shapiro 1991; Stimson 1991, 2004; Kellstedt 2003).

Political campaign strategists, journalists, and public relations professionals generally have a pretty good ear for determining whether a story, an event, issue, or revelation of some sort will have "legs" or will fade away buried into oblivion by whatever arises in the next news cycle (Arno 2009). Their judgment is intuitive and no doubt based on their experience with thousands of stories. But as media professionals they may opine, but they are not inclined formalize or test their thinking in terms of a systematic theory. That is the job of the research community.

Unfortunately the state of the literature developed by the research community on what gives an issue or news story legs—that is, ongoing public concern and attention and media coverage—is not yet convergent. A number of scholars have compiled lists of the factors that contribute to an event's sustaining viability in the public sphere but a coherent model with at least some empirical support remains elusive. Table 6.2 represents a paraphrased summary of several prominent lists. Scholars in the field of journalism and mass communication have struggled for much of the last century with developing a reliable typology of "news factors" that is qualities of an issue or event that make it newsworthy and sustain attention in the press and public (Hughes 1940; Ostgaard 1965; Galtung and Ruge 1965; Buckalew 1969; Molotch and Lester 1974; Gans 1979; Schulz 1982; Gamson 1984; Staab 1990; Heath and Heath 2007). The difference in emphasis among the sampled typologies in Table 6.2 is striking. Most of this work simply analyzed news content or journalists' "gatekeeping" behavior without attempting to assess public interest. The studies that empirically contrast journalists' judgments of newsworthiness with citizens find sur-

Table 6.2 Models of Issue Endurance in the Public Sphere

Heath and Heath (2007)	Brosius and Kepplinger (1995)	Gans (1979)	Buckalew (1969)
1. Simplicity	1. Personal consequences	1. Rank of official	1. Significance
2. Unexpectedness	2. Danger and threat	2. Collective impact on nation	2. Conflict
3. Concreteness	3. Change in knowledge	3. Impact on largest number of people	3. Prominence/celebrity
4. Credibility	4. Feedback	4. Significance for future	4. Proximity
5. Emotional involvement	5. Relatedness		5. Timeliness
6. Storylike structure	6. Symbolic value		6. Visual interest

prisingly low correlations and sometimes no correlation at all (Martin, O'Keefe, and Nayman 1972; Neuman, Just, and Crigler 1992; Jones 1993; Boczkowski and Peer 2011). Further, Doris Graber (2007, 265), for example, reminds us that in the Pew Research Center surveys over the past two decades only 7 percent of the stories tested attracted "a great deal of attention" from at least half those interviewed. So we can conclude that research on what makes a particular issue resonate over time in the public sphere remains at an early stage, and that the extensive over-time big data on media content and audience attention, audience commenting, and audience forwarding/linking offers great promise to move theory building and testing forward.

Table 6.3 represents my effort to integrate the key dimensions in the developing news factors literature with an emphasis on the question not just of newsworthiness but of longer-term continuity. So returning to the original paradox of Anthony Downs's incorrect prediction that the ecology issue was likely to disappate in media and in public attention over time we see that each of these proposed elements in Table 6.3 was at least partially engaged and served to sustain and evolve the public issue. First, new scientific evidence became public about the serious health consequences of environmental neglect so the importance and the number of people affected was understood to be larger than originally perceived. Second,

Table 6.3   Revised Model of Issue Endurance in the Public Sphere

Dimension	Key Elements	Sources
1. Perceived issue salience/ importance	Fundamental newsworthiness in eyes of media and of audience in terms of importance and human interest	Galtung and Ruge 1965 Shoemaker and Vos 2009
2. Ongoing developments, unanswered questions	Additional "energy" to sustain issue interest from new developments	Brosius and Kepplinger 1995
3. Public feedback	Public opinion, protest, commentary	Brosius and Kepplinger 1995
4. New dimension to recognized issue	Shift in issue framing by media or audience	Carmines and Stimson 1989
5. New players in existing issue or event	Turnover among political or economic leaders	Zaller 1999

new events from the revelation of new superfund sites to nuclear reactor disasters keeps environmental issues in the public mind. Third, a full-fledged environmental movement of activists and supporters arose to draw public attention to these issues. Fourth, new dimensions to environmental concerns such as, for example, the matter of global climate change, draw new attention and energy to this issue domain. And fifth, not so much new elites, but in this case an entire generation of concerned younger citizens adopted the issue as their own (Van Liere and Dunlap 1980; Guber 2003; Nisbet and Myers 2007; Hansen 2009).

The accumulated research in several research traditions convergently conclude that rather than a simple and perhaps mechanical model of agenda setting, it appears that the relationship between the media and the public is highly dynamic and variable. Sometimes issues wear out; other times they grow into mass movements. We are just beginning to understand the conditions that distinguish the two.

Perhaps the ultimate question here could be characterized as a variant of Shanto Iyengar's famous distinction between the more concrete "episodic" framing and the more historically socially grounded "thematic" framing of public issues. Iyengar's analysis focused on how such differences in media framing could affect audience perceptions at one point in time. Drawing on Iyengar's insight is likely to be useful as well in the analysis of longer-term trends in the public sphere. It appears, for reasons we do not yet fully understand, that both the media and the public evolved from a largely episodic to a predominantly thematic interpretation of the environmental issue in the decades following Downs's failed prediction.

Spirals of Dominant Opinion

Finally, we turn to Noelle-Neumann's provocative thesis about spirals of silence. When I introduced this work earlier in the chapter it was noted that Noelle-Neumann's theorizing added several new dimensions to the analysis of media–audience interactions over time. First of all, the very notion of self-reinforcing spiraling dynamics is a commendable step forward from rudimentary stimulus-response causal models. Second, although her own data were primarily short-term electoral polling studies, she positioned her theory in a much richer historical grounding of longer-term social change with examples from Europe and North America. Third, although she focused on the prospect of negative spirals, as those holding minority opinions would become increasingly reluctant to speak out, it is clear these go-with-the-crowd psychodynamics could also work as well in the opposite direction as fads and fashions drew larger numbers into supporting a particular position or framing of an issue. Fourth, the key underlying idea in this work is that in addition to the distribution of opinion, there is, importantly, a second distribution of the perceived climate of opinion that may or may not accurately reflect community feelings. It is presumed that the media could be particularly influential in influencing the latter distribution. Fifth, these dynamics highlight the question of trends in opinion polarization, political comity, and cultural tolerance that are hypothesized to be linked to the changing structure of media technologies. All of these issues raised in the spiral of silence tradition are quite resonant with the broader body of work concerning potential hegemonic influences in, for example, Gramsci, Habermas, and Hall, but with an emphasis on over-time dynamics and empirical investigation of the success or failure of potential influence efforts by elite individuals or institutions.

As with the cultivation analysis research tradition, which often focused quite narrowly on expectations of violent crime and perceptions of the numbers employed in law enforcement, spiral of silence research has focused on a relatively small set of issues originally identified by

Noelle-Neumann. The challenge now is to move beyond the seminal studies to a broader array of issues and research designs to assess the dynamics of both positive and negative spirals and the potential role of the media in these processes. Her challenge to the field in the 1970s remains relevant and resonant today: "Mass media are part of the system which the individual uses to gain information about the environment. For all questions outside his immediate personal sphere he is almost totally dependent on mass media for the facts and for his evaluation of the climate of opinion. He will react as usual to the pressure of opinion as made public (i.e., published). Research will have to be increasingly concerned with questions about how the prevalence of opinion on specific topics or persons originates in the media system, and which factors promote or inhibit it" (1974, 50–51).

Noelle-Neuman was a pollster, and most of the researchers working in the tradition she established are also survey research specialists. As a result, most of the research has focused on opinion trends rather than the institutional character of the media system. So it may be useful to step back and ask the most fundamental question—given that the media systems of modern industrial democracies are primarily privately owned commercial enterprises, would we expect these institutions to primarily emphasize and reinforce the prevailing climate of opinion or to pay particular attention to minority or deviant points of view?

There may not be a full consensus on this question, but my reading of the literature leads me to conclude that the answer is that the commercial media tend to be bipolar—that is, they are very conservative and because of their commercial marketing orientation are careful not to offend any significant group by violating conventional norms, and they are also drawn to highlighting deviant behavior because it attracts audience attention. On issues such as the depiction of racial diversity and professional women outside the home in commercial television in the United States, for example, the industry was slow to adjust to changing cultural norms. On the other hand, on the issues of antiwar protest and environmentalism in the 1960s and 1970s the media was generally quite attentive. Like the media, the research literature here is a bit bipolar as critics from the Left and from the Right critique the media for being captured by the other side. As a result a balanced assessment may need to be excavated from the polemics.

There are two important findings from the more recent spiral of silence literature that deserve special attention and may help integrate this tradition in with a broader model of the evolving structure of political communication. The first finding is drawn from a carefully conducted meta-analysis of empirical studies in this tradition. The conclusion is that although the effect is evident in most studies and most often statistically significant, the size of the effect is actually quite small (Glynn, Hayes, and Shanahan 1997; Shanahan, Glynn, and Hayes 2007). So the reluctance to speak out that would otherwise be clearly inimical to an open marketplace of ideas is limited to some citizens and some circumstances and as a result still important but perhaps less critical. The second finding is that reluctance to speak out appears to be most evident among personal acquaintances rather than a more generalized abstract community of public opinion (Moy, Domke, and Stamm 2001). This draws attention to the fact that the original empirical studies Noelle-Neumann conducted were in Germany, a country noted for its (until recently) relative ethnic and cultural homogeneity and national identity. In more culturally and ethnically diverse nation states, it is likely that subcommunities of opinion expression and activism may represent an important counterbalancing process.

There is an analytic concept from a neighboring arena of research that could and should be drawn into this sphere of scholarship. The concept is referred to as indexing (Bennett 1990) or bracketing (Hallin 1984), a dynamic defined basically as "by defining the boundaries of acceptable controversy, the media define the range of legitimate opinions that the public may adopt" (Mutz 2006, 239). In other words, although minority views may be covered by mainstream media, the coverage may characterize those views as deviant or without legitimacy. It is a promising area for more sophisticated textual analysis of the media combined with expanded survey research not just on opinions but perceived climates of opinion.

So the final question to be addressed here is what evidence we have developed so far on how the changing digital environment might influence spirals of silence or of issue engagement. The literature leads us to five tentative conclusions.

1. The profusion of channels of electronic communication generates a high-choice environment. Markus Prior (2007), among others, has pointed out that those with little interest in political life, when given the option will opt for nonpolitical entertainment content and as a result spiral from limited to perhaps even more limited

political information and engagement. Basically inadvertent exposure to politics is reduced. Inadvertent exposure may not have been the site of significant political education but still important at the margin. So, for some we may witness a spiral of disengagement.

2. In contrast, for the politically engaged the environment of profusion is a virtual candy store of public information and political debate. Pippa Norris (2000) identifies this as a virtuous circle. Accordingly, in other circumstances we may witness a positive spiral of political engagement and possibly both positive and negative trends for different groups at the same time.

3. The profusion leads to an environment of what might be termed niche news—highly specialized, more detailed, and often much more opinionated and adversarial news and commentary. There are two important qualifiers to this observation. Although niche news is available, the overwhelming majority of online information seekers continue to rely on the mainstream media, typically established newspapers and broadcast networks (Hindman 2009). And further, we cannot be sure these niche sources of information are actually a new phenomenon given the ongoing importance of interpersonal conversations, group participation, and two-step flows of information and interpretation (Katz and Lazarsfeld 1955; Huckfeldt and Sprague 1995; Mutz 1998, 2006; Huckfeldt, Johnson, and Sprague 2004). It may be simply that some portion of interpersonal discussions has shifted to an online format.

4. Importantly, the extensive research on increasing isolation from and lack of knowledge about individuals and communities different from oneself hypothesized by Turow (1997) and Sunstein (2001) is not supported in the data (DiMaggio and Sato 2003; Mutz 2006; Garrett 2009; Hindman 2009). Republicans may not be happy with the policies the Democrats propose, but they are certainly well aware of them.

5. And, similarly, although there is dramatic evidence of increased polarization of elites, this polarization, so far at least, is not also reflected in the mass population (DiMaggio, Evans, and Bryson 1996; Evans 2003; Baldassarri and Gelman 2008; Fiorina, Abrams, and Pope 2010). The issue of bracketing and indexing remains an

important one, but unfortunately, there is simply not adequate analysis to date to confirm a trend or continuities in that dimension as the public sphere becomes increasingly electronic.

A Digital Marketplace of Ideas

The focus of this chapter has been on the trials and tribulations of sustaining cultural and political pluralism in the public sphere as we move from a system based on traditional print and broadcast media to a vastly larger and more complex electronic structure of communication. An ongoing theme has been the question—will the natural human impulse to identify with others like ourselves and to be invidious in our dealings with out groups be transported from the level of individual psychology to the organizational structures of the public sphere? To answer such a difficult question I started first with the relevant research literatures of the past century that theorized about what was and was not working in protecting the public presentation of minority views from various mechanisms of majority censorship in the traditional media systems with an emphasis on the American case. I put six popular theories to work, each positing a particular social mechanism that restricts openness or distorts market mechanisms in favor of established interests. Three of these theories dealt primarily with possible patterns of elite domination—agenda setting, the Iron Law of Oligarchy, and the Matthew Effect. Two others dealt with the limitations of public attention—collective attention space and the issue-attention cycle. And the final theoretical tradition—the spiral of silence—added an important new dimension to the analysis of participation in the public sphere—the perception of the legitimacy of expressing unpopular or minority views.

A pattern has emerged in our analysis. And the pattern reflected a general weakness in the state of theory building and theory testing in the social sciences and those empirically oriented reaches of cultural studies that are closer to the social sciences. The pattern was this: the seminal theorists would posit a mechanism such as, for example, agenda setting and then proceed to try to find evidence to support the existence of the mechanism. Because over time measurement is expensive and time-consuming most of the work involved short-term surveys and content analyses or experiments. Counterevidence was generally ignored, and dramatic variance in indicators of the strength of the mechanism was also ignored. All that was required was some measure of statistically significance and publication followed with celebration of purported evidence that the posited mechanism does indeed exist.

The argument of the chapter then shifted from analysis to exhortation in an attempt to make the case that the dramatic variations in the strength of effects (and absence of effects and reverse effects) were actually the most interesting contribution to theoretical refinement—under what conditions and for what types of issues do these mechanisms of control of and distortion of an otherwise open marketplace of ideas take hold?

The movement from a mechanical model of effect and true–false dichotomous notion of theory testing to a more dynamic and refined one is still at an early stage. The older and simpler models are more intuitive and resonant with normative groundings of particularly the political left. Hopefully a movement to more conditional models of agenda setting and the like will not be taken as an abandonment of political first principles.

Do we have a firm conclusion about whether the digital public sphere will be more conducive to a vibrant and open marketplace of ideas than the industrial print and broadcast media of the twentieth century? A conclusion, yes. A firm conclusion, well, not yet. The conclusion is that the diversity, openness, flexibility, and global span of the electronic public sphere all represent an important step forward. There are exceptions and the forces of various versions of cultural and political uniformity have not diminished as the centuries turned. Many would, as China and some other nation-states have tried, to reprogram the Internet. So we conclude with a tip of the hat to Benjamin Franklin, who, as he left Independence Hall on the final day of deliberations at the close of the Constitutional Convention of 1787, was approached by a woman who asked, "Well, Doctor, what have we got—a republic or a monarchy?" "A republic," replied Dr. Franklin, "if you can keep it." That is the challenge to the electronic public sphere. Largely off to a promising start. The future in this regard, however, will hinge not on the evolution of technology but of public policy. To address such questions we need to draw on a different literature, a somewhat different style of argument and analysis, and even a different form of interaction between the normative and the demonstrably empirical. This is the challenge of Chapter 7.

7

Public Policy

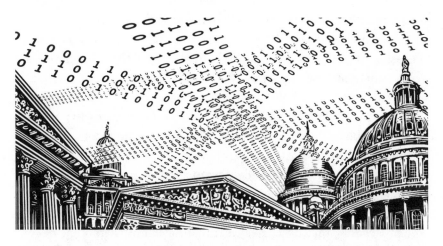

We need to recognize that the entire information sector—from music to
newspapers to telecoms to Internet to semiconductors and anything
in-between—has become subject to a gigantic market failure in slow motion.

—ELI NOAM (2004)

A culture without property, or in which creators can't
get paid, is anarchy, not freedom.

—LAWRENCE LESSIG (2006)

Unless publics and counter-publics are able to reach and influence
institutional power, online networks risk the fate of previous
attempts to establish democratic media: that they wither under the
shadow of their own political insignificance.

—STEPHEN COLEMAN AND JAY BLUMLER (2009)

I HAVE ATTEMPTED in Chapter 6 to make the case for the marketplace of
ideas as a guiding normative principle of communication scholarship. I
turn now to explore the marketplace-of-ideas notion as an equally good
grounding principle for public policy more broadly defined. There exists
a broad array of criminal and civil laws that guide the practice of business
and finance—why would we need a special category of public policy and

regulation for the domain of human communication? Good question. But the answer is unambiguously yes; public speech is a special category.

Mark Fowler, who was appointed by President Reagan as chairman of the principal regulatory body in this domain, the Federal Communication Commission (FCC), had an answer: "Television is just another appliance. It's a toaster with pictures. Let the people decide through the marketplace mechanisms what they wish to see and hear. Why is there this national obsession to tamper with this box of transistors and tubes when we don't do the same for *Time* magazine?" (Fowler 1981). His remark enraged and energized liberals, who continue to quote that remark to this day as a sign-post of conservative cluelessness about the special needs that require ongoing public scrutiny and at times regulation to protect the vibrancy, openness, and fairness of the public sphere.

Fowler's remark, now three decades old, is so iconic that all one needs to do in a gathering of American communications policy specialists is mention the word *toaster* to provoke a round of knowing chuckles. But before proceeding, I need to make clear that the simple stereotype of conservative and liberal views on government regulation—conservatives want less; liberals want more—that underlies Fowler's metaphor can be misleading in this domain of policy. Policy activists and lobbyists who rely on all the traditional conservative rhetoric are more often than not actually seeking government intervention. They are delighted with government intervention in the marketplace when it serves the economic interests of the particular entities they represent. It is a point that has been made powerfully by economist Bruce Owen and his associates in a series of case studies (Owen and Braeutigam 1978; Noll and Owen 1983). There are few philosophical purists in the lobbying business. I make this point emphatically at the outset because it is an important component of this chapter's central theme. The groups of active and resourceful policy players in the United States and around the world each seek a digital marketplace of goods and of ideas that gives them and their associates a sustainable competitive advantage. A truly open marketplace could only be the goal of naive dreamers or a useful rhetorical strategy for those who seek precisely the opposite. So it is to the good fortune of the policy process in the United States and globally that there actually are some philosophical purists in the lobbying business, groups like the Electronic Freedom Foundation, Free Press, and Public Knowledge that struggle along on public contributions and mostly volunteer labor from a mix of enthusiastic students and veteran attorneys. They describe their reason for being as fighting for a "free and open Internet" (Declaration of Internet Freedom 2012). They are resourceful and committed, but they are overwhelmingly outgunned by the various large corporations and industrial groups. It is David versus Goliath. Thus the theme in this chapter—*as various policy decisions are made in the decades ahead that will shape the structure of public communication, it is unlikely that an open marketplace of ideas will be the central motivating value*. It will instead be motivated by goals of protecting the profitability of intellectual property and law enforcement and protecting children from pornography. Sociologist Robert Merton famously warned that the "empirical observation is incontestable: activities oriented toward certain values release processes which so react as to change the very scale of values which precipitated them" (1936, 903). In other words, as we seek various other, perhaps even desirable, ends, without sustained vigilance, a vibrant and open digital marketplace of ideas may be slowly but effectively eroded. Table 7.1 outlines some of the political forces that occupy our attention in the pages ahead.

Regulating the Marketplace of Ideas

Mark Fowler's provocative metaphor is, unfortunately, not particularly exceptional in American policy debates. The characteristic and some would say defining American impulse is to rely on a market mechanism rather than an administrative one whenever possible (Lindblom 1977; Williamson 1985). Most of the world's nations until quite recently contentedly relied on the government to run not only the post office but the telephone system and broadcasting system as well (Neuman, McKnight, and Solomon 1998). One has to step back from the accepted American institutional traditions to realize that the system is actually more than a little bizarre. In the 1930s the government essentially gave the public broadcast spectrum for radio and later for television to commercial applicants who were simply asked to serve "the public interest, convenience, and necessity" a curious phrase with historical roots in utility regulation that was never defined with any clarity and virtually never enforced in eight decades of routine license renewal (Pool 1983; Neuman, McKnight, and Solomon 1999). An analysis of judicial and regulatory rulings reveals that it was simply assumed that the

Table 7.1   Political Forces in Internet Policy

Interest Group	Motivation	Prototypical Legal Initiative—United States
Internet service providers	Exploit price discrimination	Network neutrality policy
Intellectual property owners	Limit potential infringement	Digital Millennium Copyright Act 1998
		Stop Online Privacy Act (proposed)
Security and law enforcement	Intercept criminal communication	Communications Assistance for Law Enforcement Act 1994
Political authorities	Prevent/monitor seditious speech	No explicit initiative to date in United States, numerous examples globally
Tax authorities	Enforce local sales tax online	U.S. federal legislation proposed
Moral authorities	Limit sexually explicit content	Children's Internet Protection Act 2000
Moral authorities	Limit hate speech	No explicit initiative to date in United States, some examples in Canada and globally
Marketers	Develop detailed profiles of consumer behavior	Extensive data gathering, some privacy protection legislation pending

discipline of the commercial marketplace tied to audience ratings would suffice (Krasnow, Longley, and Terry 1982; Robinson 1989; Brock 1994; Einstein 2004). As a result, when candidates wish to share their policy views and aspirations with potential voters at election time, they are required to pay commercial rates for adverting time. Despite the growth of online electioneering, television is still the main forum of the electoral public sphere, and commercial time is expensive. The 2012 campaign season witnessed the expenditure of more than two billion dollars in advertising for the presidential race and another two billion for the races in the House and Senate—again primarily for television time (Center for Responsive Politics 2012). The corrupting influence of commercial advertising expenses on the political process is widely acknowledged (Ackerman and Ayres 2004). Europeans (campaigning in Europe is largely publicly funded and airtime provided as a public service) observe this process of constantly raising and spending billions of dollars for access to the public spectrum with curious fascination (Nassmacher 1993). If the president wanted to speak to the American people, and the press, for some reason, chose to ignore the request, the presumptively most powerful individual in the world would be helpless and without the resources to effectively communicate save perhaps for the occasional visitor to Whitehouse.gov. Such a scenario, of course, is unlikely, but it serves nonetheless to starkly highlight the curious structural arrangements of the American public sphere.

The present chapter reviewing how public policy affects the structure of communication, gives us the opportunity to explore the underlying debate about digital political economy in a little more detail. *The Internet provides the opportunity in the public sphere for all who had only the opportunity to listen the opportunity now to speak, as well.* But the Internet has had a dramatic, some would say disastrous, impact on the financial support for independent journalism, especially the critically important and expensive tradition of investigative journalism (Downie and Schudson 2009). And setting up a website or blog does not guarantee an audience. Hindman notes, for example, "Yes, almost anyone can put up a political Website. But this fact means little if few political sites receive any visitors. Putting up a political Website is usually equivalent to hosting a talk show on public access television at 3:30 in the morning" (2009, 56). So in addition to a discussion of the public policies themselves guiding public communication, I review the dynamics of audience behavior and associated media economics of the digital environment. We turn first to the four key paradoxes identified in the Chapters 3–6 as each suggests a somewhat different take on the public policy challenges ahead.

Profusion. An abundance of information and channels for communication, of course, is the principal rationale behind the deregulatory movement advocated among the conservative ranks. Market-championing libertarians like Peter Huber (1997) propose abolishing the FCC and his conservative compatriot George Gilder (2002) would agree pointing to limitless bandwidth and limitless communication. This rejection of regulation is more than a little ironic because much of what has made the media so profitable over the past century has been the limitation on market entry of potential competitors in telecommunications through common carriage regulations that mandate monopoly provision of regulated wireline telephone service as a natural monopoly and the spectrum scarcity that effectively limits broadcasting to those largely commercial entities that already have licensed frequencies.

American political culture celebrates the First Amendment and the tradition of a free press as a counterpoint to centralized government power. A particular irony of the digital profusion is that the advertising-based business model for independent newspapers (and to a certain extent broadcast news) is being threatened by the profuse competition. For a series of complex reasons the advertising income for online-distributed news is only one-seventh the income from traditional newsprint (Edmonds et al. 2012). Media economics is in flux, so it is not yet clear whether a new commercial model will replace the old one, but there is reason for concern about the economics for supporting high-quality professional and independent journalism, with sufficiently deep pockets to stand up to both government and various vested interests. Highly profitable media enterprises evolve when demand exceeds supply, and the digital profusion is wreaking havoc in that department.

Critics of the commercial public sphere have traditionally and appropriately wrung their hands over the potentially stultifying gatekeeper function of the media industries—the core of the concern with diversity of media ownership and agenda setting. Profusion turns the gatekeeper notion on its head—but raises the prospect of a whole new gatekeeper/agenda-control problem. When information is abundant, the problem is not the existence of information but finding the wanted informational needle in a potentially massive digital haystack. So control of the search process, rather than the production process, becomes an increasingly serious concern. In the United States, Google famously controls about two-thirds to three-quarters of the market for web searches (Auletta 2009; Segev 2010). They profess that they would not consider steering search to serve their interests for fear of alienating users, but concerns appropriately remain—biases can be subtle and the difference in visibility of being listed on the first of subsequent pages of search are significant. Listings on the first page of Google search results are thirty times more likely to engender a response than those on the second page (iProspect 2006). Furthermore, there is even a more serious and subtle contender for digital gatekeeper—the internet service providers (ISPs) that have the technical capacity to filter the flow of information and, if inclined, to effectively censor any number of sources. This has been evident in some authoritarian regimes at the authorities' behest and not yet evidently in Western industrialized democracies—but, as before, vigilance is appropriately motivated and a potential delicate question for public policy. In the American case this has thus far been manifested in a debate about "network neutrality," a topic I address in some detail shortly.

Finally, profusion inherently raises the question of the citizen's capacity to control access to the private details of their lives. Virtually every communicative and financial exchange leaves a complex digital footprint. The contested politics of privacy in a networked world could have consumed the attention of this entire manuscript, but I attempt only a brief review in the pages ahead.

Polysemy. Human communication's poetic strength and its critical weakness derives from the fact that what one says and what another hears engages the fundamental and inevitable ambiguity of words and images. And, indeed, depending on the circumstances, contrasting, say, an art gallery with a surgeon's operating room, we might alternatively celebrate or agonize over the prospect of communicative ambiguity. How might such a fundamental and defining characteristic of collective communication inform the design of public policy? Here, perhaps a relatively modest insight from the domain of content regulation. Take, for example, the long-standing attempt to constrain the propagation of obscenity and pornography. United States Supreme Court Justice Potter Stewart famously remarked, in his attempt to describe a threshold test for pornography in *Jacobellis v. Ohio* (1964): "I shall not today attempt further to define the kinds of material I understand to be embraced within that shorthand description [of hard-core pornography]; and perhaps I could never succeed in intelligibly doing so. But I know it when I see it, and the motion picture involved in this case is not that." Such judicial candor is rare, but it reveals the difficulty of trying to apply hard-and-fast bureaucratically enforced rules to censor one category of speech and protect another. The impulse to censor and regulate public speech can be strong. Precisely because of the paradox of polysemy, however, *the better public policy is to counter bad ideas in the public sphere with more speech rather than try to identify ill-conceived utterances* *and aspire to extinguish each and every one.* Put in general terms, policies reliant on censorship appear to be evidently ill conceived, and fears of evil propaganda can motivate strange public policy.

I have noted recurrently that when explicitly confronted with the question, virtually all observers acknowledge the fundamentally polysemic character of human communication. Yet when it comes to crafting policy and responding to the existence of highly unpopular and threatening speech, the third-person effect takes hold, and thinking regresses to a mechanistic and simplistic notion of cause and effect. This appears to be the propaganda problem reborn. The challenge to public policy is to sustain faith in the open-market principle and to empirically verify that the best counterpoint to error-ridden or hurtful speech is corrective speech rather than attempts at suppression.

Polarization. A recurring theme I have addressed in multiple chapters is a concern epitomized by legal scholar Cass Sunstein (2001) and others about increasing political and cultural polarization particularly in the United States but also globally. In a world of push media, there are structural incentives associated with maximizing audience size and not alienating potential advertising targets that reinforce the journalistic principles of moderation, balance, and fairness. Sunstein in particular goes further to argue that the identity-driven impulse toward homophily leads to a self-serving isolation, a *Daily Me* that leads to increasing ignorance of the views and logic of those with whom we disagree or of different cultural traditions. It turns out Sunstein got it half-right. People do indeed behave homophilically and like to stick with their own kind and to expose themselves to opinions reinforcing their own views, but, importantly, they do not seem to exercise the predicted avoidance behavior (Garrett 2009). In fact, advocates and opinionated pundits spend more than a little time critiquing and making fun of rather than avoiding the views of "the other side." Democracy is messy. Advocates may disagree on both means and ends. I review the prospects for protecting a public sphere that is vibrant but as civil and deliberative as possible.

Pluralism. Sustaining a lively and open pluralism is precisely the challenge at hand if we are to think about how communication flows could be self-consciously structured to counter the various individual-level and institutional-level distortions we have reviewed. The thesis of critical communication scholarship is that because of the historical dominance of the capitalist elite, the media have hegemonically distorted the flow of information and public discussion to reinforce and reproduce their position of dominance. One need not imagine smoke-filled rooms and a secret society of media executive conspirators. The critique merits further attention even if mechanisms involved turn out to be partial, not particularly self-conscious, and frequently unsuccessful. Powerfully articulating one's self-interest in the public sphere is not in itself a bad thing—the trick is making sure everybody gets their chance and that it is a reasonably level playing field.

Technologies of Freedom

In 1983, quite early in the digital revolution, the MIT political scientist Ithiel de Sola Pool published a prescient volume entitled *Technologies of Freedom*, which brought together his thinking about how the forces of technology would likely change the nature of broadcasting and publishing. The title is an engaging turn of phrase, and it introduces what turned out to be an important and influential book. The manuscript was published a full decade before the modern public Internet would first start to take shape. But Ithiel, who started his career focusing on political philosophy, had become increasingly concerned that the shifting technical basis of communication he witnessed firsthand from interaction from his more technically inclined colleagues at MIT might be used to weaken the First Amendment tradition in the United States. Although he had dabbled in Marxism as a young man, he had become disillusioned with the Left and the new Left and at this point was not particularly concerned with potential commercial abuses of the changing media environment. His focus was government's inclination and the increasing ability to constrain or discourage open criticism. His overview of the American regulatory traditions in communication is masterful and merits careful attention, but I have space only to summarize these issues briefly here. What made his overview particularly unusual is his inclusion of the history of telephonic regulation as well as publishing and broadcasting. A schematic of his model is summarized in Table 7.2.

The central technologies-of-freedom argument posits that our regulatory traditions for public speech are deeply rooted in different historical

Table 7.2   Three Regulatory Traditions in American Communication Policy. See Pool (1983).

Technical Basis of Communication	Regulatory Tradition	Basic Rationale
Publishing	First Amendment— free press	A celebrated principle of American democracy since 1791 and supported by numerous court decisions in the ensuing centuries—since there is no limit to the number of printing presses, counteract potentially false speech with more free speech rather than some form of government censorship, regulation or taxation.
Public telephone network	Common carriage	Since 1913 the then privately owned and competitive telephone companies became a single system, in effect a monopoly, regulated by the principles of common carriage—guaranteed access with charges regulated by public utility commissions and no constraints on speech.
Broadcasting	Public trustee	Because of the limited availability of electromagnetic spectrum for radio and television broadcasting the FCC in a formal process since 1934 makes channels available in exchange for broadcasters to serve as public trustees of the spectrum and serve the "public interest, convenience and necessity" a vague and seldom enforced requirement that is believed by many simply to be served best by the discipline of the commercial marketplace though advertising demands and audience ratings.

eras, different technologies, and different assumptions about whether the character of those technologies make possible open-market entry for competition or might require a stronger regulatory hand because of spectrum scarcity or natural monopoly properties. Pool's review of the history and his prescient prediction about the electronic convergence that was to take place dramatically over the twenty-five years following his book's publication raises the corresponding question—if these three separate traditions of regulation are no longer appropriate, what model should we adopt for the future—a mix of the three, one of the three, an entirely new formulation, or complete deregulation and a proud public policy of communicative laissez-faire? Furthermore, perhaps we can build on what we have learned from the successes and failures of these traditions to date to meet the ideals of a vibrant public sphere: respect and even encouragement for minority voices and a resultant open market-place of ideas.

Pool's answer to this question was unambiguous—rely on First Amendment principles, protect the public from government censorship, and rely on market dynamics to keep journalists and public advocates on their toes. Pool's historical positioning of the question is nuanced and effective. His answer is characteristically crisp and coherent. But, in my view, his answer is incomplete. It is incomplete because his concern about distortions in the open marketplace of ideas focused on government censorship and effectively ended there. But, as we have seen, there are multiple other powerful forms of market distortion and self-interested manipulation and market failures that require ongoing vigilance—vigilance by the public, by the diverse array of policy advocates, both paid and unpaid, and by the government itself. This is especially true when the centuries-old business model for advertiser-supported independent journalism is in peril and the capacity of artists, authors, and other creators to protect the intellectual property they have created is also in peril because of technological upheavals. So we will take Pool's model as our starting point, the First Amendment, a platform on which to build a more complete public policy for protecting openness in communication structure.

Pool's analysis focused on the American case, so he did not comment in any depth on an alternative to the public trustee/commercial broadcast model that has been widely accepted elsewhere around the world—the public service broadcasting model as exemplified, for example, by the BBC in England and the CBC in Canada. The public service model is supported by a special-purpose tax to media users and the programming decisions are made by an arm's-length agency supported by the fees but largely independent of direct government control. With no commercial financing the mission of public service broadcasting emphasized informing and educating the public, as well as providing entertainment (Scannell 1990; Blumler and Gurevitch 1995; Coleman and Blumler 2009). Public broadcasting in the United States came much later historically, has a much smaller level of funding, has a smaller sliver of public attention, and its very existence continues to attract controversy (Ickes 2006). In fact in the American case there continues to be a statutory "domestic dissemination ban" that expressly forbids the internal broadcast of any government-created programming by, for example, the State Department's Office of Public

Diplomacy (Manheim 1994). The fear, of course, is that the government's official voice might powerfully overwhelm the public sphere in an era of limited publishing and broadcast outlets. Such evolved traditions need to be freshly addressed in the digital era.

Commercial Hegemony

I attempt to address a critically important question at this point. In its most straightforward formulation the question is this: *Has the digital revolution and the attendant shift from push to pull audience dynamics significantly challenged the long-standing commercial dominance of the media marketplace?* In my view the most straightforward formulation of an answer to this question is that it will depend on how we as a community of concerned public citizens respond to this critical juncture in public policy and commercial practice. Thus, I argue, such conclusions speak to the urgency of the issues this book attempts to address. In the pages ahead I adopt a two-stage strategy for addressing this question. The first stage is to interrogate how the now well-established community of critical communication scholars has responded to the digital revolution. Also in this first stage I examine whether the principal arguments of the critical community hold up well in the new media environment. The second stage is to review some new directions for critical commentary and research especially appropriate and reenergized by the altered conditions of the media/audience environment. I conclude that an entirely fresh paradigm of critical evaluation and empirical scholarship will be required, perhaps a "NeoCritical Perspective." (The term derives from the commonly used epithet *neoliberal,* which is generally taken to mean, in effect, conservative and market oriented. The term *neoconservative,* of course, also means conservative and market oriented, so terminological options are limited.)

The Critical Community Responds to the Digital Revolution. Is there a central theme and consistent voice discernible among the responses of the various academic critics of the hegemonic commercial media system? I believe there is, and it would be summarized by the admonition "Not so fast!" By that I mean to suggest that the critical community feels the optimistic voices proposing that the Internet and attendant digital novelties solve many of our long-standing problems with the dominance of big corporations, a one-way public sphere, and banal popular culture are being naive. The critics, it would appear, are enjoying the opportunity to poke fun at the optimists. Among them one might include Vincent Mosco. His 2005 volume is entitled *The Digital Sublime: Myth Power and Cyberspace*. Perhaps he is having too much fun quoting business consultant Don Tapscott (who has written now thirteen books about revolutionary change in the business world and makes his living selling these books, giving speeches and, of course, consulting). Mosco notes, "Here is a common refrain from one of the song books of the cyberspace revolution: 'Today, we are witnessing the early turbulent days of a revolution as significant as any other in human history. A new medium of human communications is emerging, one that may prove to surpass all previous revolutions—the printing press, the telephone, the television, the computer—in its impact on our economic and social life'" (2005, 18).

Mosco feels such hyperbole is in fact representative of a long-standing American celebration of genius inventors and material innovation and warns that such messianic rhetoric serves to distract us from continuing concerns about media ownership and dominant ideologies. Dan Shiller's *Digital Capitalism* (2000) makes many of the same points and emphasizes how the new technologies boost the reach of the increasingly large and transnational corporations. The bottom line shared among virtually all within this community is that although some celebrated institutions of twentieth-century communication capitalism may be in decline—notably the revered and once dominant metropolitan newspapers—the fundamental dynamics of political economy are still with us.

A second point frequently made, also a note of caution, is that we have had open technologies of the people before—notably citizens band radio, amateur radio, community access cable television channels, and the like, and it is difficult to sustain an independent public voice without financial resources and institutional support (McChesney and Scott 2004). The most impressive demonstration of the paradoxical failures of public access without institutional support took place in the United States in the 1970s when local franchise authorities and the federal government mandated that cable operators make channels available to local citizens to program whatever they cared to, as long as it did not violate copyright, pornography, and libel laws. Cable companies were required to provide video-recording equipment and/or studio space to those who requested it, but nothing more. After a flurry of experiments by high school groups and aspiring filmmakers, both the audience and the citizen programmers drifted away to other pursuits. In the mid-1970s the FCC required larger systems to make as many as twenty channels available, but they were largely unused. A few interview shows and coverage of local community meetings survived. In 1979 the Supreme Court ruled that these local origination content requirements violated the free speech rights of the cable companies, and the empty channels were replaced with commercial broadcast retransmissions (Pool 1977; Englemann 1990; Boyle 1997; Linder 1999; Lewis 2000; Mullen 2003).

A third point emphasized by Edwin Baker and Robert McChesney, among others, is that the rhetoric extolling the expansion of free market competition in the digital age ignores the important fact that the market never was free but was in fact a creation of federal subsidies and protections of incumbent players. According to McChesney, "It may appear that the profit-driven nature of the U.S. media system generates an inexorable logic that requires businesses to act as they do, for better or for worse. There is an element of truth to such a position, but taken in isolation it is also misleading. The larger truth is that the current media market's nature is set by explicit government policies, regulations, and subsidies" (2004, 210).

McChesney goes on to point out that the American broadcast television industry was extraordinarily successful in engineering what he calls a "corrupt" government giveaway of seventy billion dollars' worth of spectrum to keep broadcasters on the air with a digital signal at newly assigned frequencies at no cost to the broadcasters, which also managed to make sure no new spectrum for TV broadcasting was available to increase competition (2004, 214). *Corrupt,* however, implies it was a shady, behind-the-scenes deal or bribe. In fact it was much more interesting, a fascinating in-front-of-the-scenes case study of brilliant legal and technical maneuvering in a relatively open policy process, a case study in powerful vested incumbents winning the day (Brinkley 1997; Neuman, McKnight, and Solomon 1998). But it does not always work out that way, and the over-the-air broadcasters a few years later found their spectrum under fire again as other powerful interests, including the cellular telephone and computer industries, coveted their spectral resources (Genachowski 2011).

A fourth argument that arises frequently in this literature makes the point that the impulse toward commodification and consumerism remains strong and in many ways characterizes the evolving culture of the web. Pop-up windows may be said to represent commercial culture's attempt to put "push" back into the media mix. Indeed there was fad in 1996 when a hot Internet start-up called PointCast attracted a lot of attention by offering a "free" news and entertainment service supported by advertising formulated in a particularly obtrusive interruptions and flashing banners. It labeled its initiative a "push" technology and filed a patent using that language. PointCast discovered web users even in these early days had different expectations; annoyed by the interruptions, users migrated elsewhere, leaving PointCast as an early example of a dot-com flop. It is remembered now as one of the ten worst ideas in Internet history (Himelstein and Siklos 1999; Meyer 2005). The push concept lives on in spam email and variations of the pop-up window, but spam filters and pop-up blockers limit their effectiveness. So although the character of the interaction between the potential consumer and the marketer has changed, the commercial consumerist character of American public culture has not. Interestingly, analyses of how individuals use search engines day to day reveal that 30–40 percent of typical searches involve queries about products and services. So the materialism in this case comes from pull rather than push behavior (Neuman and Gregorowicz 2010). Whether the celebrated culture of materialist consumption in America is exaggerated by advertising-based media or simply exploited by them is a difficult question for further research. We know that pornography is a prominent element in the content associated with newly developed media technologies, and then it gradually recedes to the background as other content categories grow, of course, not disappearing (Theroux 2012). Perhaps we will see the same with pull inquiries about goods and services.

A fifth argument focuses on how the legal structure evolving out of older media technology becomes a significant impediment to the new, a phenomenon we might tag as the revenge of the old media (Marvin 1988; Chadwick 2013). The key example here is traditional intellectual property law designed originally to incent invention and cultural creativity that has been awkwardly grafted onto the new media environment (Litman 2000; Lessig 2004, 2006). Protecting the rights of creators to benefit financially from their creations is a significant challenge in a world where peer-to-peer copying and sharing is both costless and convenient. These issues are addressed in more detail ahead, but it is paradoxical that laws designed to promote innovation and creativity are being exploited largely by the receding older media interests, such as record labels and book publishers, to resist innovation and creativity.

Finally, many in the critical community, and most prominently Robert McChesney, recognize that if the public is going to become engaged with these issues and make a difference, the time is ripe as confidence in old business models and regulatory paradigms is shaken and new technologies raise fresh issues or more often old issues in new circumstances. McChesney uses the phrase *critical juncture* to identify this window of opportunity. I enthusiastic agree. He professes:

> Critical junctures [are] those historical moments when the policymaking options are relatively broad and the policies put in place will set the media system on a track that will be difficult to reroute for decades, even generations. Critical junctures are another way to say that society holds a "constitutional convention" of sorts to deal with the problem of the media. At these points there tends to be much greater public criticism of media systems and policies and much more organized public participation than during less tumultuous periods. Critical junctures can come about when important new media technologies emerge, when the existing media system enters a crisis, or when the political climate changes sufficiently to call accepted policies into question or to demand new ones. When two or all three factors kick in, there is a high probability of a critical juncture; at these historical moments, opportunities to recast the media that would be nearly impossible under normal circumstances can materialize. (McChesney and Scott 2004, 24)

McChesney is unusual among academic critics because he has a history of getting directly involved with the policy process. He organized a successful advocacy group called Free Press, which lobbies on media policy issues and organizes a large annual conference of academics, community organizers, foundation staff, independent media activists, and artists—the National Conference for Media Reform.

The Critical Community's Response to the Digital Revolution Remains Incomplete. One must admire the energy and perseverance of the media reform community. It is clearly characterized by a David and Goliath dynamic, and there should be no doubt about who is playing the David role in this scenario. The big media Goliath employs a phalanx of well-paid and sophisticated lobbyists bearing significant campaign contributions. Many of these lobbyists are former members of Congress and congressional staff specialists in the media field. Although the impulse to root for David is strong, I proceed by being critical of the critics for clinging to old slogans and appeals no longer relevant in the digital age and for failing to embrace and pursue the many positive elements of the digital difference that swirl around them.

The literature of media reforms celebrates a golden age of broadcasting when an activist FCC insisted that broadcasters meet their public trustee obligations to educate and inform and to promote open public discussion of controversial issues—the era of the Fairness Doctrine. The fact that numerous regulatory constraints on broadcasters associated with ascertainment of community interests, public access requirements, and prohibition of multiple station ownership have been weakened or abandoned is a source of great frustration. There are, however, two problems with this golden-age thinking in my judgment. First, the regulations never worked that well, and, second, the broadcasters were notoriously successful at not letting any of them cramp their commercial style. Public service and educational programming frequently found itself scheduled for the less-than-high-profile midnight to 6:00 a.m. slot. If a full-scale reinstatement of the Fairness Doctrine were instituted (albeit, an extremely unlikely scenario), it would represent a tiny blip in the torrential flow of information and entertainment to the public. The action now is online, and the online environment actually does overcome the gateway of access problem of the era of large-scale publishing and commercial broadcasting. It may sound melodramatic, but it is true that we have returned to an era that encourages open pamphleteering, not unlike the era of the American Revolution, and we await the next John Peter Zenger and Thomas Paine. It strikes me that some use their blogs and websites to scold the media for ignoring independent and alternative voices and visions, rather than using them to put forward the alternative voices and visions. Thomas Kuhn (1962) is famous for arguing that old paradigms are never abandoned by their champions and only decline when the old champions are eventually replaced by younger champions of new paradigms. Perhaps such a dynamic will characterize the literature of media reform, as well.

The good news is that the institutional, financial, and technical barriers to public speech have receded dramatically. The blogosphere is vibrant.

The number of open discussion venues is large and growing. The topical agenda is boundless. This could be a cause for celebration. But the critical community is quick to point out that although political diversity abounds, the most popular websites, as we have shown, continue to be dominated by familiar commercially oriented old-media corporations like ESPN, CNN, and the *New York Times.* This is an important point, and it should be the subject of serious research on the dynamics of attention in an online media environment—but it is not a matter that could or should be addressed by the FCC or regulatory invocations from Congress.

Perhaps the most exciting structural innovation in the new media environment is the explosion of collaborative and social media that permit new forms of networking and information sharing and the structured aggregation of content. The classic exemplars are Wikipedia, Facebook, Twitter, YouTube, Flickr, and Digg. This more participatory, interactive, and flexible style of online behavior has been dubbed Web 2.0, acknowledging a transformative generational shift in how the web is being used. Web 1.0 was based on websites—basically a broadcast/publishing model had been more or less moved wholesale from broadcasting and printing to Internet transmission and screen display. Some of the players in this process of innovation, such as the Wikipedia Foundation, have made a point of remaining a nonprofit. Others, and that would be most of the others, anticipate income from advertising, notably targeted advertising and new forms of recommendation-based marketing. Facebook had revenues exceeding five billion dollars in 2012, so despite initial travails concerning the firm's stock price there remains widespread anticipation that Zuckerberg and his colleagues will continue to successfully monetize the popularity of their social networking platform. But it is far from a picture of uninterrupted capitalist profiteering. Most start-ups lose money for years, and only a few turn the corner to profitability. And success in terms of popularity does not necessarily reflect success in terms of profitability. The widely used YouTube open-access platform for sharing videos engages such extensive technical costs associated with storing and streaming video that one estimate calculated that the enterprise until recently reportedly lost a half-billion dollars a year and survived with ongoing patronage from its parent company, Google (Spangler 2009). I return to the dynamism of these new forms of social interaction and the mix of commercial and noncommercial institutions later in this chapter. My point of emphasis here is that these matters are seldom addressed by the critical community as it continues with a now decades-old diatribe against corporate mergers and a lack of regulatory oversight of the old media public sphere.

Does the classic reformist critique need to be updated? Yes, of course. *The fundamental concern that elites will use their resources to skew the flow of information and interpretation to reinforce their position of advantage is, of course, as worthy of attention as ever.* But the reliance on the traditional critiques of corporate media and the need for a beefed-up regulatory regime may be counterproductive strategies. To strengthen that point, I return to a classic critique of the corporate media published originally during the Reagan era by Edward Herman and Noam Chomsky (1988). They dubbed their analysis the "Propaganda Model" and developed six themes in their introductory chapter that guided their analyses in the subsequent chapters and case studies. The themes are summarized in Table 7.3, along with a commentary on what has changed in the structure of public communication. The emphasis in the Herman-Chomsky volume is on media bias in the coverage of international events rather than inequities within the U.S. system of political economy. The motivations for any presidential administration to spin-doctor the interpretation of events to put them in the most favorable light is as strong as ever, but in virtually all of the six themes the original critique is either irrelevant, or more often, incomplete, ignoring the increasing potential for individuals to consult news from global sources. In 2011, for example, the uprisings in the Middle East drew increasing attention to the Al Jazeera English cable channel, which featured extensive coverage. Because, predictably, few cable systems carried the channel when it was introduced, viewers took advantage of the online video stream and, more recently, a mixture of video streaming and social media (Weprin 2011).

State Hegemony

Critics of the American media system, as noted above, focus on the tight linkages and mutual support between the corporate media and the federal government as both share a natural interest in protecting the status quo. In the American case, the state does not actually do the writing, editing, and producing of the media fare itself. Further, direct censorship of media content remains extremely rare outcome almost always related to special

Table 7.3   Herman-Chomsky Propaganda Model

Elite-Controlled Information Filters	Old Media Pattern	New Media Pattern
1) Media access	Critical and minority voices cannot afford to own major print and broadcast media outlets.	Major media still out of reach, but now plentiful alternatives online.
2) Audience maximization	Advertising economics skew content toward apolitical and lowest-common-denominator entertainment.	Lower production and distribution costs lower dependence on advertising, expansion of long tail, completely noncommercial alternatives viable online.
3) News reliance on official sources	Demand for reliable newsworthy news sources leads to dependence on official spokespersons reflecting elite perspectives.	Still largely true, but amateur blogosphere, alternative international sources, and amateur video capture of events in process offer possible counterpoint.
4) Flak against contrary voices	Institutionalized criticism of antielite views by right-wing think tanks and foundations.	Still largely true, but increased opportunity for commentary from Left.
5) Anticommunist ideology	Vilification of threat of communism justifies support of anticommunist dictators and distorts flow of international news.	Largely irrelevant, although vilification of other threats, notably Islamic elements, may lead to similar kinds of distortion.
6) Elite propaganda campaigns	Government elites draw attention to events and issues carefully crafted to strengthen their ideological position.	This is a variation of the long-standing belief that the Right in the United States is better at public relations than the Left. This may be true but has not yet been subject to systematic research or a model of why that would be necessarily true.

cases of national security. But in other state systems around the world, state control and censorship is much more direct and prominent, so a brief review of direct state hegemonic practices is in order.

The term *hegemony,* of course, focuses attention on some of the more subtle forms of filtering, interpreting, labeling, framing, narrative creating, and agenda setting that characterize the flow of communication in indus-trialized capitalist democracies with some measure of a free press. But in many putative democracies, the exercise of state hegemony, well, it might be said, lacks subtlety. The threats, imprisonment, and sometimes outright murder of government critics and independent journalists continue to be a major concern around the world. The occasional murder or imprison-ment of a high-profile journalist in one of the more prominent nation-states may find its way into the middle pages of American newspapers, but the precariousness of the journalism profession worldwide may not be evident to the casual observer. The International Federation of Journalists (IFJ), based in Brussels, has been issuing annual reports on the intimida-tion of journalists for two decades. It reports in a recent annual tally: "It is a poignant roll call of tragedy and loss recording the death of 2,271 colleagues, including . . . 94 journalists and media staff [that] were killed, victims of targeted killings, bomb attacks and crossfire incidents" in 2010. "Despite many promises, governments continue to fail in their duty to hunt down the killers. Many deaths are recorded without serious investi-gation. The absence of the rule of law, whether due to police corruption, judicial incompetence or political indifference not only endangers jour-nalists, it imperils democracy and compromises hopes for peace and de-velopment" (White and Sagaga 2011, 2). Not all of these tragic incidents were orchestrated by that state apparatus; it is likely other elements such as criminal gangs are involved—it is paradoxical that in most cases of attempts of this sort to silence an inquisitive press, it is not clear whose fingerprints are associated with the crimes.

Table 7.4 attempts to sketch out several continua of state intervention, both positive and negative, in efforts to protect or disrupt free and at times critical speech. The nonprofit Freedom House, much like the IFJ, has tracked the dynamic chemistry of state–press relationships. The story they tell is an interesting one—after the dramatic gains in the late twen-tieth century with a wave of authoritarian states evolving into appar-ently successful electoral democracies and the corresponding liberation of the states of the former Soviet Union, there has been a disappointing re-versal and backpedaling, perhaps most noticeably in Russia itself as the state and judicial system appear to be reverting to their former ways under the continuing guise of a free press and open electoral democ-racy (Puddington 2011).

Table 7.4   Elements of State Hegemony

	Negative Government Actions	Positive Government Actions
Censorship	Direct control of media content	
	Censorship of nongovernment media	
	Blocking websites	
Intimidation	Physical attacks on journalists/critics	
	Imprisonment of journalists/critics	
	Threats against journalists/critics	
Economic	Taxation of media	Finance public broadcasting
		Finance political campaigning
		Finance public Internet
Law and regulation	Libel, defamation, sedition laws	Legal protection journalist sources
	Control of access to spectrum/net	
		Legal protection free speech
		Access to government information

Of particular interest in our case is the special case of free speech on the net. Again, a fascinating story:

> When the internet first became commercially available in the 1990s, very few restrictions on online communications and content were in place. . . . Even the authorities in China, which today has the most sophisticated regime of internet controls, exerted very little oversight in the early days. However, as various dissident groups in the late 1990s began using the internet to share information with audiences inside and outside the country, the government devoted tremendous human and material resources to the construction of a multilayered surveillance and censorship apparatus. Although China represents one of the most severe cases, similar dynamics are now becoming evident in many other countries. . . . In 23 of the 37 countries assessed, a blogger or other internet user was arrested for content posted online . . .
>
>   In 12 of the 37 countries examined, the authorities consistently or temporarily imposed total bans on YouTube, Facebook, Twitter, or equivalent services. Moreover, the increased user participation facilitated by the new platforms has exposed ordinary people to some of the same punishments faced by well-known bloggers, online journalists, and human rights activists. Among other recent cases, a Chinese woman was

sent to a labor camp over a satirical Twitter message, and an Indonesian housewife faced high fines for an email she sent to friends complaining about a local hospital . . .

Of the 37 countries examined, the governments of 15 were found to engage in substantial blocking of politically relevant content. In these countries, instances of websites being blocked are not sporadic or limited in scope. Rather, they are the result of an apparent national policy to restrict users' access to dozens, hundreds, or most often thousands of websites, including those of independent and opposition news outlets, international and local human rights groups, and individual blogs, online videos, or social-networking groups. . . . Two of the countries categorized by Freedom House as electoral democracies—Turkey and South Korea—were also found to engage in substantial political censorship. . . . The widespread use of circumvention tools has eased the impact of content censorship and at times undermined it significantly. Such tools are particularly effective in countries with a high degree of computer literacy or relatively unsophisticated blocking techniques. For example, YouTube remained the eighth most popular website among Turkish users despite being officially blocked in that country for over two years, and the number of Vietnamese Facebook users doubled from one to two million within a year after November 2009, when the site became inaccessible by ordinary means. Users need special skills and knowledge to overcome blockages in countries such as China and Iran, where filtering methods are more sophisticated and the authorities devote considerable resources to limiting the effectiveness of circumvention tools. Still, activists with the requisite abilities managed to communicate with one another, discuss national events in an uncensored space, and transmit news and reports of human rights abuses abroad. . . .

Rather than relying exclusively on technological sophistication to control internet content, many governments employ cruder but nevertheless effective tactics to delete and manipulate politically or socially relevant information. These methods are often ingenious in their simplicity, in that their effects are more difficult to track and counteract than ordinary blocking. One common method is for a government official to contact a content producer or host, for example by telephone, and request that particular information be deleted from the internet. In some cases, individual bloggers or webmasters are threatened with various reprisals should they refuse the request. (Kelly and Cook 2011, 1–7)

There is, in my view, a clear bottom line to this brief review of the direct state involvement side of the hegemony problem. It is simply that in the absence of a strong and deeply rooted free press/free speech tradition (and sometimes despite its presence) the impulse to censor criticism is extremely strong among incumbent authorities, and maintaining an open marketplace of ideas is a constant battle requiring extensive energy and resources to sustain it. In the traditional media world of the twentieth century, those energies and resources came from a mixture of foundations and academics but primarily from the culture of professional journalism and the profitable print and broadcast news industries that supported that professional culture. The basis of these resources in the digital age is far from guaranteed. It is a cause for concern.

Beyond Ideology

The field of political economy is organized around a singular, delicate, fundamental, and dynamic question: How best do we organize the boundary between the polity and the economy? The increasingly strident debates between the partisans of the Left and Right pivot on their essentially polar opposite impulses on that question. When in doubt, the liberal imperative dictates we rely on public institutions to rein in the rapacious impulses and destructive cycles of the unfettered marketplace. Correspondingly, the conservative perspective celebrates the primacy of the free market as an economic engine and vilifies the delays, self-serving bureaucracies, inefficiencies, and taxes necessary to support the regulatory apparatus. Both points of view have a legitimate point to make, of course. But, again paradoxically, it is not clear that the best way to dynamically define the boundary between the polity and the economy is rhetorical team sports with squads of jousting ideological combatants trying desperately to knock the others off their horses with frenzied adversarial crowds cheering them on. The polarization is reinforced in the American case by the historically evolved economic alignments of the Democratic and Republican Parties. It may be reinforced as well by the human propensity for tribal identities, charismatic rhetoricians, and brutal tournament combat.

Ultimately, of course, the optimal balance of and capacity for undistorted communication between market and public mechanisms in supporting social, cultural, and economic productivity is a question subject to empirical verification. It appears historically that many, perhaps most, political entities from towns to nation-states cycle through an awkward sequence of excesses in each direction followed by an equally predictable overcorrection. The cycle is not unlike the celebrated boom-and-bust cycle of unregulated markets. Framing the question as an either-or binary of the market versus the government that is so common in political rhetoric obscures the more fundamental underlying question. Although the rhetoricians may find it useful to ignore the fact, almost every allocative process in modern societies is a complex mix of public policies and market mechanisms in interaction. Powerful players are equally incented to manipulate the structure of markets as well as the structure of public policy to serve their vested interests. The challenge to political economy is keeping the administrative and market structures in some reasonable balance and both as free as possible from the distortions introduced by the various self-interested parties.

If the optimal balance of market and administrative mechanisms in various domains such as, for example, agriculture, national defense, transportation, or communication is subject to empirical verification, why all the ideological fuss? Of the multiple answers to such a question, four resonate particularly with the central themes of this book. First, measuring productivity is both difficult and itself subject to numerous subtle and not so subtle distortions. Second, some dimensions of productivity are valued differently by different observers. Classically, for example, some are more concerned about equity, while others are more concerned about efficiency of a structural process. Third, for many observers the issue of states versus markets is a matter resembling religious orthodoxy and, accordingly, is unlikely to be influenced by empirical feedback of any sort, especially complex and potentially biased measurements. Fourth, the policy process is a historically messy one. Most policy processes simply do not have a tradition of experimentation, assessment, and adjustment or one of experimentation with systematic controls. So in all but a few cases of policy debate in the public sphere, the available empirical feedback on programmatic success or failure is ambiguous, and ideological interpretations tend to dominate.

If the argument being made here is a reasonable one, the ramifications for public policy in public communication should be fairly straightforward. And, not surprisingly, such an approach is resonant with the free marketplace of ideas model. Communication in the public sphere should be structured to make feedback as transparent as possible. Since the self-interested framing and filtering of feedback will be unavoidable, the best strategy is probably to let a thousand flowers bloom with efforts to keep the cost of the production and distribution of information as low as possible and to use sustained vigilance to protect against any systematic filtering of the flow of public communication. More on systematic filtering shortly.

The interaction of the psychology of ideology and the impact of technology on the evolving public sphere offers a promising opportunity for the communication research community to be relevant and useful. We have some limited understanding of the conditions under which partisans will attend to the ideas and information provided by those they oppose and particularly the conditions under which partisans will actually be responsive to such information. We have some understanding historically, structurally, and psychologically of how the ideological divide between liberals and conservatives came to be. We know something about why the great majority of citizens pay limited attention to any component of the political and public policy domain. It is this limited attentiveness that explains why paid political advertising, particularly advertising interspersed with entertainment content, remains critical to reaching potential voters. Most citizens most of the time are not particularly inclined to seek out political websites or policy brochures. But sometimes they do, and research on these evolving dynamics continues to be important.

Digital Economics and Public Policy

I have noted with some emphasis that if there were historical conditions ripe for a self-aware restructuring of the institutions and norms of public communication, the disruptive forces of the digital revolution offer special promise. The various players in the modern media system from the publishers and broadcasters to the consumer electronics manufacturers and telecommunications conglomerates, as well as the associated labor unions, professional associations, music labels, copyright administrators, and the like, and their sophisticated lobbyists know a lot about how the existing system works and will undoubtedly devote their considerable talents and resources to protecting an equivalently sustainable and profitable position in the resultant evolving digital media system. So if recognition of public values such as reasonable costs, open competition, and a vibrant public sphere are also to be part of the process, it may be useful to

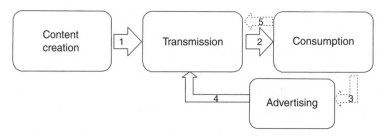

Figure 7.1.   Fundamental Economics of the Media Industries

explore in some additional detail how it is that the technical developments are actually impacting traditional media economics.

The basic economic model in Figure 7.1 may strike some as awkwardly simplistic, but there are some subtleties in the details of how that model has traditional worked that are commonly misunderstood and critically important to how the digital revolution is impacting media economics and audience behavior.

The three dominating elements in the center of the figure—content creation, transmission, and consumption—of course, represent the traditional push model of commercial media production. What becomes clear when depicted this way is that the transmission technologies of industrialized printing and broadcasting play a gatekeeper role. Professional journalists, authors, musicians, and creative artists of various traditions do not, for the most part, deal directly with their audiences. The publishing houses, recording labels, newspaper chains, movie studios, and broadcast, cable, and satellite systems represent critically important gatekeeper intermediaries who try to predict audience tastes as measured by ratings, best seller lists, and box office tallies. It is they who contract with the creative community and provide the financial resources for creation and, in effect, deliver an audience to the creator and the creator to an audience via arrows 1 and 2. The power of the gatekeepers, their orientation toward profits rather than creativity, and the large bureaucratic character of these media conglomerates have been unending targets of frustration and criticism within the creative community for generations.

The key financial engine for this process in most modern industrialized democracies, of course, is the advertising industry, which most often pays the bulk of the costs (arrow 4). In traditional over-the-air broadcasting advertising supports 100 percent of production and transmission, for newspapers it is 70 percent, magazines 50 percent, and cable television 50 percent. Motion pictures, books, and recordings are traditionally financed directly by audience payments (arrow 5). The system has been relatively stable and profitable for more than a century with sufficient demand to keep the older industries such as newspapers and radio in business as newer technologies like television (1940s) and cable television (1970s) came along. The size of the media industry in the United States alone is estimated to be between one and one and a half trillion dollars, depending on how industrial boundaries are defined for the purposes of the statistics (Veronis, Suhler, and Stevenson 2011).

The logic of the system, if *logic* is the right word, is that consumers ultimately pay for the cost of production and distribution either directly (arrow 5) or indirectly through the increased cost of consumer goods necessitated by manufacturing firms' advertising budgets (arrow 3). So, for example, General Motors spends between one thousand and two thousand dollars per car (estimates vary) on advertising (Goetz 2011) and McDonald's famously spends annually about twenty-three cents per person on advertising for every human on planet Earth (Nelson 2010).

The key to the profitability of the system, particularly the transmission media, is limiting the supply to keep prices up. And for a series of complex institutional, historical, and technical reasons, most metropolitan markets had only one or perhaps two newspapers that dominated local retail advertising, and because of scarce broadcast spectrum, they had a limited number of radio and television stations. As a result broadcasters and newspaper publishers traditionally had profit margins of between 20 percent and 25 percent, which is four to five times the American industrial average of about 5 percent (Compaine and Gomery 2000). The digital revolution, however, has been wreaking havoc in the media sector as digital distribution increasingly weakens the oligopolistic control of the traditional media. Unless you are an oligarch, this is very important and positive news.

Careful observers will note that the arrows labeled 3 and 5 in Figure 7.1 are depicted with a dotted rather than a solid line. The reason is that those are the economic transactional interfaces that are most directly influenced by the digital difference. Arrow 5 depicts direct payment for content rather than indirect payment through advertising. Economic theorists are champions of direct payment because in their judgment markets work much better that way—consumers evaluate alternatives and make judgments on quality that ultimately determine demand and price. Indirect markets introduce all sorts of distortions as advertisers prefer particular audience demographics and treat media content as commodity vehicles for advertising rather than potentially valuable elements of public culture in their own right. More important, perhaps, *the digital transmission of intellectual property offers the prospect of cutting out the middleman transmission/gatekeeper corporations entirely and putting consumers more directly in interaction with the actual journalists, authors, and artists whose work they value.* Again, an important, promising element of the digital difference.

But there is another development signaled by the dotted-line arrow number 3 in Figure 7.1 that is just now becoming understood. It might be termed the *Google Revolution* in advertising and marketing. It has two elements. First it replaces push advertising with something approaching a concept of pull advertising—the use of keyword-based promotion of products when potential consumers are searching in a product category. If you think about it for a moment, it makes a lot of sense in terms of targeting promotional messages to those who are most likely to be in the market for a product. Looking for some Italian high-heel shoes? Search for that phrase in the Google search engine and in addition those websites determined by the Google search algorithm as what most people want to find. There are eleven paid ads and a twelfth from Google's own shopping system. Online advertising spaces on pages associated with such search terms as *cars, insurance, shoes,* and *perfume* are worth a lot of money, and Google lets the marketplace work its magic by auctioning off that limited promotional space to the highest bidder, providing Google with the large majority of its, for example, fifty billion dollars in revenues in 2012 (U.S. Security and Exchange Commission 2012).

Second, the engineering mentality at Google leads to a series of auction- and market-based mechanisms for pricing the advertising. It signals what is likely to be a significant movement in the industry from impression-based advertising based on the cost-per-thousand exposures (known as CPM) to cost-per-action (CPA) an "action" typically being a click-through to seek further information or an actual online purchase. Advertisers had acknowledged for years that the return on their advertising dollar was hard to pin down—an acknowledgment captured in the frequently quoted aphorism "Half the money I spend on advertising is wasted; the trouble is I don't know which half," a lament attributed to nineteenth-century department store magnate John Wanamaker. Actually the best data available on yearlong television advertising campaigns for established brands suggests that only 17 percent of them have a statistically significant effect on sales (Lodish et al. 1995).

The potential significance of these two developments to the media industries is immense. If advertisers move their promotional budgets entirely or even largely from impression-based to search-based mechanisms, it would represent a major financial loss to the media industries. Further, consider if marketers are better able to determine when their ads are actually working because online the medium that delivers the message is also often the medium that closes the sale. If that is the case, they would be able to abruptly abandon the unsuccessful or less successful advertising investments. Accordingly, instant feedback on the marketing–sales connection represents a further and potentially significant loss of income for the creation and distribution of intellectual property.

The traditional analog transmission media of publishing, broadcasting, motion pictures, and the like, have their fixed quasi-monopoly position in their local markets. Take comic strip artist Garry Trudeau, for example. He has been publishing his syndicated strip *Doonesbury* in local newspapers since the 1970s through the Universal Press Syndicate and developed quite a following of fans. Until recently there was simply no obvious way to get a daily strip to its audience without going through and adding value to the local newspaper monopoly and aggregator. Now most strips, including Trudeau's, are available online, supported by direct subscriptions and/or advertising and typically available as well in newspapers for the still-substantial ink-on-paper readership. But from such an example, one can see how the relative power of the gatekeeper has declined. It is true, of course, that awareness of the *Doonesbury* strip and its subsequent independent popularity was initially enhanced by its newspaper-based visibility. It is relatively easy for any aspiring artist to publish a strip online. Finding an audience of critical mass may be another issue. But it may be that viral recommendations and social media pass-alongs will become the ultimate arbiters of popularity and in time come to replace the official imprimatur of big media as a key to public visibility and economic sustainability.

Figure 7.2 captures this power shift as an increasingly universal digital distribution system diminishes the power of the formerly requisite mid-

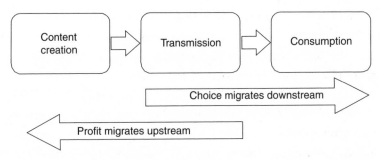

Figure 7.2.   Economic Impact of the Transition from Analog to Digital Transmission

dlemen as analog transmitters, promoters, and profit recipients. *In a digital environment, profitability migrates upstream to the actual creators and owners of intellectual property rights. And choice migrates downstream as pull replaces push.* For example, in 1960 when the typical American home had access to 3.4 television stations, 8.2 radio stations, 1.1 newspapers, 1.5 recently purchased books, and 3.6 magazines—that represents a human-scale choice (Neuman, Yong, and Panek 2012). In contrast, the web provides instant access in recent estimates to more than forty billion web pages (worldwidewebsize.com). True, the prominent search engines and particularly Google are not without the power to steer attention, which could potentially be used in a self-interested or censorial manner. So a critical and entirely new research question comes to the fore: *Will the dynamic, transparent, and socially networked character of the public sphere on the web constrain the exercise of such potential power?* It represents a fresh incarnation of the familiar media gatekeeper/agenda-setting hypothesis. The era of the big media gatekeepers and big media profit centers would appear to be in jeopardy. These well-financed corporations are highly incented to find a way to reestablish their quasi-monopolistic position in the new digital environment, and that likely battle for exploitation rights occupies much of the rest of this chapter.

Network Neutrality

Network neutrality is a slogan, and, like most slogans, it has political purpose. And, like most effective slogans, it has some inherent potential persuasive symbolic power. Who wouldn't be in favor of network neutrality? The rhetorical alternative would appear to be a "network bias," which does not sound very appealing. The arguments for and against network neutrality dominate attention in the domain of Internet policy and economics in the United States and to some extent globally, so it is useful briefly explore the origins of the debate and the broader significance of the issue. It turns out that beneath the layers of rhetorical flourish and technical obfuscation, the future of an open digital marketplace of ideas may be at risk (Farrell and Weiser 2003; Wu 2003).

It is said in the corridors of power in Washington, DC, these days that there is no independent lobbyist or consultant who is free to comment independently on the issue of network neutrality. The reason for this, as the story goes, is that the two sides in the battle have already hired everybody there is with any expertise and experience with these issues. So then, how are these two sides in the battle over what turns out to be economically important but rather arcane technical policy issues defined? On one side in the debate, those proclaiming to favor network neutrality are the big firms that use the Internet to conduct business, notably such companies as Google, Facebook, Amazon, and Netflix. In the opposite corner are the big telephone and cable ISPs such as AT&T, Verizon, and Comcast. It is a battle of the corporate giants—the big network users versus the big network providers. It must bring a smile to the otherwise grim professional policy propagandists of K Street in Washington (Herman 2006).

The big users adopted the phrase *network neutrality* because they want to make sure the providers do not charge them extra for their extensive use of the network. Comcast, for example, could charge an extra fee to Netflix because Netflix's movie streaming requires a great deal of bandwidth. From Netflix's point of view, such a practice would be a form of extortion and a violation of the fundamental Internet design principle that a bit is a bit and every bit has equal access to the network. From Comcast and the other carriers' points of view, they need to "shape traffic" to efficiently optimize their networks and to limit the behavior of what they consider "bandwidth hogs." The carriers have the technical capacity to block or simply slow down traffic from any web address they might choose. Since there are only two practical options for broadband connectivity in any local market for most American households—the local cable and the local telephone line—such power in the hands of a de facto duopoly concerns the other big players (Pasquale 2010).

Why should we be concerned about the relative profitability of the various corporate players? The reason is that the issue of network neutrality, although it is currently defined quite narrowly in terms of competing corporate interests, represents a deeper and much more enduring question about the architecture of the technical network and how potential distortions in that architecture impact the promise of an open marketplace of ideas. At this point it may be useful to return briefly to historical origins of the Internet and how it came to evolve its famously open design.

The Internet, originally the ARPANET, was based on a then newly invented flexible system of "packet switching." The design developed by academics and government researchers for the U.S. Defense Department's Advanced Research Projects Administration (ARPA, sometimes DARPA) and was optimized to sustain military communication under potential enemy attack. The telephone and most other analog systems of that era had centralized hub-based switching systems that were potentially vulnerable to targeted attack. The Internet was based on a geodesic grid of routers that forwarded packets to their destination, and if any route was blocked or overloaded, each router had a set of alternative routes for the traffic—and thus there existed no vulnerable central communicative hub, just a broad distribution of smaller routers each determined to get the packets of bits through as quickly and efficiently as possible. The design principle was to keep the network itself as simple as possible with all the fancy intelligence of interpreting audio, video, and encryption at the "edges" of the network—that is, the users' computers (Blumenthal and Clark 2001). A bit is a bit. No high-priority bits. No first-class section in the bit streams. No charging by the bit—this is a government system. It did not occur to the designers to build in a billing system or to install speed bumps for those with especially large files. Of course, no capitalists in full possession of their faculties would have designed a system without a way to fully control it and charge with precision for its use in time or volume, but these engineers could not have anticipated the ARPANET's spectacular growth and now nearly universal use as system of broadband connectivity. What a wonderful historical accident (Edwards 2010).

The Internet is much more complex now than in the 1960s and 1970s, but its basic design principle remains the same—a bit is a bit, much like the original concept of a common carrier. It is not the network's job to judge the political correctness, economic value, or moral standing of any packet of bits. In fact, under ordinary conditions the routers just read the addresses on the digital packets and pass them on to the next router in the chain and ignore the contents. But it is technically possible to read the contents of the packets; it is called "packet sniffing." And there are various interests who are uncomfortable with the Wild West openness of how the network works and who have proposed at various times a diverse array of potential roadblocks, gatekeepers, speed bumps, bit police, and censors. *So it is reasonable to predict that if the open Internet based on network neutrality was a happy historical accident, it will be a significant challenge to keep it that way.*

I am using *network neutrality* here in a somewhat broader sense that typically arises in technical and policy debates in American and international fora, and it engages the behavior of not just the heavy users and ISPs but also the portals, the social media platforms, the information aggregators, and particularly the search engines. The iconic case study in the current network neutrality debate is Madison River Communications, a midsize ISP and telecommunication company that blocked its customers' access to voice-over-Internet-protocol (VOIP) competitive telephone services until it was caught and fined for anticompetitive practice by the FCC (Yoo 2006). Such cases of economic self-interest are relatively straightforward and easy to spot, and in this case the blocked competitors were very quick to recognize the technical situation and even quicker to complain about it. But if search engines, online encyclopedias, or social networks adjust their complex algorithms to favor their own products or their favored political position perhaps in possibly a subtle way, it may be much more difficult to detect and to address (Crawford 2016). Further, as is the case in China and other authoritarian regimes, there is extensive pressure on commercial firms and ISPs to censor what is seen to be undesirable speech (MacKinnon 2012).

There is an unfortunately extensive and sustained tradition in human history of killing the messenger. It may be consoling for the recipient of bad news, but it wreaks havoc on the messenger business. If the mavens who manage the Internet backbone believe that they need to exercise more control over differentiated types of Internet traffic in order to exploit particularly profitable bit streams, they may soon find themselves in the unwanted position of digital traffic cops—in Negroponte's (1995) term of appropriate derision, the "bit police." When criminals use the Internet in some way as a contributing element to their evil designs, in the public eye these become Internet crimes. The impulse is to regulate the ISPs. This is particularly true in cases of online sexual predation and exhortations to violence and terrorism. If for the past century the use of a telephone in committing a crime did not make it a telephone crime—perhaps that is a tradition that could be sustained in the digital era.

Digital Property Rights

The rich and celebrated tradition of copyright acknowledges the unique character of intellectual property, which, unlike physical property, requires special handling. There it is in article 1, section 8 of the U.S. Constitution as an explicit assignment to the newly created federal government: "The Congress shall have power . . . to promote the progress of science and useful arts, by securing for limited times to authors and inventors the exclusive right to their respective writings and discoveries"—tucked in between the responsibility for the post office and the federal court system. Copyrights and patents are important because once an inventive idea exists, it might easily be copied without appropriate compensation to the author or inventor. The fundamental idea of copyright is uncontroversial and globally acknowledged. It has roots in the 1710 British Statute of Anne, which begins with the acknowledgment: "Whereas Printers, Booksellers, and other Persons, have of late frequently taken the Liberty of Printing . . . Books, and other Writings, without the Consent of the Authors . . . to their very great Detriment, and too often to the Ruin of them and their Families." A series of international agreements from the original Berne convention of 1886 now overseen by the UN's World Intellectual Property Organization in Geneva symbolize global acknowledgment of the special case in principle if not always in practice (Johns 2009).

Since the popularization of Gutenberg's press and through the development of a diversity of recording and broadcasting technologies, the predictable assemblages of entrepreneurial middlemen have emerged that facilitate the movement of ideas from creators to audiences for a fee, and equally predictably they have made it their business to exploit the copyright tradition to protect their business model. (American copyright, it is sometimes noted, is based in significant measure on legal battles over the copyrightability of piano rolls for that celebrated digital technology, the player piano. The Supreme Court ruled unanimously in 1908 that piano rolls were not copies of a musical work under current law. Congress changed the law in 1909.) As a result, in the long tradition of the iron law of oligarchy, the legal structure has become increasingly designed to benefit music labels, publishers, and trade groups and has lost some connection to its original motivating purpose of motivating creators and disseminating useful arts (Litman 2000; Lessig 2001).

Note that in the last sentence, there are two implicit purposes that motivate copyright—that new ideas and creative arts get created and that they get "used." It is the getting used part that has receded to the background in all the hubbub about artists and their many representatives getting paid. The importance of this point has been captured in economist Stanley Besen's particularly insightful series of papers on intellectual property economics. For example, in his essay with legal specialist Leo Raskind he offers the following language to describe the foundational purpose of copyright: "the objective of intellectual property protection is to create incentives that maximize the difference between the value of the intellectual property that is created and used and the social cost of its creation" (Besen and Raskind 1991, 5). An economist right down to his DNA, Besen cannot resist using language that highlights the benefits of efficiency, that is, maximizing the ideas created and used and minimizing the costs. Accordingly, Besen and his associates are skeptical of the layers of middlemen and administrators and use prohibitions that the evolved system has inadvertently spawned. It is refreshing to have an analyst point out that the original motivation in the constitution was to incent creation and facilitate use rather than to focus only on penalizing unauthorized use.

One egregious example of overreach is the current length of copyright enforcement in the American case. The term of copyright enforcement in the original American legislation was fourteen years and eligible for a fourteen-year extension upon formal renewal. That would seem to be sufficient to motivate creativity. But the length of copyright enforcement in the American case has been expanded in five subsequent legislative initiatives to its present interval of life of the author plus 70 years or 120 years from time of creation (see Figure 7.3). Grandchildren are indeed a delight, but it is hard to imagine an author declining to undertake authorship in the absence of a statutory guarantee that their grandchildren and great-grandchildren will be adequately compensated. Of course, it was not

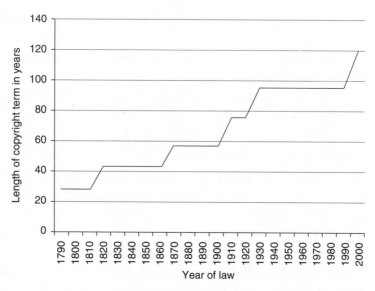

Figure 7.3. Extension of Copyright Duration in Response to Political Pressure from Industry

a cluster of grandchildren in short pants that shadowed the halls of Congress in each of these legislative initiatives; it was the familiar cluster of corporate attorneys in dark suits. Another egregious example is the prohibition of publication or promulgation of so-called orphan works, those whose authors are unavailable or unknown (Litman 2000).

The presumptive logic is that if previously unavailable authors come forward after publication by another party, they need to be able to sue for damages and enjoin publication. The net result is that such liabilities effectively prevent beneficial use of orphaned works. A legal structure that would encourage publication and develop funds for compensation in the rare cases that lost authors come forward has been proposed but not yet approved in a compromise with the various industry representatives. The proposal was initiated by Google as part of its effort to digitize virtually all books in existence (Samuelson 2010).

The technical capacity of private individuals to make copies of protected intellectual property with relative ease is not new. Through the second half of the twentieth century the existence of photocopiers and audio and video recorders made it possible and with each year increasingly convenient and inexpensive. Interestingly, the entertainment industry, although uncomfortable about dual cassette players that made audio copies easy and VCRs that allowed for time shifting and sharing, accepted the fact that it was not in a position to legislate these technologies out of existence. In fact the Audio Home Recording Act of 1992 explicitly acknowledges the legality of making copies for personal use (Fisher 2004). In the Betamax case before the Supreme Court in 1984, the justices ruled definitively that using a VCR in the home for personal use such as time shifting was not infringing (Samuelson 2006).

What made the industry particularly nervous was that the newer digital technologies, unlike their analog forebears, could make flawless copies and do so from copies of copies. Their concern with that high-quality copies might lower demand for the actual recordings, but it is unlikely that youthful music enthusiasts who were playing cassettes on boom boxes would even notice the difference. *The real technology problem for the industry turned out to be not the quality of the copies but the networked capacity to share copies of any quality.* It is easy to identify the historical moment that this became clear—in 1998 in a dorm room at Northeastern University in Boston as nineteen-year-old Shawn Fanning was listening to his roommates complain about how hard it was to find music online. Fanning had a hunch. He figured first that there must be a lot of music out there on people's hard drives; second, that he could probably combine Windows API protocols, Internet Relay Chat, and Unix server commands to make a friendly peer-to-peer music-sharing system for every individual who might choose to become part of the network; and three, that given the chance people would indeed share. Fanning quit school and worked around the clock in some space in his uncle's office back in Hull, Massachusetts, a particularly remote strip of land in Boston's Massachusetts Bay. He worked around the clock because he had to assume others had come up with the same idea. He called his sharing program Napster after his own school nickname, and his idea and the start-up company he founded with a few friends took off like a shot in June 1999, a mere six months later. Each of his hunches was on the mark. Needless to say, the media corporations and trade associations took notice and before the year was out had filed suit. A series of court hearings finally resolved in the spring of 2001 in a ruling that concluded that peer-to-peer sharing did not qualify under the fair use doctrine and that even if Napster's servers were capable of blocking 99.4 percent of the exchange of protected intellectual property, that was not sufficient and the system needed to be shut down, which it was that September (Zittrain 2006; Gillespie 2007).

It is difficult to overestimate the importance both of the creation of the Napster peer-to-peer sharing system and the definitive court ruling to shut it down. In the Betamax case two decades before the Supreme Court had basically concluded that home copying technologies were legal. Judge Marilyn Patel of Northern California's Ninth Circuit Court concluded—not when they are networked. Her decision was upheld upon appeal and in turn has had a continuing and significant impact on the structure of communication. Judge Patel's ruling did not put the very concept of peer-to-peer communication in legal jeopardy, but influence of her ruling on its structure is more than subtle. The television industry survived the Betamax ruling and the existence of video recorders. It is interesting to speculate on what might have happened if, like Betamax, the Napster system was deemed fair use of intellectual property if the originating copy were legally obtained (a counterfactual scenario explored in Carrier 2012).

Intellectual property should thrive in the digital age. The new technologies lower the cost of production, make it easier for audiences to find authors, and make it easier for authors to transmit their works to audiences. Further, with a little creativity the networks can provide a variety of ways for audiences to compensate authors. The potential losers in such a scenario are the historical middleman/gatekeepers who derived their considerable profits from control of the transmission technology. These are, for the most part, rather sophisticated players who recognize that the digital transition is here to stay. The historically evolved complexities of intellectual property law offer them a chance to create their middleman role and profitability in a networked world. Accordingly, we can expect that the domain of digital rights management will be a battlefield where the fresh ideas of entrepreneurs will do battle with the vested interests of incumbent corporations.

That said, it is important to note that none of this argument is intended to be inimical to the compensation of authors—such an enterprise is entirely praiseworthy. When the Recording Industry Association of America (RIAA) realized that online peers were sharing music, it did what comes natural to a trade association—it initiated suit. It initially targeted 261 individuals for egregious downloading, including a twelve-year-old honors student in New York and a seventy-one-year-old grandfather from

Richardson, Texas. RIAA threatened a liability of up to $150,000 per song. Most of these symbolic targets settled for small sums, and the few who went to court ended up facing major damages. Suing one's customers into submission does not offer the prospect of representing a sustainable business plan. The RIAA felt it was successful in sending a message to the online public in general and proposed to drop the mass litigation strategy. One analyst reported the RIAA spent sixty-four million dollars in legal fees and recovered a little over one million in damages (Beckerman 2008; Electronic Freedom Foundation 2008).

Contrast that response to a record company that discovered that the song "Forever" by Chris Brown was being used in a viral wedding dance video downloaded by a record number of twelve million viewers in a single week in 2009. The record company was Sony, and after requesting that the video be disabled on YouTube, the company changed its mind and permitted the video, adding a link on the video to purchase the song at iTunes or the Amazon music store. The song reached number four on the iTunes most downloaded chart and number three on Amazon (Parfeni 2009).

Legal scholar Michael Carrier (2012) has estimated that the litigious strategies of the gatekeepers have had a significant and sustained negative effect on innovation and experimentation in the development of new approaches to the creation, aggregation, and distribution of intellectual property. The story is not new. The Western Union Telegraph Company was not happy about the invention of the telephone and might have been successful in sidetracking the invention and tying Bell up in patent disputes. Fortunately, the executives at Western Union dismissed the telephone as an electronic toy (Wu 2010). The good news is that as the new media start-ups grow into a mature industry, the inevitable legal battles will be less appropriately described as David versus Goliath. For the record, a few decades after Western Union dismissed the telephone as a toy, Western Union was purchased by the then much larger and more successful Bell Telephone Company and renamed Western Electric as one of Bell's manufacturing subsidiaries (Brock 1981).

Lessons of the Long Tail

An open marketplace of ideas benefits from a diversity of ideas. If everybody has a book in their hands, that's a good thing. If all those books are identical, say, Mao's Little Red Book, not so good. The key to having the best ideas rise to the top is establishing the most diverse collection to start with. Scott Page (2007), among others, has demonstrated that empirically with more successful idea generation among groups of increasing diverse backgrounds as contributors not so likely to all think alike. It is an important insight. So when Chris Anderson, as editor-in-chief of *Wired*, published his now famous essay "The Long Tail" in 2004, followed by the book-length version in 2006, there was reason to celebrate. He begins with some impressive statistics that appear to contradict the familiar 80-20 Pareto Principle, the law of the vital few, the idea that roughly 80 percent of the effects come from 20 percent of the causes. In the public sphere that translates into a few intensely popular movies, books or recordings and a very large number of those largely ignored. He achieves his effect by formulating the question a little differently: "What percentage of the top 10,000 titles in any online media store (Netflix, iTunes, Amazon, or any other) will rent or sell at least once a month? Most people guess 20 percent. . . . Only 20 percent of major studio films will be hits. Same for TV shows, games, and mass-market books—20 percent all. The odds are even worse for major-label CDs, where fewer than 10 percent are profitable, according to the Recording Industry Association of America. But the right answer . . . is 99 percent. There is demand for nearly every one of those top 10,000 tracks" (Anderson 2004).

He introduces his famous graphic, reproduced here as Figure 7.4, which arrays the different titles in a particular media domain such as movies, books, or video games from the most popular to the least. We see the familiar dramatic head of the curve denoting the blockbusters and best sellers and a gentle sloping tail of modest sales or popularity for the great majority of titles available. Anderson's argument, with a focus on Amazon and Netflix as provocative exemplars, is twofold. First, unlike a local bookstore, national online providers of intellectual property have warehouses with virtually "infinite" shelf space. They can stock a much larger number of the less popular titles so the section of available titles moves profitably much farther down the curve and thus "the long tail." Second, these online retailers are motivated to promote the attention of their customers to the less popular titles. That is especially true of Netflix's mailing of physical DVDs in red envelopes to subscribers. By pushing demand down the curve they need only stock fewer of the most popular titles, especially for

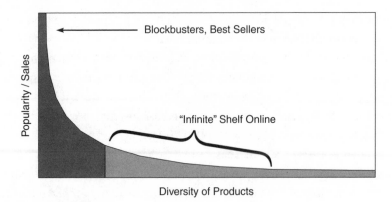

Figure 7.4.   The Long Tail

the predictable moments of peak popularity at time of release. So Anderson's story of digital diversity is a promising one and has been well received. There is only one problem. It is highly misleading.

Let's go back to the original premise. His point is not that the shape of the curve has changed, only that firms like Amazon and Netflix can profitably sell titles farther down the curve than local brick-and-mortar retailers can. In fact the viral dynamics of online popularity may actually lead to a steeper curve of attentional inequity between the most popular and the least. It is a point frequently made by those who study online network dynamics. Physicist and network theorist Albert-László Barabási (1999, 2002) calls it "preferential attachment"—the more connected a network node is, the more likely it is to receive new links. Political scientist Matthew Hindman calls it "Googlearchy" and reminds us that the very essence of Google's popularity as a search engine is based on its original insight manifested in the famous PageRank algorithm that attributes importance in proportion to the number of links that a page receives from other sites (Hindman, Tsioutsiouliklis, and Johnson 2003; see also Menczer et al. 2006). The current algorithm is more sophisticated, but that logic remains the core of its success, and it is mimicked by virtually all present-day competitors in online search. So by definition, the results of a semantic search reinforce the winner-take-all/rich-get-richer attentional dynamics. Hindman applied the classic formulas used by economists to measure industrial concentration and found that in terms of attention online, the popular and mostly corporate websites take a bigger share of attention than the equivalent competitors in radio, newspapers, and magazines. The top ten radio stations in the top fifty radio markets garner only 7 percent of the listenership. In magazines the equivalent concentration is 27 percent. For the web, the top ten websites dominate 62 percent of the attention of the entire top one hundred websites (Hindman 2006, 338).

So we need to put Anderson in perspective. Yes, that the Amazons and Netflixes of the world have unprecedentedly extensive shelf space for diverse content and making titles from farther down the popularity curve conveniently (and profitably) available to the public is unambiguously good news for the marketplace of ideas. But the actual shape of the popularity curve is not likely to change anytime soon, and because of online attentional dynamics it may actually get steeper. Rutgers's media policy analyst Philip Napoli (2008) explains why. Three factors, he argues, reflect the continuing forces of massification. First, the basic economics of intellectual property has not changed. It costs a great deal to produce the first copy of a book, record, or movie. The cost of making the second and subsequent copies is minimal. So once the sunk cost investment of creating a work has been spent, all energy is devoted to marketing it to the largest possible audience. This largest-possible-audience logic is reinforced by advertising economics, which relishes the efficiency of mass rather than niche media adverting exposure. Second, he notes, "the distribution of audience attention in most media contexts tends to cluster around high-budget, mass appeal content, which of course also tends to be the content produced by the traditional institutional communicators (with the resources to expend on big-budget content)." And, third, he draws attention to what he labels a matrix of institutional forces, including the legacy of traditional thinking within the media industries and biases in the technologies of audience measurement (Napoli 2008, 59).

Napoli's analysis raises an interesting question. Are the first generations of online culture heavily constrained by the traditional culture of one-way industrial commercial media? The terms *Web 1.0* and *Web 2.0* highlight the distinction between the first historical stage of basically putting magazines and newspapers on the computer screen in static web page format (Web 1.0) contrasted with the interactive, dynamic, and networked social media that evolved in the years following (Web 2.0).

Hindman's calculations on the top-heavy concentration of public attention on a few, mostly commercial, websites and web portals represent cause for concern. But as new generations of users who grow up on on-line interactivity, electronic mobility, and gaming emerge, there is reason to believe new patterns of audience behavior and institutional response will evolve. It is widely acknowledged in the study of technological adoption that the initial uses of innovative technology are not a reliable predictor of the use at maturity (Rogers 2003).

Pondering Policy

I began this chapter by reviewing the question of whether the domain of public (and private) communication requires special attention and perhaps a unique set of practices and legal principles. I concluded that, yes, indeed it does. A television is not just a toaster with pictures. The marketplace for kitchen appliances engages public policy addressing public health and safety, environmental impacts such as energy use, and, of course, the usual rules about truth in advertising and market competition rather than monopoly. But the marketplace of ideas has special properties and special importance because it is the venue in which all the other policies and collective initiatives are debated and determined.

The American case that occupied most of our attention here has historically emphasized individual initiative, private enterprise, and a relatively limited role for public institutions in the domain of public culture, tele-communications, broadcasting, and publishing. For the industrial era the United States was something of a global outlier in this respect. But concurrently with the digital revolution, the global pattern looks more and more American with shrinking public institutions, deregulation, and an expansion of private enterprise. So lessons from the American case may prove to be broadly relevant.

The American model is based on the proposition that less public sector involvement is best. The premise is that left pretty much alone, the marketplace of ideas will take care of itself. Such a philosophy has been successful in avoiding authoritarian statism, but less successful in avoiding corporate hegemony of various forms influencing the way policies are framed and ultimately decided. It is difficult to conclude that the playing field is level and the marketplace of ideas unfettered.

The digital difference in the transition from the industrial to the information age on the surface promises a revolutionary movement toward diversity, equity, and an openness of opportunity in the digital commons. In the industrial age the voices were few, the barriers to entering the marketplace of public speech were daunting, and the gatekeepers were intimidatingly powerful. Now the voices and choices are abundant, the cost of entry minimal, and, if anything, it is the diversity of choice that is daunting. But in our discussion of digital economics, I addressed the case that the industries of culture and information are under tremendous economic pressure to find a way to recapture their position of power and profitability that the era of scarcity provided them. The future of network neutrality, intellectual property rights, personal privacy, and the delicate dynamics of access to public attention are policy domains in flux. These issues are, as they say in Washington, "in play." If the Internet was a happy accident, the economic forces motivated to try to reverse the results of that accident are incredibly strong. Recognition of such perils motivates book covers like Harvard legal scholar Jonathan Zittrain's recent cover picture of train tracks leading off the face of a cliff with his book title in large capital letters: *The Future of the Internet and How to Stop It.*

There will be many voices participating in these decision processes, including the voices of some independent academic researchers. Most of these academic voices will be economists and legal scholars. Few communications scholars will take the initiate to rise and address these policy questions, and likely fewer will be asked. It could be otherwise. It should be otherwise. That is the subject of Chapter 8.

8

Praxis

Media theory is essentially an informed consciousness of what is happening
when systems of public communication change. . . . For policy application,
the main emphasis should be on problem identification and the
evaluation of alternative courses of action.

—DENIS MCQUAIL (1986)

Communication science is not yet an established, operating reality.

—STEVEN CHAFFEE AND CHARLES BERGER (1987)

The state of research on media effects is one of the most notable
embarrassments of modern social science.

—LARRY BARTELS (1993)

THE BREADTH of our exploration of the digital difference has, perhaps
inevitably, encountered something akin to the forest and trees conundrum—so many facts, figures, examples, and counterexamples it is easy
to lose the central thread of the argument. Further, even if the arguments
are reasonably persuasive, what next? How would we put these observations to work in communication scholarship and the connection of communication scholarship to professional practice and to policy? This is the
frequently overlooked issue of praxis. Frequently overlooked, perhaps,

because praxis is the hard part. I attempt to identify eight principles of practice in communication scholarship and scholarly outreach around which a consensus may come to emerge. Each should be familiar and resonant with the arguments and analysis above. Again, these are ideals and long-term goals, understandably not easily achieved.

The quotations at the beginning of this chapter, particularly the quotes from Chaffee and Berger and from Bartels, resonate with my sense of the state of communication scholarship. Such a view may be in the minority. It is likely that the majority of scholars in the field consider the research corpus to be strong and the theories and methods dramatically improved in recent decades. I am not in disagreement on that point, but my judgment is that we may be able to move further and faster using the digital revolution as the critical fulcrum for considerable leverage and, importantly, that we still have a long way to go.

A Multilevel Field of Study

Anthropologists and psychologists do not have much in common. Although both devote themselves to the study of human behavior, they do so at different levels—the anthropologist at the level of cultural and collective phenomena and the psychologist at the level of the individual. Similarly, those in the field of communications who study persuasion and perception draw on different literatures, theories, and methods than those who study media institutions, media economics, communication policy, and media culture do. The same might be said of economics and the distinct traditions of macro and micro economics. But there is a critical difference. In the case of economics there are key theoretical and paradigmatic linkages connecting research at the different levels. I think such a connection is not only possible in communication; I believe it would be highly productive.

Attentive readers of this book will note that the discussion of polysemy and polarization dealt primarily with the individual level of analysis, focusing on individual differences in interpretation and social identification. In turn, the discussion of pluralism and public policy dealt primarily with the social, institutional, and collective levels. The two sets of literatures overlap only at the margins. This is, in my view, because the theoretical connective tissue is not yet fully developed. What is the nature of this elusive theoretical linkage?

The potential connection might be articulated as follows. The psychology of individual humans is tribal in nature. Humans are prone to polarization, to the interpretation of polysemic speech in different ways resulting from their diverse social identities and lifeworld experiences. Miscommunication and noncommunication result. The challenge to the establishment of collective norms and institutions in the structuring of communication and the public sphere at the social level is to take that component of human nature self-consciously into account—to institutionally and culturally compensate in response to the impulse toward tribal polarization.

It is a simple idea. It may strike some as elementary and naive. At first blush it might appear to align communication scholarship with such enterprises as conflict management, peace negotiations or labor relations. But perhaps more appropriately it can be seen as an enthusiastic acknowledgment of Habermas's (1979, 1981) historically grounded concepts of communicative action and the ideal of universal pragmatics. Habermas (1988) called for a reunification of empirical social science and social philosophy, although his own predilections kept his writings at a highly abstract and philosophical level. Habermas's lifelong project was working out how social conditions might enhance the natural human capacity for communicative competence and in turn mutual understanding in the public sphere. Because of the deeply philosophical flavor of his intellectual style, his oeuvre has not yet inspired much empirically grounded work on systematically testing the details of these social conditions and their prophesized impacts on consensus building and mutual understanding. But it could, and in my view it should (Neuman, Bimber, and Hindman 2011).

My call for integrative multilevel research resonates with a number of analyses and recommendations in the communication literature (Paisley 1984; Chaffee and Berger 1987; McLeod, Kosicki, and McLeod 2009). McLeod and associates have been particularly forceful on this point. They point out that inference across levels is particularly delicate given the problems of the widely discussed ecological fallacy but note that the concepts of emergent properties and independent auxiliary (that is, connective) theories have proven helpful (McLeod, Kosicki, and McLeod 2009). They emphasize the importance of new data analytic tools, such as hierarchical analysis, which are explicitly structured to address these issues (Pan and McLeod 1991). They argue further that such theoretical traditions as cultivation analysis and the spiral of silence are inherently multilevel in their logical structure. Intriguingly, sociologist James S. Coleman (1987) has tackled the macro–micro disjuncture and concluded that the market model of economics that links the micro and the macro can be generalized and draws attention to parallel dynamics in sociology. I believe the same is true for communication. It is my view that the marketplace-of-ideas metaphor and the notion of public resonance (rather than media causal effects) offer special promise for connecting the individual and the social. Finally, McLeod and Blumler (1987) emphasize that the linkage between levels is critically important for communication research to be taken seriously in the corridors of power and public policy.

New Opportunities for Connection of Research with Policy and Practice

Elihu Katz has walked the corridors of power (founding director, Israel Television) as well as those of the academy (recently Hebrew University and the University of Pennsylvania), and his experiences have engendered a particularly nuanced ear for the delicate connection between independent/ nonpartisan scientific research and the complexities of policy decision making. Because of his broad experience with international broadcasting he was asked to participate in a review of British broadcasting as a social scientist and wrote about his experiences under the revealing title "Looking for Trouble" in the *Journal of Communication* (Katz 1978). The communication industries have no problem funding and working with proprietary research when they control the questions asked and are at liberty to make research findings public selectively as it suits their interests. But working with independent scholars and scholarly organizations is asking for trouble. It is the modern reflection of the original response of industry executives to John Marshall and his enthusiastic band of first-generation communication researchers in the 1940s when one executive explained with unusual candor that they did not want to "rock the boat" and saw no upside in getting involved with or funding an independently derived research agenda.

Katz's essay is a sober reflection on the difficulties of connecting research with policy and practice. He quotes his mentor Paul Lazarsfeld, who remarked that for an industry that makes a point of being critical of others, the media are remarkably sensitive to criticism of themselves. The BBC, of course, reflects the culture of public service broadcasting, which in contrast to commercial broadcasting includes in its mandate a concern about social impacts beyond the counting the viewership in demographic categories of interest to advertisers. He notes that in the American case the connection between the research community and the domains of industry and public policy are not so much troubled but rather nonexistent.

The digital difference of our era, however, offers the prospect of change in the connection between research and practice along three intertwined dimensions. The first dimension is digital disruption of traditional business models in broadcasting, publishing, and telecommunications. The executives in broadcasting and publishing during the industrial era managed extraordinarily profitable enterprises. They were hardly inclined to seek outside advice or to explore new business models. But the pace of technologically disruptive change in these industries is rapid as new players replace old ones and, as discussed in Chapter 7, the fundamental structure of the industry is yielding to tectonic pressures for change. Accordingly there is a currently a remarkable openness to new ideas and new perspectives, perhaps an openness tinged with desperation. Google, Microsoft, Yahoo!, and a number of other high-tech firms have their own labs hiring anthropologists, psychologists, economists, and data scientists and conducting an intriguing mix of pure and applied research and sponsoring work in the academy. It will be interesting to see how the chemistry of interacting research cultures develops.

A second important dimension of change is the development of entirely new dimensions and applications of public policy. The public trustee tradition of broadcasting and the first amendment tradition of publishing persist but require new definitions and applications to practice. The Federal Communication Commission (FCC) has been struggling with a set of regulatory issues associated with the idea of network neutrality. In the American case, the Internet service providers (ISPs) have not yet systematically exercised their power to filter or delay access of their customers to particular content or particular content providers, but it is clearly within their technical capacity. The courts have made it clear that the FCC has the legal authority to regulate the ISPs if they do so under common carriage authority normally associated with the regulation of traditional telephony. The FCC has been reluctant to take that route but decided to proceed, hesitantly, along those lines in 2015. Many predict a delicate interaction of regulatory rulings, new legislation, and court rulings over the next decade as new guidelines are developed to try to guarantee equal access to an open marketplace of ideas. As reviewed throughout the book, the incentives to try to self-interestedly distort what would otherwise be a level playing field are strong and persistent. It is clearly a policy debate that would benefit from diverse scholarly participation and open public discussion. In addition there are new dimensions of intellectual property law and privacy law that need to be rethought in the light of the new digital realities.

The third dimension is a change of context, an as yet modest but important shift toward the global. The regulation of speech, publishing, and broadcasting has traditionally been the domain of the nation-state. But in an era of satellite broadcasting and a global Internet, the marketplace of ideas becomes an issue of global rather than national policy. It is intriguing to note that this obvious and compelling observation is seldom acknowledged. It seems to be convenient for nation-states to conduct business as usual as if the flow of public communication were still strictly the dominion of traditionally regulated local publishers and broadcasters. The World Intellectual Property Organization has been organized to deal with global questions of copyright enforcement. The Internet Corporation for Assigned Names and Numbers and the International Telecommunication Union have been chartered to address questions of technical standards and interconnectivity, and there are about a dozen small nongovernmental organizations that attempt to keep an eye on the free flow of information in the global public sphere. But as yet there is no agreed-on global forum for discussion or the development of policy. It may never happen. Many observers question whether the United Nations is well equipped to address such issues, although in the early 2000s in cooperation with the ITU it conducted a World Summit on the Information Society, drawing some of the world players in this space together (Dutton and Peltu 2009). In a field dominated thus far by rhetorical flourish rather than serious and systematic research, it is not yet clear that ongoing research will be supported and a meaningful research–policy connection sustained. As they say, we live in interesting times.

A New Historical Grounding

In the review of the origins of systematic communication research in the mid-twentieth century in Chapter 1, I noted that the onset of World War II and the Cold War that followed generated a strong sense of urgency and historical and political relevance among the founding practitioners of the field of communication research. Propaganda was an urgent problem. The bad guys were easily delineated. The energy and focus of the research community was palpable. They were inventing a new field.

But over time energy and focus tend inevitably to diffuse as does the sense of urgency and historical connection. The thesis of this book is that the digital difference could and should serve as a pivotal energy source for a renewed sense of challenge, urgency, and relevance. Instantaneous inexpensive broadband global communication is at hand, but there appears to be no obvious corresponding changes in despotic politics, political demonstrations (involving varying levels of violence), terrorism (state sponsored or otherwise), participatory open democratic elections, ethnic and religious prejudice, economic inequality, or economic growth. If anything, it would appear that polarization is simply increasing on multiple fronts of cultural, geographic, religious, and economic invidiousness. This represents yet another paradox. *The capacity for communication can diminish the propensity for violent conflict. But the capacity for communication and actual meaningful communication are not the same thing.* The optimist might argue that we simply have not yet figured out how to use these new technologies we find at hand to address questions of inequity and productivity and to protect an open marketplace of ideas. The pessimist might argue that the incumbents and the powerful will be more skillful in their use of these new tools of communication and will employ them to reinforce the dynamics of the Iron Law of Oligarchy. It is interesting to note that these two propositions are not mutually exclusive.

Perhaps the strongest and most influential voice of recent years to make the case that we are witnessing a critical historical juncture is Manuel Castells. In his trilogy *The Information Age* (1996–1998), he elaborates how network structures and global communication influence not just the character of the nation-state but the very character of political power and of social identity in our era. Over the next decade in a series of books, particularly in *Communication Power* (2009) and *Networks of Outrage and Hope* (2012), he develops and refines these themes with updated research and new case studies. He draws on the language of political sociology focusing on the dynamics of social protest and the framing of issues of political and economic inequality, the polarity of the network versus the self, and what he terms the space of flows and mass self-communication. Although his vocabulary is somewhat different from my own, the underlying concerns are the same.

Castells grew up in fascist Spain and was active in the anti-Franco movement. As a result he had to leave his home country and complete his studies in Paris, so it is not surprising to see that after proclaiming the importance of an open political forum he enters the forum himself as a progressive advocate. The immediacy of his case studies and the global breadth of his analysis are particularly compelling. My argument, however, is directed more toward the academic audience and an attempt to influence the direction of the research enterprise rather than the political one. But I hope the case is, in its own way, also compelling.

There are other voices like Castells's that celebrate the prospect of a historically grounded paradigm shift of research and theory and a historically grounded connection to policy and practice. New forms of social organization made possible by ubiquitous networking, including open source, peer production, and collaborative models, are attracting public and scholarly attention (Raymond and Young 2001; Benkler 2006; Tapscott and Williams 2006; Shirky 2008). New approaches to intellectual property have also been stimulated by the fact that public copying and sharing digital media is effortless and virtually without cost (Litman 2000; Lessig 2006). Some observers note that the fundamental relationship between digital authors/performers and their audiences is undergoing change (Jenkins 2006; Zuckerman 2013).

Thus far in communication scholarship the study of "new media" is defined as a specialized field for those researchers who study social media, digital economics, or mobile technologies, and the academic job listings typically designate these rarified specialties. In my view such a vocabulary represents simply a curious rhetorical rendering of what is the study of human communication in our historical era (Chadwick 2013).

A New Form of Data

The corpus of quantitative communication research evolving over the past three quarters of a century has been built primarily on a platform of surveys and experiments. The fundamental paradigm in this tradition assesses to what degree an individual or group has been exposed to a persuasive message/perspective and accordingly to assess statistically significant differences in attitudes and behaviors that might be associated with exposure. Chapter 2 reviewed these issues in some detail. The persistent problem for this research tradition has been that the measurement typically relies on individual self-reports of exposure and of behavior or attitudes, a practice fraught with unreliability and frequently systematic distortion. The digital difference of our era is that what is seen in the media and in response what is said in the social media can be assessed independently and without the distortions of partial recollections and motivated self-presentation to interviewers.

Real-world data collected from extant digital social and traditional media draw the analyst's attention to a more holistic media ecology as various media and messages rise and fall, representing continuous variation of public attention over time. The diversity and complexity of messages in social context leads the analyst to shift from message "x" potentially affecting attitude "y," to a question of which of the multitude of messages appear to resonate with public consciousness, which attract attention or comment, which are passed on, and which are sustained over time? It complements rather than replaces models of traditional media effects. Real-world big data is particularly well attuned to analyzing agenda dynamics, self-reinforcing mechanisms, damping effects, thresholds, and spiral phenomena in historical context (McCombs and Shaw 1972; Noelle-Neumann 1974; Neuman, 1990; Page and Shapiro 1991; Erikson, MacKuen, and Stimson 2002; Slater 2007).

The emphasis on exposure/persuasion effects made historical sense in the early days of communication research in the mid-twentieth century in part because of the limited number of published and broadcast sources of information for the average audience member. But in an age for which each broadband connection to the web is equally technically empowered to speak as well as to listen, our theoretical lens broadens from the dynamics of attitude change to include the dynamics of attention—of the many voices in the digital cacophony, what attracts attention, and which framings of public issues seem to resonate most strongly in the ongoing public discussion?

The phrases *big data* and *computational social science* are becoming generic labels for data and analyses of this general type. The breathless enthusiasm of early advocates is giving was to the acknowledgment that these computational techniques will complement and expand rather than replace more traditional methods (Shah, Cappella, and Neuman 2015).

Early research on new media typically contrasted those who reported using the Internet frequently with those who did not as researchers examined possible differences in knowledge, attitudes, and behavior, and experimentalists contrasted paired comparisons of online and traditional media representations of content (DiMaggio et al. 2001). This research question was McLuhanesque in its focus on the medium of transmission. Such research designs were arguably appropriate, because in the earlier days of Internet diffusion, now dubbed Web 1.0, online newspapers and other media were essentially a re-creation of their original content and format on a video screen. Other researchers in the tradition of organization research had a similar model asking "are the new media different" by contrasting face-to-face communication with computer-mediated communication, which in the early days was largely email (Walther 1996).

But as increasingly large majorities are online and ubiquitously so, and as a result the online-offline contrast makes less sense as an analytic approach. Further, with the younger generation sometimes texting as frequently as talking (teenage girls, eighty texts per day, boys a mere thirty, according to the Pew Internet Project), the notion of digitally mediated communication as some rarified or special type of communication becomes less and less appropriate. The digital difference is that digital communication *is* communication. When the digital generation meets for drinks or dinner the smartphones come out and are strategically placed on the table. Norms vary on the acceptability of glancing down at texted updates or responding to the device's ringtone demands.

The promises and perils of big data are addressed in more detail in related works (Neuman et al. 2014; Shah, Cappella, and Neuman 2015). Clearly what is texted, tweeted, posted, and emailed is not a representative sample of what a community is thinking and saying, but it represents an exciting new window into the dynamics of the public sphere. Obviously, as with other forms of research on human subjects, appropriate norms and procedures to protect the privacy of individuals need to be developed and carefully and consistently implemented. The evolving definition of what is private and what is public in the domain of social media is itself an important focus of further research.

Communication scholars have studied Nielsen ratings, box office trends, and best-seller lists for a better understanding of what issues and ideas seem to resonate with the collective consciousness of communities and countries. Such data continue to be of interest, but the analysis of aggregated search engine queries provides a much richer and fine-grained picture of what is on the public's mind and how these questions evolve over time (Segev 2010; Ripberger 2011).

The primary new form of data I have been emphasizing is the naturally occurring media content and the digitally mediated public response in social networks and other media now accessible through commercial aggregators. The key is "naturally occurring," rather than in response to the artificially constructed environment of an experiment or the formality of a survey questionnaire. Eugene Webb and colleagues (2000) have emphasized the importance of supplementing traditional surveys through the use of creative unobtrusive indicators of social behavior such as the study of graffiti or the wear of floor tile for different exhibits in a museum. They were particularly successful in drawing attention to the reactive biases in traditional social science tools, but their creative inventions were sometimes more amusing than serious alternatives. The digital footprints of our era, however, provide a much better resource for research than the physical footprints of Webb's early provocative invention do.

But there are other important developments in experimental and survey research that lower the level of artificiality and obtrusiveness by taking advantage of the digital infrastructure to assess responses in more natural field settings and various forms of natural experiments. One particularly promising technique known as experience sampling methodology asks subjects to respond to a few questions in real time in natural environments (Kubey, Larson, and Csikszentmihalyi 1996; Hektner, Schmidt, and Csikszentmihalyi 2006). Originally researchers used existing technologies such as beepers or preset timers to request subjects to record behaviors such as media exposure and concurrent attitudes or moods. Modern smartphones, of course, allow for more sophisticated assessments of ecology of real-time assessment in natural environments.

Further, the experimental method is poised, when appropriate, to move out of the laboratory. The ubiquity of broadband cellular connectivity permits increasingly sophisticated field experimental designs including the all-important element of random assignment to different conditions (Yaros 2006; Gerber and Green 2012; Ryan 2012).

New Models of Data Analysis

Consider again the fundamental paradigm for communications effects research. In an appropriate sample of a given population, the analyst assesses exposure to a persuasive or informative message and corresponding levels of attitude or behavior and typically posits that any apparent correlations between the two may be evidence of communication effects. The persistent and critically important difficulty, of course, is that the attitudes or beliefs may have led to the exposure rather than the other way around. In a onetime survey or any other single assessment, it is extremely difficult to determine causal direction or to rule out spuriousness resulting from an unmeasured third variable. In the experimental tradition, exposure is most often randomized so self-selection is ruled out and causal direction more easily assessed. But experiments take place in artificial environments over short periods of time. The reciprocal over-time spiral phenomena of increased exposure and increasingly strong belief identified by Noelle-Neumann (1984), Price and Allen (1990), and Slater (2007), among others, may prove to be particularly characteristic of how communications effects manifest themselves in real-world settings that are not typically accessible in the laboratory.

Enter big data. In this case, it is a new form of data that makes a new form of data analysis possible. Patterns of message exposure and recipients' potentially resultant attitudes, beliefs, and behaviors over time is routinely collected and aggregated by commercial firms. The primary commercial interest is in tracking the ups and downs of brand reputations and potential effects of advertising and promotion on purchases. That is about it. The commercial researchers focus on their brand and a few key competitors. All the rest of the aggregated data in the traditional media and the social media about public issues, health concerns, economic concerns, foreign policy, and cultural trends are available for analysis usually at quite reasonable cost after the brand managers have paid for the heavy lifting and data aggregation. In addition, importantly, this is over-time data, currently about a decade's worth—the most sophisticated data for the past five years—a gold mine of interacting trends for the curious data scientist.

The fundamental logic is powerful. If variable x precedes variable y in time, x is likely to be the causal source rather than the other way around. Further, if there is a reciprocal effect over time; that too can be assessed in the trend data. Economists, of course, have been using highly sophisticated mathematical techniques for sorting out causal patterns in trend data as over-time economic data have been available for centuries. This field of statistical analysis is generally referred to as econometrics, but the techniques are perfectly appropriate for over-time data from any source. It isn't magic; it is hard work. The potential for errors in measurement and errors in inference remains significant. And, importantly, it is not put forward as a substitute for but rather a complement to traditional tools of measurement and analysis.

Renewed Attention to the Active Audience

The issue of audience activity has been characterized as "one of the longest-lived controversies in the short history of communication research" (Levy and Windahl 1985, 109). In the pages above and in an independent bibliometric analysis of the research literature (Neuman and Guggenheim 2011) I have characterized the progression of communication theorizing as largely a reaction to the passive and anomic characterization of audience members within the propaganda-centered research literature emerging from the Second World War. In one sense it is not actually much of a controversy if since 1960 pretty much all the scholarship identified itself as on the active-audience side. Uses/gratification, selective attention, and social identity theories emphasized that different audience members exhibited highly differentiated motivations and expectations in attending to media content. The two-step flow model and the network analysis and diffusion traditions drew attention to opinion leadership and the social structure of information diffusion. Concepts such as the social construction of reality, constructivism, and sense making emphasized that media content interacts with the audience member's preexisting understandings rather than mechanically and unidirectionally affecting beliefs and attitudes.

The digital difference, however, adds important new elements to theorizing about audience activity. There are two fundamentally new dimensions of audience activity in the two-way capacities of digital broadband. First the "audience" members pull the content desired rather than having it pushed out to them as in the traditional one-way broadcast and publishing logic of the industrial age—the widely acknowledged profusion of choice. Second, the "audience" members can speak as easily as they can listen, to comment, to reedit, to retweet, to click the "like button." The active audience, as before, is able to dynamically interpret and perhaps selectively ignore much of what is pushed out or pulled their way, but these two new dimensions of interactivity influence the complex processes that bring new meaning to the notion of an active audience or perhaps suggest we need to move beyond the fundamental concept of an "audience" in the digital age altogether.

Part of the original conceptualization of the active audience was the famous two-step flow of media content engaging opinion leadership and group discussion. In our era of social media, it is clearly a multistep flow with complex patterns including the evolution of subcultures, sometimes splitting off, sometimes combining, patterns of seemingly viral contagion, and smart mobs blurring the distinction between virtual connection and physical presence. The electronic publication of traditional news stories and, of course, blogs, routinely invites commentary and, naturally, commentary on the commentary. We are only beginning to understand how the dynamics of multistep flow, multistep flow with editing, and multistep flow with commentary actually typically work (Garrett 2011; Bakshy et al. 2012; Boczkowski and Mitchelstein 2012). Researchers are developing graphic and mathematical tools to analyze patterns of retweeting, the editing and resolving disputes within Wikipedia, polarization versus convergence in political blogs, and the diffusion of information in email networks (Haythornthwaite 2005; Gaines and Mondak 2009; Wu et al. 2011).

There is more at work here than new technological affordances, networking data, and new analytics. The digital difference is significant enough to prompt the rethinking of old theories as well as entirely new theoretical approaches. Many examples have already been discussed, but it may be appropriate here to point to some additional new directions.

One is a perspective known as the constitutive role of communication in organizations. Traditionally in the study of organizations, analysts would look for patterns of communication within a given organization structure. Following some seminal work by Robert McPhee and Pamela

Zang (2000) and by Linda Putnam and associates (Putnam and Nicotera 2009) analysts have been turning that original formulation on its head, arguing that organization can be seen as being constituted by patterns of communication rather than the other way around. This is particularly important because it draws attention to the prospect that even unintended or accidental inclusions or exclusions in, for example, email notification lists and electronic document access lists can have profound structural consequences for social organization. It is broadly acknowledged in organizational studies that communication technologies and data system linkages weaken organizational boundaries and increase "outsourcing" of functions and even decisions (Brynjolfsson and McAfee 2014).

These observations have relevance for the broader questions of the public sphere and marketplace of ideas we have been reviewing. Take, for example, Michael Schudson's provocative essay "Why Conversation Is Not the Soul of Democracy" (1997). Schudson argues that democratic theorists in the tradition of Dewey and Habermas celebrate the ideal of public conversation perhaps even to a level of obsession. But, he argues, typical human conversation follows rules of informality and sociability that fall short of what is actually required. "Democratic talk is not essentially spontaneous but essentially rule-governed, essentially *civil,* and unlike the kinds of conversation often held in highest esteem for their freedom and their wit, it is essentially oriented to *problem-solving*" (298). One might recall the formalities of *Robert's Rules of Order* as designed to address such questions. The question for research, and indeed for actual practice, is how to design such rules of conversational engagement for the digital public sphere. Many websites and discussion forums have evolved such rules and even designated trusted participants to enforce them. Wikipedia has one of the most elaborate and sophisticated systems of structured editing and dispute resolution (Leskovec, Huttenlocher, and Kleinberg 2010). Other entities like the technology-oriented discussion forum Slashdot have evolved peer moderation systems that permit collaborative and collective decisions on the qualities and appropriateness of submissions (Poor 2005). These dynamic systems are the Robert's Rules of online democracy. They are likely, in a very real and practical way as they evolve, to either permit or frustrate an open marketplace of ideas. Lawrence Lessig (2006) was one of the first to draw public concern to these issues with his elaboration of the double meaning of code—code of law and computer code—and his clarion call that both forms of code may constrain our freedoms and steer our capacity for interaction and that the latter was becoming increasingly powerful and increasingly independent of public scrutiny and control. The active audience of the industrial age could gather at the salon or coffeehouse or town hall, attend protests, and write letters to editors. Rules and norms evolved for such participation. The active audience of the digital age is perhaps equally constrained by rules and norms. Research needs to address the questions of who designs these rules of engagement, how they are enforced, how aware the public is of their existence, and how they might be contested.

A Rethinking of a Taken-for-Granted Research Paradigm

The academic field of communication research is about sixty years old. It is, I argue, time for a midlife crisis. It may even be a little late for such an exercise in soul searching and reevaluation. As the term *midlife crisis* is normally used in psychology and in popular culture, there is a period of highly predictable and taken-for-granted routine and then a triggering event of some sort followed by an energized questioning of first principles and reconsideration of both goals and strategies. The triggering event in my scenario is the digital difference.

My concern is that although there is broad acknowledgment that that the new technologies are, well, new and different, the scholarly reaction will be to double down on business as usual and perhaps hire somebody new to "do" new media in the department. My argument is that to define new media as a new and separate subfield is to subvert its importance and to ignore the extremely important opportunity for paradigmatic reformation these developments represent.

The previous paragraph uses the phrase *paradigmatic reformation*. I have used the terms *paradigm* and *paradigm shift* frequently in these pages. It is not without irony. Many would argue that the field of communication scholarship has no paradigm. It is simply an odd collection of diverse forms of scholarly inquiry and notions borrowed from other disciplines united only by the term *communication* in the departmental designation and a shared coffee break room. The social science and humanistic traditions of scholarship may not even be entirely comfortable with sharing the same coffee break room. In Chapter 1 I quote Robert Craig's (1999, 120) lament: "The conclusion that communication theory is not yet a coherent field of study seems inescapable." I have always appreciated his thoughtful wording and his emphasis on "not yet." My view is that the central elements of a fundamental and unique paradigm of communication science and communication scholarship have been there all along. For various historical and structural reasons the core elements have been concealed and distorted.

My proposal is for a ten-step program of paradigmatic reform. All of these points have been addressed in some detail the pages above but summarized briefly might serve as an agenda for reformation.

1. Abandon a propaganda-like mechanical concept of communication effects. Audiences are active. Human communication is fundamentally polysemic. Meaning as received is frequently at odds with meaning as intended.

2. Instead of mechanical/causal communication effects consider that some ideas expressed resonate with listeners while others do not. Different sorts of ideas resonate with different sorts of listeners.

3. Instead of a single hypothesized communication effect, consider a distribution of responses usually ranging from the intended response to its very opposite.

4. Abandon the fruitless distraction of trying to determine whether communication effects are minimal or not so minimal.

5. Take as a central paradigmatic question—what are the conditions under which meaning as received corresponds to meaning as intended? When is communication successful? That is the central question of communication in policy and professional practice. It is the unique question that distinguishes communication scholarship from all its sister disciplines.

6. A profound abundance of communication is the mark of our times. The character of this profusion and how elites and audiences respond to it represents a central question for communication research. An abundance of communication emphasizes the importance of selective attentional dynamics in addition to persuasive dynamics. Search engine dynamics supplement traditional journalistic norms and at times may replace them.

7. Publics are predisposed to polarization. Most communication is valenced. This derives from the central role of social identity in

human existence. Tribal hatreds in global networks represent a particular challenge to meaningful dialogue. Research in this area could represent to most important contribution of the field to policy and professional practice.

8. Political, economic, and cultural elites are prone to use their powers to protect their powers. This is equally true in the domain of public communication and represents a persistent and systematic challenge to the maintenance of an open marketplace of ideas and public dialogue. This iron law of oligarchy is closely related to the persistence of polarization and research on these issues represents an equally important potential contribution to policy and professional practice.

9. Scholars inclined to study texts in the tradition of the humanities could benefit from including the study of audience responses within their paradigm. Scholars in the social science tradition could benefit from addressing the complex and polysemic character of the texts they label as simply persuasive.

10. The birth of the field of communication research during and following World War II was marked by a historically rooted sense of urgency and relevance much of which has now diffused and dissipated. The digital difference offers an important opportunity for a reenergization and reconnection.

The Marketplace of Ideas Redux

Despite the expansive diversity of issues and literatures addressed in these pages, it is my contention that this volume can be seen as one long argument. In its most abbreviated formulation the argument is this—*the revolution of digital networks and media provides us with a most welcome opportunity to rethink how we systematically study and how, accordingly, we might self-consciously structure the practices, institutions, and norms of the public sphere to better serve important values and ideals.* The digital difference, then, is a double-edged sword—a difference in how we communicate and a difference in how we understand the complex processes by which we communicate.

Scholars of communication from the outset have been articulately critical of weaknesses and biases of the public sphere. They note that many segments of the public are neither well informed nor politically engaged and that the two factors are, in all likelihood, closely related. All but a tiny minority of political observers would agree that a well-informed and well-engaged citizenry represent uncontroversially important and positive ideals. The disagreements arise when it comes to assessing which structural arrangements might best serve such ideals. Americans, for example, typically favor privatized market mechanisms (Streeter 1996), while others emphasize the contributions of public sector models, such as public service broadcasting (Blumler and Gurevitch 1995). The digital difference influences this classic debate in three important ways. First, the economics of privatized public communication are in flux, some would say under siege—most notably thus far the case in point is the newspaper industry. Notions of privatized independent investigative journalism as a counterpoint to state power, corporate power, and other incumbents exercising the iron law of institutional self-preservation need to be carefully evaluated (Downie and Schudson 2009). Second, new models of networked communication offer intriguing novel opportunities for networked engagement and spontaneous public mobilization and organization (Benkler 2006; Bennett and Segerberg 2013). Third, systematic research can contribute to the process of answering the difficult question of which institutional innovations actually contribute to the agreed-on values and ideals. Take the public service broadcasting issue, for example. It turns out that there is actually systematic evidence that public models are associated with higher levels of hard news knowledge among citizens, but, importantly, it depends on the character of and conditions in which the public service institutions operate (Soroka et al. 2012). All such conclusions are time bound, of course, given the dynamic changes in public and private institutions and in audience expectations. But that is the point. Such institutional change could be informed, in some cases in real time from evolving big data and field research on the impacts of these changes on public beliefs, public participation, and the dynamics of the public sphere. Such research is not likely to be determining, but it could be part of the process. These are not arcane academic issues. The network neutrality facet of the public–private debate is evident daily in energized public policy debates and in the strategic calculations of private enterprises in the United States and around the world.

The story told here turned out to be frustratingly complex with paradox building on paradox as we worked through and contrasted the literatures of media economics, media psychology, political communication, and cultural studies. The most promising ideal for structuring the public sphere, I have suggested, is the concept of an open and vibrant marketplace of ideas.

Most observers acknowledge the appeal of such a notion in theory. It turns out to be very difficult to engage such a notion in practice, however, because so many of us have cherished beliefs associated with our religious, ethnic, social class, or ideological identity or our gender, and we are emotionally and politically engaged in correcting what we see as the cultural and structural biases that run counter to that valued identity. So the classic exemplification of communication research scholarship for the past half century has attempted to demonstrate two things—that mediated communication has big effects and those effects are prejudicial, that is, likely to reinforce political bias, perhaps various forms of sexism, racism, ageism, or to overemphasize violence or commercialism. The work is well motivated, and the argument is not without merit. But this stance of scholarship has led to a complete disconnect between media research and professional practice. The disjuncture is unfortunate because if there were ever an opportunity for systematic scholarly attention to be put to constructive use, it is the current disruptive transition from traditional to networked digital media we have termed the digital difference.

I remain convinced that the paradigm of modern communication scholarship continues to resonate with its deep roots in the moral aversion to the threat of European fascist propaganda from the mid-twentieth century. It is a noble legacy. Active researchers may scoff at such an assertion because their theories and methods and substantive foci have grown in sophistication and moved far beyond the simple persuasion models of the era of Lasswell, Lazarsfeld, and Hovland. My argument is that the field can move further—not a quantum-style revolution but more of a pivot. The big-effects mantra (focusing on effect size) and the bad-effects mantra (focusing on countering negative persuasive appeals and depictions), in my view, have become a serious constraint on the field's capacity to address the new and critical challenges and opportunities of a new era.

Communication has structure. The structure of communication makes a difference. The goal can and should be an open marketplace of ideas that thrives on a level digital playing field as new ideas compete with old.

Keeping an economy open and healthy is hard work. Our colleagues in economics are busy. The importance of their work is widely acknowledged.

Keeping a democracy open and healthy is hard work. Our colleagues in political science and public policy are busy. The importance of their work is widely acknowledged.

Keeping the structure of communication open and healthy is hard work. Researchers in communication thus far have only indirectly and obliquely addressed this important challenge. The research community has been busy with other worthy normative concerns and research challenges. The digital difference provides an opportunity for the communication research community to rise to the challenge and for the urgent importance of this work to be widely acknowledged within the domains of both scholarship and public policy.

REFERENCES

ACKNOWLEDGMENTS

INDEX

References

Abdelal, Rawi, Yoshiko M. Herrera, Alastair Iain Johnston, and Rose McDermott, eds. (2009a). "Identity as a Variable." In *Measuring Identity: A Guide for Social Scientists*, 17–32. Cambridge: Cambridge University Press.

———. (2009b). *Measuring Identity: A Guide for Social Scientists*. Cambridge: Cambridge University Press.

Aborn, Murray. (1984). "The Short and Happy Life of Social Indicators at the National Science Foundation." *Items (Social Science Research Council)* SSRC Items 38, no. 2/3 (September): 32–40.

Abrams, Dominic, and Michael A. Hogg, eds. (1999). *Social Identity and Social Cognition*. New York: Blackwell.

Abrams, Jessica R., and Howard Giles. (2007). "Ethnic Identity Gratifications Selection and Avoidance by African Americans: A Group Vitality and Social Identity Gratifications Perspective." *Media Psychology* 9 (1): 115–134.

Achen, Christopher. (1975). "Mass Political Attitudes and the Survey Response." *American Political Science Review* 69:1218–1231.

Ackerman, Bruce, and Ian Ayres. (2004). *Voting with Dollars*. New Haven, CT: Yale University Press.

Adamic, Lada A., Rajan M. Lukose, Amit R. Puniyani, and Bernardo A. Huberman. (2001). "Search in Power-Law Networks." *Physical Review* 64 (4): 1–8.

Adams, Laura L. (2009). "Techniques for Measuring Identity in Ethnographic Research." In *Measuring Identity: A Guide for Social Scientists*, edited by Rawi Abdelal, Yoshiko M. Herrera, Alastair Iain Johnston, and Rose McDermott, 316–344. Cambridge: Cambridge University Press.

Adomavicius, Gediminas, and Alexander Tuzhilin. (2005). "Toward the Next Generation of Recommender Systems: A Survey of the State-of-the-Art and

Possible Extensions." *IEEE Transactions on Knowledge and Data Engineering* 17 (6): 734–749.

Adorno, Theodor W. (1969). "Scientific Experiences of a European Scholar in America." In *The Intellectual Migration: Europe and America, 1930–1960*, edited by Donald Fleming and Bernard Bailyn, 337–370. Cambridge, MA: Harvard University Press.

Adorno, Theodor W., Else Frenkel-Brunswik, Daniel J. Levinson, and R. Nevitt Sanford. (1950). *The Authoritarian Personality*. New York: Harper and Row.

Alessi, Christopher. (2011). "Occupy Wall Street's Global Echo." *Council on Foreign Relations*. www.cfr.org/united-states/occupy-wall-streets-global-echo/p26216.

Alexander, Jeffrey C., ed. (1998). *Neofunctionalism and After*. Malden, MA: Blackwell.

Allen, Jodie T. (2010). *How a Different America Responded to the Great Depression*. Washington, DC: Pew Research Center.

Allport, Gordon W. (1954). *The Nature of Prejudice*. Garden City, NY: Doubleday.

Alterman, Eric. (2004). *What Liberal Media?: The Truth about Bias and the News*. New York: Basic Books.

Alwin, Duane. (2010). "How Good Is Survey Measurement? Assessing the Reliability and Validity of Survey Measures." In *Handbook of Survey Research*, edited by Peter V. Marsden and James D. Wright, 405–436. Bingley, UK: Emerald.

Amar, Vikram David, ed. (2009). *The First Amendment, Freedom of Speech: Its Constitutional History and the Contemporary Debate*. New York: Prometheus Books.

Anderson, Benedict. (1983). *Imagined Communities*. London: Verso.

Anderson, Chris. (2004). "The Long Tail." *Wired*, October.

———. (2006). *The Long Tail: Why the Future of Business Is Selling Less of More*. New York: Hyperion.

Anderson, Doug, and Radha Subramanyam. (2011). *The New Digital American Family: Understanding Family Dynamics, Media and Purchasing Behavior Trends*. New York: A. C. Nielsen.

Anderson, James A. (2011). *Media Research Methods: Understanding Metric and Interpretive Approaches*. Thousand Oaks, CA: Sage.

Andersson, Lynne M., and Christine M. Pearson. (1999). "Tit for Tat? The Spiraling Effect of Incivility in the Workplace." *Academy of Management Review* 24:452–471.

Ang, Ien. (1985). *Watching Dallas: Soap Opera and the Melodramatic Imagination*. New York: Methuen.

———. (1996). *Living Room Wars: Rethinking Media Audiences for a Postmodern World*. New York: Routledge.

Ansolabehere, Stephen, Shanto Iyengar, Adam Simon, and Nicholas Valentino. (1994). "Does Attack Advertising Demobilize the Electorate?" *American Political Science Review* 88 (4): 829–838.

Ansolabehere, Stephen, and Brian F. Schaffner. (2014). "Does Survey Mode Still Matter? Findings from a 2010 Multi-Mode Comparison." *Political Analysis* 33 (2): 285–303.

Arendt, Hannah. (1951). *The Origins of Totalitarianism.* New York: Harcourt Brace.

Aristotle. (2009). *Nicomachian Ethics.* New York: Oxford University Press.

Arno, Andrew. (2009). *Alarming Reports: Communicating Conflict in the Daily News.* New York: Berghahn Books.

Arthur, W. Brian. (1994). *Increasing Returns and Path Dependence in the Economy.* Ann Arbor: University of Michigan Press.

Ascher, William, and Barbara Hirschfelder-Ascher. (2005). *Revitalizing Political Psychology: The Legacy of Harold D. Lasswell.* Mahwah, NJ: Lawrence Erlbaum.

Ashley, Richard, Clive W. J. Granger, and Richard Schmalensee. (1980). "Advertising and Aggregate Consumption: An Analysis of Causality." *Econometrica* 48:1149–1168.

Association for Psychological Science. (2008). "Are Humans Hardwired for Fairness?" *ScienceDaily,* April 18. www.sciencedaily.com/releases/2008/04 /080416140918.htm.

Auletta, Ken. (1992). *Three Blind Mice: How the TV Networks Lost Their Way.* New York: Vintage.

———. (2009). *Googled: The End of the World as We Know It.* New York: Penguin.

Babbie, Earl R. (2010). *The Practice of Social Research.* Florence, KY: Cengage Learning.

Baird, Jay W. (1975). *The Mythical World of Nazi War Propaganda, 1939–1945.* Minneapolis: University of Minnesota Press.

Baker, C. Edwin. (1994). *Advertising and a Democratic Press.* Princeton, NJ: Princeton University Press.

———. (2007). *Media Concentration and Democracy: Why Ownership Matters.* New York: Cambridge University Press.

Bakshy, Eytan, Itamar Rosenn, Cameron Marlow, and Lada Adamic. (2012). "The Role of Social Networks in Information Diffusion." In *Proceedings of the 21st International Conference on World Wide Web,* International World Wide Web Conference Committee (IW3C2), 519–528. Lyon, France: ACM.

Baldassarri, Delia, and Andrew Gelman. (2008). "Partisans without Constraint: Political Polarization and Trends in American Public Opinion." *American Journal of Sociology* 114 (2): 408–446.

Bandura, Albert. (1977). *Social Learning Theory.* Englewood Cliffs, NJ: Prentice Hall.

———. (2001). "Social Cognitive Theory: An Agentic Perspective." *Annual Review of Psychology* 52:1–26.

———. (2009). "Social Cognitive Theory and Mass Communication." In *Media Effects: Advances in Theory and Research,* edited by Jennings Bryant and Mary Beth Oliver, 94–124. New York: Routledge.

Barabási, Albert-László. (1999). "Internet: Diameter of the World-Wide Web." *Nature* 401 (6749): 130.

———. (2002). *Linked: The New Science of Networks.* Cambridge, MA: Perseus Publishing.

Barkow, Jerome H., Lisa Cosmides, and John Tooby, eds. (1992). *The Adapted Mind: Evolutionary Psychology and the Generation of Culture.* New York: Oxford University Press.

Baron, Sabrina A., Eric N. Lindquist, and Eleanor F. Shevlin, eds. (2007). *Agent of Change: Print Culture Studies after Elizabeth L. Eisenstein.* Amherst: University of Massachusetts Press.

Bartels, Larry M. (1993). "Message Received: The Political Impact of Media Exposure." *American Political Science Review* 87:267–285.

———. (2005). "Homer Gets a Tax Cut: Inequality and Public Policy in the American Mind." *Perspectives on Politics* 3:15–31.

———. (2008). *Unequal Democracy: The Political Economy of the New Gilded Age.* New York: Princeton University Press.

Baum, Matthew A., and Philip B. K. Potter. (2008). "The Relationships between Mass Media, Public Opinion, and Foreign Policy: Toward a Theoretical Synthesis." *Annual Review of Political Science* 11 (1): 39–65.

Bayley, Edwin R. (1981). *Joe McCarthy and the Press.* New York: Pantheon Books.

Beckerman, Ray. (2008). "Large Recording Companies v. The Defenseless: Some Common Sense Solutions to the Challenges of the RIAA Litigations." *Judges' Journal* 47 (3). http://shop.americanbar.org/eBus/Store/ProductDetails.aspx ?productId=218521&sponsor=Judicial%20Division.

Behr, Roy L., and Shanto Iyengar. (1985). "Television News, Real-World Cues, and Changes in the Public Agenda." *Public Opinion Quarterly* 49 (1): 38–57.

Bell, Daniel. (1973). *The Coming of Post-Industrial Society: A Venture in Social Forecasting.* New York: Basic Books.

———. (1979). "The Social Framework of the Information Society." In *The Computer Age,* edited by Michael L. Dertouzos and Joel Moses, 163–211. Cambridge, MA: MIT Press.

Beniger, J. R. (1978). "Media Content as Social Indicators." *Communication Research* 5 (4): 437–453.

———. (1986). *The Control Revolution: Technological and Economic Origins of the Information Society.* Cambridge, MA: Harvard University Press.

———. (1987). "Toward an Old New Paradigm: The Half-Century Flirtation with Mass Society." *Public Opinion Quarterly* 51:S46–S66.

Benkler, Yochai. (2006). *The Wealth of Networks: How Social Production Transforms Markets and Freedom.* New Haven, CT: Yale University Press.

Bennett, W. Lance. (1990). "Toward a Theory of Press-State Relations in the United States." *Journal of Communication* 40 (2): 103–127.

———. (2011). *News: The Politics of Illusion.* New York: Longman.

Bennett, W. Lance, and Shanto Iyengar. (2008). "A New Era of Minimal Effects? The Changing Foundations of Political Communication." *Journal of Communication* 58 (4): 707–731.

Bennett, W. Lance, and Alexandra Segerberg. (2013). *The Logic of Connective Action.* New York: Cambridge University Press.

Benoit, William L., and R. Lance Holbert. (2008). "Empirical Intersections in Communication Research: Replication, Multiple Quantitative Methods, and Bridging the Quantitative–Qualitative Divide." *Journal of Communication* 58 (4): 615–628.

Berelson, Bernard. (1952). *Content Analysis in Communications Research*. New York: Free Press.

Berger, Arthur Asa. (2000). *Media and Communication Research Methods: An Introduction to Qualitative and Quantitative Approaches*. Thousand Oaks, CA: Sage.

Berger, Charles R., Michael Roloff, and David Roskos-Ewoldsen, eds. (2009). *Handbook of Communication Science*. Thousand Oaks, CA: Sage.

Berger, Peter L., and Thomas Luckman. (1966). *The Social Construction of Reality: A Treatise in the Sociology of Knowledge*. Garden City, NY: Anchor.

Bernard, H. Russell, Peter Killworth, David Kronenfeld, and Lee Sailer. (1984). "The Problem of Informant Accuracy: The Validity of Retrospective Data." *Annual Review of Anthropology* 13:495–517.

Bertrand, Ina, and Peter Hughes. (2005). *Media Research Methods: Audiences, Institutions, Texts*. New York: Palgrave Macmillan.

Besen, Stanley M., and Leo J. Raskind. (1991). "An Introduction to the Law and Economics of Intellectual Property." *Journal of Economic Perspectives* 5 (1): 3–27.

Bettinger, Robert L. (1991). *Hunter-Gatherers: Archaeological and Evolutionary Theory*. New York: Plenum Press.

Bikhchandani, Sushil, David Hirshleifer, and Ivo Welch. (1992). "A Theory of Fads, Fashion, Custom, and Cultural Change as Informational Cascades." *Journal of Political Economy* 100 (5): 992–1026.

Billig, Michael, and Henri Tajfel. (1973). "Social Categorization and Similarity in Intergroup Behavior." *European Journal of Social Psychology* 3:27–52.

Bimber, Bruce. (2003). *Information and American Democracy: Technology in the Evolution of Political Power*. New York: Cambridge University Press.

Bimber, Bruce A., Andrew J. Flanagin, and Cynthia Stohl. (2012). *Collective Action in Organizations: Interaction and Engagement in an Era of Technological Change*. New York: Cambridge University Press.

Bineham, Jeffery L. (1988). "A Historical Account of the Hypodermic Model in Mass Communication." *Communication Monographs* 55:230–249.

Blalock, Hubert M., Jr. (1989). *Power and Conflict: Toward a General Theory*. Newbury Park, CA: Sage.

Blogpulse. (2011). "Blogpulse.Com." NM Incite. www.nielsen.com.

Blumenthal, Marjorie, and David Clark. (2001). "Rethinking the Design of the Internet: The End-to-End Arguments vs. The Brave New World." In *Communications Policy in Transition: The Internet and Beyond*, edited by Benjamin M. Compaine and Shane Greenstein, 91–140. Cambridge, MA: MIT Press.

Blumler, Jay G. (1980). "Information Overload: Is There a Problem?" In *Human Aspects of Telecommunication*, edited by Eberhard Witte, 229–263. New York: Springer.

Blumler, Jay G., and Michael Gurevitch. (1995). *The Crisis of Public Communication*. New York: Routledge.

Blumler, Jay G., and Elihu Katz, eds. (1974). *The Uses of Mass Communications*. Newport Beach, CA: Sage.

Boczkowski, Pablo J., and Eugenia Mitchelstein. (2012). "Clicking, Emailing, and Commenting: How Users Take Advantage of Different Forms of Interactivity on Online News Sites." *Human Communication Research* 38 (1): 1–22.

Boczkowski, Pablo, and Limor Peer. (2011). "The Choice Gap: The Divergent Online News Preferences of Journalists and Consumers." *Journal of Communication* 61:857–876. doi: 10.1111/j.1460-2466.2011.01582.x

Bollier, David. (2010). *The Promise and Peril of Big Data*. Washington, DC: Aspen Institute.

Boorstin, Daniel J. (1961). *The Image: A Guide to Pseudo-Events in America*. New York: Harper and Row.

———. (1989). *The Republic of Letters: Librarian of Congress Daniel J. Boorstin on Books, Reading, and Libraries*. Washington, DC: Library of Congress.

Borrero, Juan D., and Estrella Gualda. (2013). "Crawling Big Data in a New Frontier for Socioeconomic Research: Testing with Social Tagging." *Journal of Spatial and Organizational Dynamics—Discussion Papers* 12:6–28.

Bourdieu, Pierre. (1991). *Language & Symbolic Power*. Cambridge, MA: Harvard University Press.

———. (1993). *The Field of Cultural Production*. New York: Columbia University Press.

boyd, danah, and Kate Crawford. (2012). "Critical Questions for Big Data." *Information, Communication and Society* 15 (5): 662–679.

Boyle, Deirdre. (1997). *Subject to Change: Guerrilla Television Revisited*. New York: Oxford University Press.

Bradac, James J. (1983). "On Generalizing Cabbages, Messages, Kings, and Several Other Things: The Virtues of Multiplicity." *Human Communication Research* 9 (2): 181–187.

Bradley, Ian L. (1971). "Repetition as a Factor in the Development of Musical Preferences." *Journal of Research in Music Education* 19:295–298.

Brashers, Dale E., and Sally Jackson. (1999). "Changing Conceptions of 'Message Effects': A 24-Year Review." *Human Communication Research* 25 (4): 457–477.

Brewer, Marilynn B. (1999). "The Psychology of Prejudice: Ingroup Love or Outgroup Hate?" *Journal of Social Issues* 55 (3): 429–444.

Briggs, Asa, and Peter Burke. (2009). *A Social History of the Media: From Gutenberg to the Internet*. Malden, MA: Polity.

Brinkley, Joel. (1997). *Defining Vision: The Battle for the Future of Television*. New York: Harcourt Brace.

Brock, Gerald W. (1981). *The Telecommunications Industry.* Cambridge, MA: Harvard University Press.

———. (1994). *Telecommunications Policy for the Information Age.* Cambridge, MA: Harvard University Press.

Brosius, Hans-Bernd. (2008). "Research Methods." In *The International Encyclopedia of Communication,* edited by Wolfgang Donsbach. Blackwell Reference Online. www.communicationencyclopedia.com/public/.

Brosius, Hans-Bernd, and Mathias Hans Kepplinger. (1995). "Killer and Victim Issues: Issue Competition in the Agenda-Setting Process of German Television." *International Journal of Public Opinion Research* 7 (3): 211–231.

Brown, John Seely, and Paul Duguid. (2000). *The Social Life of Information.* Cambridge, MA: Harvard Business School Press.

Bryant, Jennings. (1993). "Will Traditional Media Research Paradigms Be Obsolete in the Era of Intelligent Communication Networks?" In *Beyond Agendas: New Directions in Communication Research,* edited by Philip Gaunt, 149–166. Westwood, CT: Greenwood.

Bryant, Jennings, and John Davies. (2006). "Selective Exposure Processes." In *Psychology of Entertainment,* edited by Jennings Bryant and Peter Vorderer, 19–34. Mahwah, NJ: Lawrence Erlbaum.

Bryant, Jennings, and Mary Beth Oliver, eds. (2009). *Media Effects: Advances in Theory and Research.* New York: Routledge.

Brynjolfsson, Erik, and Andrew McAfee. (2014). *The Second Machine Age: Work, Progress, and Prosperity in a Time of Brilliant Technologies.* New York: Norton.

Buckalew, James K. (1969). "News Elements and Selection by Television News Editors." *Journal of Broadcasting* 14:47–54.

Bucy, Erik Page, and R. Lance Holbert, eds. (2014). *Sourcebook for Political Communication Research: Methods, Measures, and Analytical Techniques.* New York: Routledge.

Buonomano, Dean V., and Michael M Merzenich. (1998). "Cortical Plasticity: From Synapses to Maps." *Annual Review of Neuroscience* 21:149–186.

Burnham, Walter Dean. (1970). *Critical Elections and the Mainsprings of American Politics.* New York: Norton.

Buxton, William J. (2003). "John Marshall and the Humanities in Europe: Shifting Patterns of Rockefeller Foundation Support." *Minerva* 41 (2): 133–153.

Cacioppo, John T., and Richard E. Petty. (1981). "Social Psychological Procedures for Cognitive Response Assessment: The Thought-Listing Technique." In *Cognitive Assessment,* edited by Thomas V. Merluzzi, Carol R. Glass, and Myles Genest, 309–342. New York: NYU Press.

Cacioppo, John T., William von Hippel, and John M. Ernst. (1997). "Mapping Cognitive Structures and Processes through Verbal Content: The Thought-Listing Technique." *Journal of Consulting and Clinical Psychology* 65 (6): 928–940.

Calhoun, Craig. (1993). "Nationalism and Ethnicity." *Annual Review of Sociology* 19:211–239.

———, ed. (1994). *Social Theory and the Politics of Identity*. New York: Wiley Blackwell.

Callamard, Agnès. (2005). "Striking the Right Balance." In *Words and Deeds: Incitement, Hate Speech and the Right to Free Expression*, edited by Ursula Owen, 30–31. London: Index on Censorship.

Campbell, Angus, Philip E. Converse, Warren E. Miller, and Donald E. Stokes. (1960). *The American Voter*. New York: Wiley.

Campbell, Donald T. (1965). "Ethnocentric and Other Altruistic Motives." In *Nebraska. Symposium on Motivation*, edited by Donald Levine, 283–311. Lincoln: University of Nebraska Press.

Campbell, Scott W., and Nojin Kwak. (2010). "Mobile Communication and Civic Life: Linking Patterns of Use to Civic and Political Engagement." *Journal of Communication* 60 (3): 536–555.

Cantor, Muriel G. (1971). *The Hollywood TV Producer: His Work and His Audience*. New York: Basic Books.

Cappella, Joseph N., and Kathleen Hall Jamieson. (1997). *Spiral of Cynicism: The Press and the Public Good*. New York: Oxford University Press.

Carey, James W. (1989). *Communication as Culture: Essays on Media and Society*. Boston: Unwin Hyman.

Carmines, Edward G., and James A. Stimson. (1989). *Issue Evolution: Race and the Transformation of American Politics*. Princeton, NJ: Princeton University Press.

Carr, Nicholas. (2008). "Is Google Making Us Stupid? What the Internet Is Doing to Our Brains." *Atlantic Monthly*, July–August. www.theatlantic.com/magazine /archive/2008/07/is-google-making-us-stupid/306868/.

———. (2010). *The Shallows: What the Internet Is Doing to Our Brains*. New York: Norton.

Carrier, Michael A. (2012). "Copyright and Innovation: The Untold Story." *Wisconsin Law Review* 891:891–962.

Carson, Rachel. (1962). *Silent Spring*. Boston: Houghton Mifflin.

Castells, Manuel. (1996). *The Rise of the Network Society*. Malden, MA: Blackwell.

———. (1997). *The Power of Identity*. Malden, MA: Blackwell.

———. (1998). *End of Millennium*. Malden, MA: Blackwell.

———, ed. (2004). *The Network Society*. Northhampton, MA: Edgar Elgar.

———. (2009). *Communication Power*. New York: Oxford University Press.

———. (2012). *Networks of Outrage and Hope: Social Movements in the Internet Age*. Malden, MA: Polity.

Cavin, Susan. (2008). "Adorno. Lazarsfeld and the Princeton Radio Project, 1938–1941." Presented at the American Sociological Association Annual Meeting, Boston.

Center for Responsive Politics. (2012). "2012 Election." www.opensecrets.org.

Chadwick, Andrew. (2013). *The Hybrid Media System: Politics and Power*. New York: Oxford University Press.

Chaffee, John. (1988). "Differentiating the Hypodermic Model from Empirical Research: A Comment on Bineham's Commentaries." *Communication Monographs* 55:230–249.

Chaffee, Steven H. (1975). "The Diffusion of Political Information." In *Political Communication: Issues and Strategies for Research*, edited by Steven H. Chaffee, 85–128. Beverly Hills, CA: Sage.

Chaffee, Steven H., and Charles R. Berger. (1987). "What Communication Scientists Do." In *Handbook of Communication Science*, edited by Charles R. Berger and Steven H. Chaffee, 99–122. Newbury Park, CA: Sage.

Chaffee, Steven H., and John L. Hochheimer. (1982). "The Beginnings of Political Communications Research in the United States: Origins of the 'Limited Effects' Model." In *The Media Revolution in America and in Western Europe*, edited by Everett M. Rogers and Francis Balle, 267–296. Norwood, NJ: Ablex.

Chaffee, Steven H., and Miriam J. Metzger. (2001). "The End of Mass Communication." *Mass Communications and Society* 4:365–379.

Chance, Michael R. A., and Ray R. Larsen, eds. (1976). *The Social Structure of Attention*. London: John Wiley.

Chang, Ray M., Robert J. Kauffman, and YoungOk Kwon. (2014). "Understanding the Paradigm Shift to Computational Social Science in the Presence of Big Data." *Decision Support Systems* 63:67–80.

Choi, Hyunyoung, and Hal R. Varian. (2012). "Predicting the Present with Google Trends." *Economic Record* 88:2–9.

Chomsky, Noam. (1972). *Language and Mind*. New York: Harcourt Brace.

———. (2004). *Media Control: The Spectacular Achievements of Propaganda*. New York: Seven Stories.

Chong, Dennis, and James N. Druckman. (2007). "Framing Theory." *Annual Review of Political Science* 10 (1): 103–126.

Christie, Richard, and Marie Jahoda, eds. (1954). *Studies in the Scope and Method of the Authoritarian Personality*. New York: Free Press.

Clarke, Peter, and F. Gerald Kline. (1974). "Media Effects Reconsidered." *Communication Research* 1 (2): 224–240.

Cmiel, Kenneth. (1996). "On Cynicism, Evil, and the Discovery of Communication in the 1940s." *Journal of Communication* 46 (3): 88–87.

Cohen, Bernard C. (1963). *The Press and Foreign Policy*. Princeton, NJ: Princeton University Press.

Cohen, Stephen S., and John Zysman. (1987). *Manufacturing Matters: The Myth of the Post-Industrial Economy*. New York: Basic.

Cole, Barry, and Mal Oettinger. (1978). *Reluctant Regulators: The FCC and the Broadcast Audience*. Reading, MA: Addison-Wesley.

Coleman, James S. (1987). "Microfoundations and Macrosocial Behavior." In *The Micro-Macro Link*, edited by Jeffrey C. Alexander, Bernhard Giesen, Richard Munch, and Neil J. Smelser, 153–173. Berkeley: University of California Press.

————. (1990). *Foundations of Social Theory*. Cambridge, MA: Harvard University Press.

Coleman, Stephen, and Jay G. Blumler. (2009). *The Internet and Democratic Citizenship*. New York: Cambridge University Press.

Collins, Randall. (1986). *Weberian Sociological Theory*. New York: Cambridge University Press.

————. (1998). *The Sociology of Philosophies: A Global Theory of Intellectual Change*. Cambridge, MA: Harvard University Press.

Comins, Neil F., and William J. Kaufmann. (2011). *Discovering the Universe*. New York: W. H. Freeman.

Compaine, Benjamin, and Douglas Gomery, eds. (2000). *Who Owns the Media? Competition and Concentration in the Mass Media Industry*. Mahwah, NJ: Lawrence Erlbaum.

Comstock, George, Steven Chaffee, Natan Katzman, Maxwell McCombs, and Donald Roberts. (1978). *Television and Human Behavior*. New York: Columbia University Press.

Condon, Richard. (1959). *The Manchurian Candidate*. New York: McGraw-Hill.

Converse, Philip. (1964). "The Nature of Belief Systems in Mass Publics." In *Ideology and Discontent*, edited by David Apter, 206–261. New York: Free Press.

————. (2000). "Assessing the Capacity of Mass Electorates." *Annual Review of Political Science* 3 (1): 331–353.

Cook, Elizabeth Adell, Ted G. Jelen, and Clyde Wilcox. (1992). *Between Two Absolutes: Public Opinion and the Politics of Abortion*. Boulder, CO: Westview.

Cook, Michael. (2003). *A Brief History of the Human Race*. New York: Norton.

Corner, John. (1991). "Meaning, Genre and Context: The Problematics of 'Public Knowledge' in the New Audience Studies." In *Mass Media and Society*, edited by James Curran and Michael Gurevitch, 267–284. London: Edward Arnold.

Corso, Regina. (2008). *National Anthem Survey Results Summary*. New York: Harris Interactive.

Coser, Lewis A. (1971). *Masters of Sociological Thought; Ideas in Historical and Social Context*. New York: Harcourt Brace Jovanovich.

Cowan, Nelson. (1998). *Attention and Memory: An Integrated Framework*. New York: Oxford University Press.

Craig, Robert T. (1999). "Communication Theory as a Field." *Communication Theory* 9 (2): 119–161.

Crawford, Kate. (2016). "Can an Algorithm Be Agonistic? Ten Scenes from Life in Calculated Publics." *Science, Technology & Human Values* 41 (1): 77–92.

Crick, Bernard. (1959). *The American Science of Politics: Its Origins and Conditions*. New York: Routledge.

Curran, James. (1990). "The New Revisionism in Mass Communications Research: A Reappraisal." *European Journal of Communication* 5 (2–3): 135–164.

Dahl, Robert A. (1982). *Dilemmas of Pluralist Democracy: Autonomy vs. Control?* New Haven, CT: Yale University Press.

Dahrendorf, Ralf. (1959). *Class and Class Conflict in Industrial Society.* Stanford, CA: Stanford University Press.

D'Andrade, Roy G. (1995). *The Development of Cognitive Anthropology.* New York: Cambridge University Press.

Davidson, Philip Grant. (1941). *Propaganda and the American Revolution, 1763–1783.* Chapel Hill: University of North Carolina Press.

Davis, Dennis K. (1990). "News and Politics." In *New Directions in Political Communication: A Resource Book,* edited by David L. Swanson and Dan Nimmo, 147–184. Newbury Park, CA: Sage.

Davis, James C. (2004). *The Human Story: Our History from the Stone Age to Today.* New York: HarperCollins.

Davison, W. Phillips (1983). "The Third Person Effect in Communication." *Public Opinion Quarterly* 47:1–15.

Deacon, Terrence W. (1997). *The Symbolic Species.* New York: Norton.

Dearing, James W., and Everett M. Rogers. (1996). *Agenda-Setting.* Thousand Oaks, CA: Sage.

Declaration of Internet Freedom. (2012). "Free Press." www.freepress.net.

Delia, Jessie G. (1987). "Communication Research: A History." In *Handbook of Communication Science,* edited by Charles R. Berger and Steven H. Chaffee, 20–98. Newbury Park, CA: Sage.

Delli Carpini, Michael X., and Scott Keeter. (1996). *What Americans Know about Politics and Why It Matters.* New Haven, CT: Yale University Press.

Demers, David Pearce, Dennis Craff, Yang-Yo Choi, and Beth M. Pessin. (1989). "Issue Obtrusiveness and the Agenda-Setting Effects of National Network News." *Communication Research* 16 (6): 793–812.

de Tocqueville, Alexis. (1955). *The Old Regime and the French Revolution.* Garden City, NY: Doubleday Anchor Books.

———. (1961). *Democracy in America.* New York: Schocken Books.

de Vreese, Claes H. (2005). "News Framing: Theory and Typology." *Information Design Journal* 13 (1): 51–62.

de Waal, F. B. M. (1982). *Chimpanzee Politics: Power and Sex among Apes.* New York: Harper and Row.

Dewey, John. (1925). *Experience and Nature.* New York: Dover.

———. (1927). *The Public and Its Problems.* Denver, CO: Alan Swallow.

Diamond, Jared. (1992). *The Third Chimpanzee: The Evolution and Future of the Human Animal.* New York: Harper Perennial.

DiMaggio, Paul, John Evans, and Bethany Bryson. (1996). "Have Americans' Social Attitudes Become More Polarized?" *American Journal of Sociology* 102 (3): 690–755.

DiMaggio, Paul, Eszter Hargittai, W. Russell Neuman, and John Robinson. (2001). "Social Implications of the Internet." In *Annual Review of Sociology,* 307–336. Palo Alto, CA: Annual Reviews.

DiMaggio, Paul, and Kyoko Sato. (2003). "Does the Internet Balkanize Political Attention?: A Test of the Sunstein Theory." Paper presented at the annual meeting of the American Sociological Association, Atlanta, GA.

Doob, Leonard W. (1948). *Public Opinion and Propaganda*. New York: Holt.

Dordick, Herbert S., and Georgette Wang. (1993). *The Information Society*. Newbury Park, CA: Sage.

Dovidio, John F., Anja Eller, and Miles Hewstone. (2011). "Improving Intergroup Relations through Direct, Extended and Other Forms of Indirect Contact." *Group Processes and Intergroup Relations* 14 (2): 147–160.

Dovidio, John F., Peter Samuel Glick, and Laurie A. Rudman, eds. (2005). *On the Nature of Prejudice: Fifty Years after Allport*. Malden, MA: Blackwell.

Downie, Leonard, Jr., and Michael Schudson. (2009). *The Reconstruction of American Journalism*. New York: Columbia University Press.

Downs, Anthony. (1972). "Up and Down with Ecology: The Issue Attention Cycle." *Public Interest* 28:38–50.

Doyle, Gillian. (2013). *Understanding Media Economics*. Thousand Oaks, CA: Sage.

Drucker, Peter. (1969). *The Age of Discontinuity*. London: Heinemann.

Druckman, James N. (2001). "Evaluating Framing Effects." *Journal of Economic Psychology* 22 (1): 91–101.

Druckman, James N., and Kjersten R. Nelson. (2003). "Framing and Deliberation: How Citizens' Conversations Limit Elite Influence." *American Political Science Review* 47:728–744.

Dupagne, Michel, and R. Jeffery Green. (1996). "Revisiting the Principle of Relative Constancy: Consumer Mass Media Expenditures in Belgium." *Communication Research* 23:612–635.

Durbin, E. F. M., and John Bowlby. (1939). *Personal Aggressiveness and War*. London: Routledge and Kegan Paul.

Durham, William H. (1991). *Coevolution: Genes, Culture and Human Diversity*. Stanford, CA: Stanford University Press.

Durkheim, Emile. (1964). *The Rules of Sociological Method*. New York: Free Press.

Dutton, William H., and Malcolm Peltu. (2009). "The New Politics of the Internet: Multi-Stakeholder Policy-Making and the Internet Technocracy." In *Routledge Handbook of Internet Politics*, edited by Andrew Chadwick and Philip N. Howard, 384–400. New York: Routledge.

Economides, Nicholas. (1996). "The Economics of Networks." *International Journal of Industrial Organization* 14:673–699.

Edmonds, Rick, Emily Guskin, Tom Rosenstiel, and Amy Mitchell. (2012). "Newspapers: Building Digital Revenues Proves Painfully Slow." In *The State of the News Media 2012*. Washington, DC: Pew Project for Excellence in Journalism.

Edwards, Kari, and Edward E. Smith. (1966). "A Disconfirmation Bias in the Evaluation of Arguments." *Journal of Personality and Social Psychology* 77:5–24.

Edwards, Michael. (2009). *Civil Society*. Malden, MA: Polity.

Edwards, Paul N. (2010). "Some Say the Internet Shouldn't Have Happened." In *Media, Technology, and Society: Theories of Media Evolution,* edited by W. Russell Neuman, 141–160. Ann Arbor: University of Michigan Press.

Efron, Edith. (1971). *The News Twisters.* New York: Nash.

Ehrlinger, Joyce, and David Dunning. (2003). "How Chronic Self-Views Influence (and Potentially Mislead) Estimates of Performance." *Journal of Personality and Social Psychology* 84 (1): 5–17.

Einstein, Mara. (2004). *Media Diversity: Economics, Ownership and the FCC.* Mahwah, NJ: Lawrence Erlbaum.

Eisenach, Jeffrey A., and Thomas M. Lenard. (1999). *Competition, Innovation, and the Microsoft Monopoly: Antitrust in the Digital Marketplace.* Boston: Kluwer Academic Publishers.

Eisenstein, Elizabeth L. (1979). *The Printing Press as an Agent of Change.* Cambridge: Cambridge University Press.

———. (2011). *Divine Art, Infernal Machine: The Reception of Printing in the West from First Impressions to the Sense of an Ending.* Philadelphia: University of Pennsylvania Press.

Elberse, Anita. (2006). "The Power of Stars: Do Stars Drive Success in Creative Industries?" Working paper, Harvard Business School. http://hbswk.hbs.edu/item/the-power-of-stars-do-stars-drive-success-in-creative-industries.

Electronic Freedom Foundation. (2008). "RIAA v. The People: Five Years Later." Washington, DC. https://www.eff.org/wp/riaa-v-people-five-years-later.

Eliade, Mircea. (1979). *The Forge and the Crucible: The Origins and Structure of Alchemy.* Chicago: University of Chicago Press.

Emmers-Sommer, Tara M., and Mike Allen. (1999). "Surveying the Effect of Media Effects a Meta-Analytic Summary of the Media Effects Research in Human Communication Research." *Human Communication Research* 24 (4): 478–497.

Entman, Robert M. (1989). *Democracy without Citizens: Media and the Decay of American Politics.* New York: Oxford University Press.

———. (1993). "Framing—toward Clarification of a Fractured Paradigm." *Journal of Communication* 43 (4): 51–58.

———. (2004). *Projections of Power: Framing News, Public Opinion, and U.S. Foreign Policy.* Chicago: University of Chicago Press.

Eppler, Martin J., and Jeanne Mengis. (2004). "The Concept of Information Overload: A Review of Literature from Organization Science, Accounting, Marketing, MIS, and Related Disciplines." *Information Society* 20:325–344.

Erbring, Lutz, Edie M. Goldenberg, and Arthur H. Miller. (1980). "Front Page News and Real World Cues: A New Look at Agenda-Setting by the Media." *American Journal of Political Science* 24:16–49.

Ericsson, K. Anders, and Herbert A. Simon. (1993). *Protocol Analysis: Verbal Reports as Data.* Cambridge, MA: MIT Press.

Erikson, Robert S., Michael MacKuen, and James A. Stimson. (2002). *The Macro Polity.* New York: Cambridge University Press.

Esser, Frank, and Barbara Pfetsch, eds. (2004). *Comparing Political Communication: Theories, Cases and Challenges.* New York: Cambridge University Press.

Evans, John. (2003). "Have Americans' Social Attitudes Become More Polarized?—an Update." *Social Science Quarterly* 84 (1): 71–90.

Evans, Richard I. (1989). *Albert Bandura, the Man and His Ideas—a Dialogue.* New York: Praeger.

Eveland, William P., Jr., and Sharon Dunwoody. (2000). "Examining Information Processing on the World Wide Web Using Think Aloud Protocols." *Media Psychology* 2 (3): 219–244.

Fan, David P. (1988). *Predictions of Public Opinion from the Mass Media.* Westport, CT: Greenwood.

Fanon, Frantz. (1963). *The Wretched of the Earth.* New York: Grove Press.

Farrell, Joseph, and Philip J. Weiser. (2003). "Modularity, Vertical Integration, and Open Access Policies: Towards a Convergence of Antitrust and Regulation in the Internet Age." *Harvard Journal of Law and Technology* 17 (1): 87–134.

Febvre, Lucien, and Henri-Jean Martin. (1997). *The Coming of the Book: The Impact of Printing 1450–1800.* London: Verso.

Fiorina, Morris P., Samuel J. Abrams, and Jeremy C. Pope. (2010). *Culture War? The Myth of a Polarized America.* New York: Longman.

Fisher, William W. (2004). *Promises to Keep: Technology, Law, and the Future of Entertainment.* Stanford, CA: Stanford Law and Politics.

Förster, Jens. (2009). "Cognitive Consequences of Novelty and Familiarity: How Mere Exposure Influences Level of Construal." *Journal of Experimental Social Psychology* 45 (2): 444–447.

Fowler, Mark S. (1981). "Reason Interview: Mark S. Fowler." *Reason,* November. Reason.com.

Frank, Thomas. (2004). *What's the Matter with Kansas?: How Conservatives Won the Heart of America.* New York: Metropolitan Books.

Fromm, Erich. (1941). *Escape from Freedom.* New York: Farrar and Rinehart.

Fuchs, Christian. (2008). *Internet and Society: Social Theory in the Information Age.* New York: Routledge.

Funkhouser, Gerald R. (1973). "Trends in Media Coverage of the '60s." *Journalism Quarterly* 50 (3): 533–538.

Gaines, Brian J., and Jeffery J. Mondak. (2009). "Typing Together? Clustering of Ideological Types in Online Social Networks." *Journal of Information Technology and Politics* 6 (3–4): 216–231.

Galtung, Johann, and Mari Holmboe Ruge. (1965). "The Structure of Foreign News." *Journal of Peace Research* 2:64–91.

Gamson, William. (1968). *Power and Discontent.* Homewood, IL: Dorsey Press.

———. (1975). *The Strategy of Social Protest.* Chicago: Dorsey Press.

———. (1984). *What's News.* New York: Free Press.

———. (1992). *Talking Politics.* New York: Cambridge University Press.

Gans, Herbert J. (1979). *Deciding What's News.* New York: Pantheon Books.

————. (1999). *Popular Culture and High Culture: An Analysis and Evaluation of Taste Revised and Updated*. New York: Basic Books.

Garrett, R. Kelly. (2009). "Politically Motivated Reinforcement Seeking: Reframing the Selective Exposure Debate." *Journal of Communication* 59 (4): 676–699.

————. (2011). "Troubling Consequences of Online Political Rumoring." *Human Communication Research* 37 (2): 255–274.

Gary, Brett. (1996). "Communication Research, the Rockefeller Foundation, and Mobilization for the War on Words, 1938–1944." *Journal of Communication* 46 (3): 124–147.

————. (1999). *The Nervous Liberals: Propaganda Anxieties from World War I to the Cold War*. New York: Columbia University Press.

Gates, Bill. (1996). *Content Is King*. Redmond, WA: Microsoft.

Geertz, Clifford. (1973). *The Interpretation of Cultures*. New York: Basic Books.

————. (2000). *Available Light: Anthropological Reflections on Philosophical Topics*. Princeton, NJ: Princeton University Press.

Gelman, Andrew. (2005). "Analysis of Variance—Why It Is More Important Than Ever." *Annals of Statistics* 33 (1): 1–53.

Genachowski, Julius. (2011). "Remarks on Spectrum." *The White House*, April 6.

Gerber, Alan S., and Donald P. Green. (2012). *Field Experiments: Design, Analysis, and Interpretation*. New York: Norton.

Gerbner, George. (1956). "Toward a General Model of Communication." *Educational Technology Research and Development* 4 (3): 171–199.

————. (1967). "Mass Media and Human Communication Theory." In *Human Communication Theory: Original Essays*, edited by Frank E. X. Dance, 40–60. New York: Holt, Rinehart, and Winston.

————. (1969). "Toward 'Cultural Indicators': The Analysis of Mass Mediated Public Message Systems." *AV Communication Review* 17 (2): 137–148.

————, ed. (2002). *Against the Mainstream: The Selected Works of George Gerbner*. New York: Lang.

Gerbner, George, and Larry Gross. (1976). "Living with Television: The Violence Profile." *Journal of Communication* 26 (2): 173–199.

Gerbner, George, Larry Gross, Michael F. Eleey, Marilyn Jackson-Beeck, Suzanne Jeffries-Fox, and Nancy Signorielli. (1977). "TV Violence Profile No. 8: The Highlights." *Journal of Communication* 27 (2): 171–180.

Gerbner, George, Larry Gross, Marilyn Jackson-Beeck, Suzanne Jeffries-Fox, and Nancy Signorielli. (1978). "Cultural Indicators: Violence Profile No. 9." *Journal of Communication* 28 (3): 176–206.

Gerbner, George, Larry Gross, Michael Morgan, and Nancy Signorielli. (1980). "The 'Mainstreaming' of America: Violence Profile No. 11." *Journal of Communication* 30 (3): 10–29.

————. (1981a). "A Curious Journey into the Scary World of Paul Hirsch." *Communication Research* 8:39–72.

———. (1981b). "Final Reply to Hirsch." *Communications Research* 8 (1): 259–280.

Gerbner, George, Larry Gross, Nancy Signorielli, Michael Morgan, and Marilyn Jackson-Beeck. (1979). "The Demonstration of Power: Violence Profile No. 10." *Journal of Communication* 29 (3): 177–196.

Gibson, James L. (1988). "Political Intolerance and Political Repression during the McCarthy Red Scare." *American Political Science Review* 82:511–529.

Giddens, Anthony. (1976). *New Rules of Sociological Method: A Positive Critique of Interpretative Sociologies.* New York: Basic Books.

Gilder, George. (1989). *Microcosm: The Quantum Revolution in Economics and Technology.* New York: Simon & Schuster.

———. (2002). *Telecosm: The World after Bandwidth Abundance.* New York: Free Press.

Gilens, Martin. (1999). *Why Americans Hate Welfare: Race, Media and the Politics of Antipovery Policy.* Chicago: University of Chicago Press.

Giles, Jim. (2005). "Internet Encyclopedias Go Head to Head." *Nature* 438 (7070): 900–901.

Gillespie, Tarleton. (2007). *Wired Shut: Copyright and the Shape of Digital Culture.* Cambridge, MA: MIT Press.

Ginsberg, Benjamin. (1986). *The Captive Public: How Mass Opinion Promotes State Power.* New York: Basic Books.

Gitlin, Todd. (1978). "Media Sociology: The Dominant Paradigm." *Theory and Society* 6:205–253.

———. (1979). "Prime Time Ideology: The Hegemonic Process in Television Entertainment." *Social Problems* 26 (3): 205–253.

———. (1980). *The Whole World Is Watching.* Berkeley: University of California Press.

———. (2002). *Media Unlimited: How the Torrent of Images and Sounds Overwhelms Our Lives.* New York: Owl Books.

Glander, Timothy Richard. (2000). *Origins of Mass Communications Research during the American Cold War: Educational Effects and Contemporary Implications.* Mahwah, NJ: Lawrence Erlbaum.

Glasgow University Media Group. (1976). *Bad News.* London: Routledge and Kegan Paul.

———. (1980). *More Bad News.* London: Routledge and Kegan Paul.

Glynn, Carroll. (1989). "Perceptions of Others' Opinions as a Component of Public Opinion." *Social Science Research* 18:53–69.

Glynn, Carroll J., Andrew F. Hayes, and James Shanahan. (1997). "Perceived Support for One's Opinions and Willingness to Speak Out: A Meta-Analysis of Survey Studies on the 'Spiral of Silence.'" *Public Opinion Quarterly* 61 (3): 452–463.

Goetz, David. (2011). "Total Auto Ad Spend to Grow 6.4%." *Media Daily News,* March 17.

Goldberg, Robert Alan. (2001). *Enemies Within: The Culture of Conspiracy in Modern America.* New Haven, CT: Yale University Press.

Goldstein, Kenneth, and Travis N. Ridout. (2004). "Measuring the Effects of Televised Political Advertising in the United States." *Annual Review of Political Science* 7 (1): 205–226.

Goodhardt, G. J., A. S. C. Ehrenberg, and M. A. Collins. (1980). *The Television Audience: Patterns of Viewing.* Lexington, MA: Lexington Books.

Graber, Dons A. (2007). "The Road to Public Surveillance: Breeching Attention Thresholds." In *The Affect Effect: Dynamics of Emotion in Political Thinking and Behavior,* edited by W. Russell Neuman, George E. Marcus, Ann N. Crigler, and Michael MacKuen, 265–290. Chicago: University of Chicago Press.

Gramsci, Antonio. (1933). *Selections from the Prison Notebooks.* New York: International Publishers.

Gray, Ann, Jan Campbell, Mark Erickson, Stuart Hanson, and Helen Wood. (2007). *CCCS Selected Working Papers, Volume 1 and 2.* New York: Routledge.

Grehan, John R., and Jeffrey H. Schwartz. (2011). "Evolution of Human–Ape Relationships Remains Open for Investigation." *Journal of Biogeography* 38 (12): 2397–2404.

Gripsrud, Jostein. (1992). "The Aesthetics and Politics of Melodrama." In *Journalism and Popular Culture,* edited by Peter Dahlgren and Colin Sparks, 84–95. Thousand Oaks, CA: Sage.

Groseclose, Tim. (2011). *Left Turn: How Liberal Media Bias Distorts the American Mind.* New York: St. Martin's Press.

Grossberg, Lawrence. (2010). *Cultural Studies in the Future Tense.* Durham, NC: Duke University Press.

Grossman, Dave, and Gloria DeGaetano. (1999). *Stop Teaching Our Kids to Kill: A Call to Action against TV, Movie and Video Game Violence.* New York: Crown Publishers.

Guber, Deborah Lynn. (2003). *The Grassroots of a Green Revolution: Polling America on the Environment.* Cambridge, MA: MIT Press.

Gulli, Antonio, and Alessio Signorini. (2005). "The Indexable Web Is More Than 11.5 Billion Pages." Paper presented at the Fourteenth International World Wide Web Conference, Chiba, Japan.

Gurr, Ted Robert. (1970). *Why Men Rebel.* Princeton, NJ: Princeton University Press.

Habermas, Jürgen (1979). *Communication and the Evolution of Society.* Toronto: Beacon Press.

———. (1981). *The Theory of Communicative Action, Volumes 1 and 2.* Boston: Beacon Press.

———. (1988). *On the Logic of the Social Sciences.* Cambridge, MA: MIT Press.

———. (1989). *The Structural Transformation of the Public Sphere.* Cambridge, MA: MIT Press.

———. (1990). *Moral Consciousness and Communicative Action.* Cambridge, MA: MIT Press.

Hagel, John, and John Seely Brown. (2005). "From Push to Pull- Emerging Models for Mobilizing Resources." *McKinsey Quarterly.* www.mckinsey.com/insights /mckinsey_quarterly.

Halberstam, David. (1979). *The Powers That Be.* New York: Knopf.

Hall, Stuart. (1980). "Encoding/Decoding." In *Culture, Media, Language: Working Papers in Cultural Studies, 1972–79,* edited by Stuart Hall, Dorothy Hobson, Andy Lowe, and Paul Willis, 128–138. London: Routledge.

———. (2003). "The Whites of Their Eyes: Racist Ideologies and the Media." In *Gender, Race, and Class in Media: A Text-Reader,* edited by Gail Dines and Jean McMahon Humez, 89–93. Thousand Oaks, CA: Sage.

Hall, Stuart, Dorothy Hobson, Andy Lowe, and Paul Willis, eds. (1980). *Culture, Media, Language: Working Papers in Cultural Studies, 1972–79.* London: Routledge.

Hallin, Daniel C. (1984). "The Media, the War in Vietnam, and Political Support: A Critique of the Thesis of an Oppositional Media." *Journal of Politics* 46 (1): 2–24.

Hamilton, James T. (1998). *Channeling Violence: The Economic Market for Violent Television Programming.* Princeton, NJ: Princeton University Press.

Hamilton, Richard F. (1972). *Class and Politics in the United States.* New York: Wiley.

———. (1982). *Who Voted for Hitler?* Princeton, NJ: Princeton University Press.

Hansen, Anders, ed. (2009). *Mass Communication Research Methods.* Thousand Oaks, CA: Sage.

Hare, Ivan, and James Weinstein, eds. (2011). *Extreme Speech and Democracy.* New York: Oxford University Press.

Hargittai, Eszter. (2002). "Second Level Digital Divide: Differences in People's Online Skills." *First Monday* 7 (4). firstmonday.org.

———. (2015). "Is Bigger Always Better? Potential Biases of Big Data Derived from Social Network Sites." *ANNALS of the American Academy of Political and Social Science* 659:63–76.

Hargittai, Eszter, and Yuli P. Hsieh. (2010). "Predictors and Consequences of Differentiated Social Network Site Uses." *Information, Communication and Society* 13 (4): 515–536.

Hargittai, Eszter, W. Russell Neuman, and Olivia Curry. (2012). "Taming the Information Tide: Americans' Thoughts on Information Overload, Polarization and Social Media." *Information Society* 28:161–173.

Harris, Richard J. (1994). "Forever Random (Factors): A Single-Message Treatment of Message Research Methods." *PsycCRITIQUES* 39 (5): 474–475.

Harris, Richard Jackson. (2004). *A Cognitive Psychology of Mass Communication.* Mahwah, NJ: Erlbaum.

Harrison, Albert A. (1977). "Mere Exposure." In *Advances in Experimental Social Psychology,* edited by L. Berkowitz, 39–83. New York: Academic Press.

Harwood, John. (2009). "If Fox Is Partisan, It Is Not Alone." *New York Times,* November 1.

Hauser, Marc D. (1996). *The Evolution of Communication.* Cambridge, MA: MIT Press.

Hauser, Marc D., Noam Chomsky, and W. Tecumseh Fitch. (2002). "The Faculty of Language: What Is It, Who Has It, and How Did It Evolve?" *Science* 298:1569–1579.

Hayes, Andrew F., Michael D. Slater, and Leslie B. Snyder. (2008). *The Sage Sourcebook of Advanced Data Analysis Methods for Communication Research*. Los Angeles: Sage.

Haythornthwaite, Caroline. (2005). "Social Networks and Internet Connectivity Effects." *Information, Communication and Society* 8 (2): 125–147.

Heath, Chip, and Dan Heath. (2007). *Made to Stick: Why Some Ideas Survive and Others Die*. New York: Random House.

Hektner, Joel M., Jennifer A. Schmidt, and Mihaly Csikszentmihalyi, eds. (2006). *Experience Sampling Method: Measuring the Quality of Everyday Life*. Thousand Oaks, CA: Sage.

Herman, Bill D. (2006). "Opening Bottlenecks: On Behalf of Mandated Network Neutrality." *Federal Communication Law Journal* 103:103–156.

Herman, Edward S., and Noam Chomsky. (1988). *Manufacturing Consent: The Political Economy of the Mass Media*. New York: Pantheon Books.

Herzstein, Robert Edwin. (1978). *The War That Hitler Won: The Most Infamous Propaganda Campaign in History*. New York: Putnam.

Hewstone, Miles, and Hermann Swart. (2011). "Fifty-Odd Years of Inter-Group Contact: From Hypothesis to Integrated Theory." *British Journal of Social Psychology* 50 (3): 374–386.

Hilgartner, Stephen, and Charles L. Bosk. (1988). "The Rise and Fall of Social Problems: A Public Arenas Model." *American Journal of Sociology* 94 (1): 53–78.

Himelstein, Linda, and Richard Siklos. (1999). "Pointcast: The Rise and Fall of an Internet Star." *Business Week*, April 26.

Hindman, Matthew, Kostas Tsioutsiouliklis, and Judy A. Johnson. (2003). "Googlearchy: How a Few Heavily Linked Sites Dominate Politics Online." Presented at the Annual Meeting of the Midwest Political Science Association, Chicago.

Hindman, Matthew Scott. (2006). "A Mile Wide and an Inch Deep: Measuring the Diversity of Political Content Online." In *Localism and Media Diversity: Meaning and Metrics*, edited by Philip Napoli, 327–347. Mahwah, NJ: Lawrence Erlbaum.

———. (2009). *The Myth of Digital Democracy*. Princeton, NJ: Princeton University Press.

Hirsch, Paul M. (1980). "The 'Scary World' of the Nonviewer and Other Anomalies: A Reanalysis of Gerbner et al.'s Findings on Cultivation Analysis, Part 1." *Communication Research* 7 (4): 403–456.

———. (1981a). "Distinguishing Good Speculation from Bad Theory." *Communication Research* 8 (1): 73–95.

———. (1981b). "On Not Learning from One's Own Mistakes: A Reanalysis of Gerbner et al.'s Findings on Cultivation Analysis Part II" *Communication Research* 8 (1): 3–37.

Hoggart, Richard. (1957). *The Uses of Literacy*. Piscataway, NJ: Transaction Publishers.

Holbert, R. Lance, and Erik Page Bucy. (2014). "Advancing Methods and Measurement: Supporting Theory and Keeping Pace with the Modern Political Environment." In *Sourcebook for Political Communication Research: Methods, Measures, and Analytical Techniques,* edited by Erik Page Bucy and R. Lance Holbert, 3–15. New York: Routledge.

Hoorens, Vera. (1993). "Self-Enhancement and Superiority Biases in Social Comparison." *European Review of Social Psychology* 4 (1): 113–139.

Hornik, Robert, Lela Jacobsohn, Robert Orwin, Andrea Piesse, and Graham Kalton. (2008). "Effects of the National Youth Anti-Drug Media Campaign on Youths." *American Journal of Public Health* 98(12): 2229–2236.

Housley, Meghan K., Heather M. Claypool, Teresa Garcia-Marques, and Diane M. Mackie. (2010). "'We' Are Familiar but 'It' Is Not: Ingroup Pronouns Trigger Feelings of Familiarity." *Journal of Experimental Social Psychology* 46 (1): 114–119.

Hovland, Carl. (1959). "Reconciling Conflicting Results Derived from Experimental and Survey Studies of Attitude Change." *American Psychologist* 14:8–17.

Hovland, Carl, Irving Janis, and Harold H. Kelley. (1953). *Communication and Persuasion.* New Haven, CT: Yale University Press.

Hovland, Carl I., Arthur A. Lumsdaine, and Fred D. Sheffield. (1949). *Experiments on Mass Communication.* Princeton, NJ: Princeton University Press.

Huber, Peter W. (1997). *Law and Disorder in Cyberspace: Abolish the FCC and Let the Common Law Rule the Telecosm.* New York: Oxford University Press.

Huckfeldt, Robert, Paul E. Johnson, and John Sprague. (2004). *Political Disagreement: The Survival of Diverse Opinions within Diverse Communication Networks.* New York: Cambridge University Press.

Huckfeldt, Robert, and John Sprague. (1995). *Citizens, Politics and Social Communication: Information and Influence in an Election Campaign.* New York: Cambridge University Press.

Hughes, Helen MacGill. (1940). *News and the Human Interest Story.* Chicago: University of Chicago Press.

Hughes, Michael. (1980). "The Fruits of Cultivation Analysis: A Reexamination of Some Effects of Television Watching." *Public Opinion Quarterly* 44 (3): 287–302.

Hunter, John E., Mark A. Hamilton, and Mike Allen. (1989). "The Design and Analysis of Language Experiments in Communication." *Communication Monographs* 56 (4): 341–363.

Huntington, Samuel P. (1993). "The Clash of Civilizations." *Foreign Affairs* 72:22–49.

———. (1996). *The Clash of Civilizations and the Remaking of World Order.* New York: Simon & Schuster.

Huron, David. (2006). *Sweet Anticipation: Music and the Psychology of Expectation.* Cambridge, MA: MIT Press.

Huston, Aletha C., et al. (1992). *Big World, Small Screen: The Role of Television in American Society.* Lincoln: University of Nebraska Press.

Ickes, L. R., ed. (2006). *Public Broadcasting in America*. Hauppauge, NY: Novinka Books.

Innis, Harold A. (1950). *Empire and Communication*. New York: Oxford University Press.

———. (1951). *The Bias of Communication*. Toronto: University of Toronto Press.

———. (1952). *The Strategy of Culture*. Toronto: University of Toronto Press.

iProspect. (2006). "Search Engine User Behavior Study." iProspect.com.

Isaacs, Harold R. (1975). *Idols of the Tribe: Group Identity and Political Change*. Cambridge, MA: Harvard University Press.

Iyengar, Shanto. (1987). "Television News and Citizens' Explanations of National Affairs." *American Political Science Review* 81 (3): 815–831.

———. (1990). "Shortcuts to Political Knowledge: The Role of Selective Attention and Accessibility." In *Information and the Democratic Process*, edited by John A. Ferejohn and James H Kuklinski, 160–185. Urbana: University of Illinois Press.

———. (1991). *Is Anyone Responsible? How Television Frames Political Issues*. Chicago: University of Chicago Press.

Iyengar, Shanto, and Donald R. Kinder. (1987). *News That Matters: Television and American Opinion*. Chicago: University of Chicago Press.

Iyengar, Sheena. (2010). *The Art of Choosing*. New York: Twelve.

Jackson, Sally. (1991). "Meta-Analysis for Primary and Secondary Data Analysis: The Super-Experiment Metaphor." *Communication Monographs* 58 (4): 449–462.

———. (1992). *Message Effects Research: Principles of Design and Analysis*. New York: Guilford.

Jackson, Sally, and Scott Jacobs (1983). "Generalizing about Messages: Suggestions for Design and Analysis of Experiments." *Human Communication Research* 9 (2): 169–191.

Jackson, Sally, Daniel J. O'Keefe, and Dale E. Brashers. (1994). "The Messages Replication Factor: Methods Tailored to Messages as Object of Study." *Journalism Quarterly* 71 (4): 984–996.

Jackson, Sally, Daniel J. O'Keefe, and Scott Jacobs. (1988). "The Search for Reliable Generalizations about Messages a Comparison of Research Strategies." *Human Communication Research* 15 (1): 127–142.

Jackson, Sally, Daniel J. O'Keefe, Scott Jacobs, and Dale E. Brashers. (1989). "Messages as Replications: Toward a Message-Centered Design Strategy." *Communication Monographs* 56 (4): 364–384.

Jacobellis v. Ohio. (1964). 378 U.S. 184 (1964) No. 11.

Jamieson, Kathleen Hall, and Joseph N. Cappella. (2008). *Echo Chamber: Rush Limbaugh and the Conservative Media Establishment*. New York: Oxford University Press.

Janowitz, Morris. (1976). "Content Analysis and the Study of Sociopolitical Change." *Journal of Communication* 26 (4): 10–21.

Jeffres, Leo W. (1997). *Mass Media Effects*. Prospect Heights IL: Waveland Press.

Jenkins, Henry. (1992). *Textual Poachers: Television Fans and Participatory Culture.* New York: Routledge.

———. (2006). *Convergence Culture: Where Old and New Media Collide.* New York: NYU Press.

Jennings, M. Kent, and Gregory B. Markus. (1977). "The Effect of Military Service on Political Attitudes: A Panel Study." *American Political Science Review* 71 (1): 131–147.

Jensen, Klaus Bruhn. (1987). "Qualitative Audience Research: Toward an Integrative Approach to Reception." *Critical Studies in Mass Communication* 4 (1): 21–37.

———. (1990). "The Politics of Polysemy: Television News, Everyday Consciousness and Political Action." *Media Culture and Society* 12 (1): 57–78.

———, ed. (1998). *News of the World: World Cultures Look at Television News.* New York: Routledge.

———, ed. (2002). *A Handbook of Media and Communication Research: Qualitative and Quantitative Methodologies.* New York: Routledge.

———, ed. (2011). *A Handbook of Media and Communication Research: Qualitative and Quantitative Methodologies.* New York: Routledge.

Johns, Adrian. (2009). *Piracy: The Intellectual Property Wars from Gutenberg to Gates.* Chicago: University of Chicago Press.

Jones, Robert Alun. (1983). "The New History of Sociology." *Annual Review of Sociology* 9 (1): 447–469.

Jones, Robert W. (1993). "Coorientation of a News Staff and Its Audience." *Communication Reports* 6 (1): 41–46.

Jonscher, Charles. (1999). *The Evolution of Wired Life: From the Alphabet to the Soul-Catcher Chip—How Information Technologies Change Our World.* New York: Wiley.

Jost, John T. (2009). ""Elective Affinities": On the Psychological Bases of Left–Right Differences." *Psychological Inquiry* 20:129–141.

Kahneman, Daniel, Paul Slovic, and Amos Tversky, eds. (1982). *Judgment under Uncertainty.* New York: Cambridge University Press.

Kahneman, Daniel, and Amos Tversky. (2000). *Choices, Values, and Frames.* New York: Cambridge University Press.

Kaid, Lynda Lee. (2004). "Political Advertising." In *Handbook of Political Communication Research,* edited by Lynda Lee Kaid, 155–202. New York: Routledge.

Katz, Daniel, and Floyd H. Allport. (1931). *Student Attitudes.* Syracuse, NY: Craftsman.

Katz, Daniel, and Kenneth Braly. (1933). "Racial Stereotypes of One Hundred College Students." *Journal of Abnormal and Social Psychology* 28:280–290.

Katz, Daniel, Dorwin Cartwright, Samuel Eldersveld, and Alfred McClung Lee, eds. (1954). *Public Opinion and Propaganda.* New York: Holt, Rinehart, and Winston.

Katz, Daniel, and Richard Schanck. (1938). *Social Psychology.* New York: Wiley.

Katz, Elihu. (1978). "Looking for Trouble." *Journal of Communication* 28 (2): 90–95.

————. (1987). "Communications Research since Lazarsfeld." *Public Opinion Quarterly* 51:S25–S45.

————. (2001). "Media Effects." *International Encyclopedia of the Social and Behavioral Sciences* 9472–9479. www.communicationencyclopedia.com/public/.

Katz, Elihu, Jay G. Blumler, and Michael Gurevitch. (1973). "On the Use of the Mass Media for Important Things." *American Sociological Review* 38 (2): 164–181.

Katz, Elihu, and Paul F. Lazarsfeld. (1955). *Personal Influence: The Part Played by People in the Flow of Communications.* New York: Free Press.

Katz, Elihu, John Durham Peters, Tamar Liebes, and Avril Orloff, eds. (2003). *Canonic Texts in Media Research: Are There Any? Should There Be? How about These?* Cambridge: Polity.

Keeley, Lawrence H. (1996). *War before Civilization.* New York: Oxford University Press.

Keller, Bill. (2011). "A Theory of Conspiracy Theories." *New York Times,* June 3.

Kellstedt, Paul M. (2003). *The Mass Media and the Dynamics of American Racial Attitudes.* New York: Cambridge University Press.

Kelly, Sanja, and Sarah Cook. (2011). *Freedom on the Net 2011: A Global Assessment of Internet and Digital Media.* Washington, DC: Freedom House.

Kenneally, Christine. (2007). *The First Word: The Search for the Origins of Language.* New York: Viking.

Key, V. O., Jr. (1961). *Public Opinion and American Democracy.* New York: Alfred A. Knopf.

Key, Wilson Bryan. (1974). *Subliminal Seduction: Ad Media's Manipulation of a Not So Innocent America.* New York: New American Library.

Kincaid, Cliff, Roger Aronoff, and Don Irvine. (2007). *Why You Can't Trust the News, Volume Two.* Washington, DC: Accuracy in Media.

Kinder, Donald R., and Cindy D. Kam. (2009). *Us against Them: Ethnocentric Foundations of American Opinion.* Chicago: University of Chicago Press.

Kinder, Donald R., and Lynn M. Sanders. (1996). *Divided by Color: Racial Politics and Democratic Ideals.* Chicago: University of Chicago Press.

King, Barbara J. (1994). *Information Continuum: Evolution of Social Information Transfer in Monkeys Apes and Hominids.* Santa Fe, NM: School of American Research Press.

King, Gary. (2004). "The Future of Replication." *International Studies Perspectives* 4:443–499.

Kirscht, John P., and Ronald C. Dillehay. (1967). *Dimensions of Authoritariansim: A Review of Research and Theory.* Lexington: University of Kentucky Press.

Klapper, Joseph. (1960). *The Effects of Mass Communication.* New York: Free Press.

Klingberg, Torkel. (2009). *The Overflowing Brain: Information Overload and the Limits of Working Memory.* New York: Oxford University Press.

Kornhauser, William. (1959). *The Politics of Mass Society.* New York: Free Press.

Krasnow, Erwin G., Lawrence D. Longley, and Herbert A. Terry. (1982). *The Politics of Broadcast Regulation.* New York: St. Martin's Press.

Kreml, William P. (1977). *The Anti-Authoritarian Personality.* New York: Pergamon Press.

Kriesberg, Louis. (2007). "The Conflict Resolution Field: Origins, Growth, Differentiation." In *Peacemaking in International Conflict: Methods and Techniques,* edited by William Zartman, 25–60. Washington, DC: United States Institute of Peace.

Krippendorff, Klaus. (1994). "A Recursive Theory of Communication." In *Communication Theory Today,* edited by David Crowley and David Mitchell, 78–104. Stanford, CA: Stanford University Press.

Krosnick, Jon A. (1999). "Survey Research." *Annual Review of Psychology* 50:537–567.

Krosnick, Jon A., and Laura A. Brannon. (1993). "The Impact of the Gulf War on the Ingredients of Presidential Evaluations: Multidimensional Effects of Political Involvement." *American Political Science Review* 87:963–975.

Krosnick, Jon A., and Shibley Telhami. (1995). "Public Attitudes toward Israel: A Study of the Attentive and Issue Publics." *International Studies Quarterly* 39:535–554.

Kubey, Robert, Reed Larson, and Mihaly Csikszentmihalyi. (1996). "Experience Sampling Method Applications to Communication Research Questions." *Journal of Communication* 46 (2): 99–120.

Kuhn, Thomas. (1962). *The Structure of Scientific Revolutions.* Chicago: University of Chicago Press.

Kull, Steven, Clay Ramsay, Stephen Weber, Evan Lewis, and Ebrahim Mohseni. (2009). *Public Opinion in the Islamic World on Terrorism, Al Qaeda, and US Policies.* Washington, DC: Worldpublicopinion.org.

Land, Kenneth C., and Seymour Spilerman, eds. (1975). *Social Indicator Models.* New York: Russell Sage Foundation.

Lang, Annie. (2013). "Discipline in Crisis? The Shifting Paradigm of Mass Communication Research." *Communication Theory* 23 (1): 10–24.

Lasswell, Harold D. (1927). *Propaganda Technique in the World War.* New York: Smith.
———. (1935). *World Politics and Personal Insecurity.* New York: Free Press.
———. (1941). "The World Attention Survey." *Public Opinion Quarterly* 5 (3): 456–462.
———. (1948). "The Structure and Function of Communications in Society." In *The Communication of Ideas,* edited by Lyman Bryson, 37–51. New York: Harper.
———. (1963). *The Future of Political Science.* New York: Atherton Press.

Lazarsfeld, Paul F. (1941). "Remarks on Administrative and Critical Communication Research." *Studies in Philosophy and Social Science* 9:2–16.

Lazarsfeld, Paul F., Bernard Berelson, and Hazel Gaudet. (1944). *The People's Choice: How the Voter Makes Up His Mind in a Presidential Campaign.* New York: Columbia University Press.

Lazarsfeld, Paul F., and Robert K. Merton. (1948). "Mass Communication, Popular Taste, and Organized Social Action." In *The Communication of Ideas,* edited by Lyman Bryson, 95–118. New York: Harper.

Lecheler, Sophie, Claes de Vreese, and Rune Slothuus. (2009). "Issue Importance as a Moderator of Framing Effects." *Communication Research* 36 (3): 400–425.

Lee, Alfred McClung, and Elizabeth Briant Lee. (1939). *The Fine Art of Propaganda.* New York: Harcourt Brace.

Leeds-Hurwitz, Wendy. (1993). *Semiotics and Communication: Signs, Codes, Cultures.* Mahwah, NJ: Lawrence Erlbaum.

Leskovec, Jure, Daniel Huttenlocher, and Jon Kleinberg. (2010). "Governance in Social Media: A Case Study of the Wikipedia Promotion Process." In *Proceedings of the 4th International AAAI Conference on Weblogs and Social Media.* www.aaai.org/Library/ICWSM/icwsm10contents.php.

Lessig, Lawrence. (2001). *The Future of Ideas: The Fate of the Commons in a Connected World.* New York: Random House.

———. (2004). *Free Culture: How Big Media Uses Technology and the Law to Lock Down Culture and Control Creativity.* New York: Penguin Press.

———. (2006). *Code: Version 2.0.* New York: Basic.

Levy, Mark R., and Sven Windahl. (1985). "The Concept of Audience Activity." In *Media Gratifications Research: Current Perspectives,* edited by Karl Erik Rosengren, Lawrence A. Wenner, and Philip Palmgreen, 109–122. Newport Beach, CA: Sage.

Lewis, Ian. (2000). *Guerrilla TV: Low Budget Programme Making.* Boston: Focal Press.

Lewis, Justin. (2001). *Constructing Public Opinion: How Political Elites Do What They Like and Why We Seem to Go Along with It.* New York: Columbia University Press.

Lichter, S. Robert, Stanley Rothman, and Linda S. Lichter. (1986). *The Media Elite.* Bethesda, MD: Adler and Adler.

Lieberman, Marvin, and David Montgomery. (1988). "First-Mover Advantages." *Strategic Management Journal* 9:41–58.

Liebes, Tamar, and Elihu Katz. (1990). *The Export of Meaning: Cross-Cultural Readings of Dallas.* New York: Oxford University Press.

Lifton, Robert Jay. (1961). *Thought Reform and the Psychology of Totalism: A Study of "Brainwashing" in China.* New York: Norton.

Lind, E. Allan, and Tom R. Tyler. (1988). *The Social Psychology of Procedural Justice.* New York: Plenum Press.

Lindberg, David C. (2007). *The Beginnings of Western Science: The European Scientific Tradition in Philosophical, Religious, and Institutional Context, Prehistory to A.D. 1450.* Chicago: University of Chicago Press.

Lindblom, Charles E. (1977). *Politics and Markets: The World's Political-Economic Systems.* New York: Basic Books.

Linder, Laura R. (1999). *Public Access Television: America's Electronic Soapbox.* Westport, CT: Praeger.

Lindlof, Thomas R. (1991). "The Qualitative Study of Media Audiences." *Journal of Broadcasting and Electronic Media* 35 (1): 23–42.

Lindlof, Thomas R., and Bryan C. Taylor. (2010). *Qualitative Communication Research Methods.* Los Angeles: Sage.

Lippmann, Walter. (1922). *Public Opinion*. New York: Free Press.
———. (1925). *The Phantom Public*. New York: Macmillan.
Lipset, Seymour Martin. (1960). *Political Man*. New York: Doubleday.
———. (1970). *Revolution and Counterrevolution: Change and Persistence in Social Structures*. Garden City, NY: Anchor Books.
———. (1985). *Consensus and Conflict*. New Brunswick, NJ: Transaction Books.
Lipset, Seymour Martin, Martin A. Trow, and James S. Coleman. (1956). *Union Democracy*. Garden City, NY, Anchor Books.
Litman, Jessica. (2000). *Digital Copyright: Protecting Intellectual Property on the Internet*. Amherst NY, Prometheus Books.
Liu, Xun, and Robert LaRose. (2008). "Does Using the Internet Make People More Satisfied with Their Lives? The Effects of the Internet on College Students' School Life Satisfaction." *CyberPsychology and Behavior* 11 (3): 310–320.
Livingstone, Sonia. (1998). "Relationships between Media and Audiences: Prospects for Audience Reception Studies." In *Media, Ritual and Identity*, edited by Tamar Liebes and James Curran, 237–252. New York: Routledge.
Locksley, Anne, Vilma Ortiz, and Christine Hepburn. (1980). "Social Categorization and Discriminatory Behavior: Extinguishing the Minimal Intergroup Discrimination Effect." *Journal of Personality and Social Psychology* 39:773–783.
Lodish, Leonard M., Magid Abraham, Stuart Kalmensen, Jeanne Livelsberger, Beth Lubetkin, Bruce Richardson, and Mary Ellen Stevens. (1995). "How TV Advertising Works: A Meta Analysis of 389 Real World Split Cable TV Advertising Experiments." *Journal Marketing Research* 32:125–139.
Logan, Robert K. (1986). *The Alphabet Effect: The Impact of the Phonetic Alphabet on the Development of Western Civilization*. New York: Morrow.
Lotz, Amanda. (2007). *The Television Will Be Revolutionized*. New York: NYU Press.
Lubken, Deborah. (2008). "Remembering the Straw Man: The Travels and Adventures of Hypodermic." In *The History of Media and Communication Research: Contested Memories*, edited by David W. Park and Jefferson Pooley, 19–42. New York: Peter Lang.
Lull, James. (1988). "Critical Response: The Audience as Nuisance." *Critical Studies in Mass Communication* 5 (3): 239–242.
Lyman, Peter, and Hal R. Varian. (2003). *How Much Information?* Berkeley: University of California.
Machlup, Fritz. (1962). *The Production and Distribution of Knowledge in the United States*. Princeton, NJ: Princeton University Press.
MacKinnon, Rebecca. (2012). *Consent of the Networked: The Worldwide Struggle for Internet Freedom*. New York: Basic Books.
MacKuen, Michael, and Steven Lane Coombs. (1981). *More Than News: Media Power in Public Affairs*. Beverly Hills, CA: Sage.
Man, John. (2002). *The Gutenberg Revolution: How Printing Changed the Course of History*. London: Headline.

Manheim, Jarol B. (1994). *Strategic Public Diplomacy and American Foreign Policy: The Evolution of Influence.* New York: Oxford University Press.

Mannheim, Karl. (1936). *Ideology and Utopia.* New York: Harcourt Brace.

Mansell, Robin, ed. (2009). *The Information Society: Critical Concepts in Sociology.* New York: Routledge.

Marcus, George E., W. Russell Neuman, and Michael MacKuen. (2000). *Affective Intelligence and Political Judgment.* Chicago: University of Chicago Press.

Marsden, Peter V., and James D. Wright. (2010). *Handbook of Survey Research.* Bingley, UK: Emerald.

Martin, Ralph K., Garrett J. O'Keefe, and Oguz B. Nayman. (1972). "Opinion Agreement and Accuracy between Editors and Their Readers." *Journalism and Mass Communication Quarterly* 49 (3): 460–468.

Marvin, Carolyn. (1988). *When Old Technologies Were New: Thinking about Technology in the Late 19th Century.* New York: Oxford University Press.

Masuda, Yoneji. (1980). *The Information Society as Post Industrial Society.* Bethesda, MD: World Future Society.

Mattelart, Armand. (2003). *The Information Society: An Introduction.* London: Sage.

Mayer, Martin. (1992). *The Greatest Ever Bank Robbery: The Collapse of the Savings and Loan Industry.* New York: C. Scribner's Sons.

Mazzoleni, Gianpietro, and Winfried Schulz. (1999). "'Mediatization' of Politics: A Challenge for Democracy?" *Political Communication* 16 (3): 247–261.

McCarthy, John D., and Mayer N. Zald. (1977). "Resource Mobilization and Social Movements: A Partial Theory." *American Journal of Sociology* 82:1212–1241.

McChesney, Robert W. (1999). *Rich Media, Poor Democracy: Communication Politics in Dubious Times.* New York: New Press.

McChesney, Robert W., and Ben Scott, eds. (2004). *Our Unfree Press: 100 Years of Radical Media Criticism.* New York: New Press.

McChesney, Robert Waterman. (2004). *The Problem of the Media: U.S. Communication Politics in the Twenty-First Century.* New York: Monthly Review Press.

McCombs, Maxwell, Donald L. Shaw, and David Weaver, eds. (1997). *Communication and Democracy: Exploring the Intellectual Frontiers in Agenda-Setting Theory.* Mahwah, NJ: Lawrence Erlbaum.

McCombs, Maxwell, and Jian-Hua Zhu. (1995). "Capacity, Diversity, and Volatility of the Public Agenda: Trends from 1954 to 1994." *Public Opinion Quarterly* 59 (4): 495–525.

McCombs, Maxwell E. (2004). *Setting the Agenda: The Mass Media and Public Opinion.* Malden, MA: Polity.

McCombs, Maxwell E., and Chaim H. Eyal. (1980). "Spending on Mass Media." *Journal of Communication* 30 (1): 153–158.

McCombs, Maxwell E., and Amy Reynolds. (2002). "News Influence on Our Pictures of the World." In *Media Effects: Advances in Theory and Research,* edited by Jennings Bryant and Dolf Zillmann, 1–18. Mahwah, NJ: Lawrence Erlbaum.

McCombs, Maxwell E., and Donald L. Shaw. (1972). "The Agenda Setting Function of the Mass Media." *Public Opinion Quarterly* 36:176–187.

———. (1993). "The Evolution of Agenda-Setting Research: Twenty-Five Years in the Marketplace of Ideas." *Journal of Communication* 43 (2): 58–67.

McGuire, William J. (1968). "Personality and Susceptibility to Social Influence." In *Handbook of Personality Theory and Research,* edited by Edgar F. Borgatta and William W. Lambert, 1130–1187. Chicago: Rand-McNally.

———. (1985). "Attitudes and Attitude Change." In *The Handbook of Social Psychology,* edited by Gardner Lindzey and Elliot Aronson, 233–346. New York: Random House.

McKnight, David. (2012). *Rupert Murdoch: An Investigation of Political Power.* Chicago: Allen & Unwin.

McLeod, Jack, Lee B. Becker, and James E. Byrnes. (1974). "Another Look at the Agenda Setting Function of the Press." *Communications Research* 2:131–166.

McLeod, Jack, and Jay Blumler. (1987). "The Macrosocial Level of Communication Science." In *Handbook of Communication Science,* edited by Charles R. Berger and Steven H. Chaffee, 271–322. Newbury Park, CA: Sage.

McLeod, Jack M., Gerald M. Kosicki, and Douglas M. McLeod. (2009). "Levels of Analysis and Communication Science." In *Handbook of Communication Science,* edited by Charles R. Berger, Michael Roloff, and David Roskos-Ewoldsen, 183–200. Thousand Oaks, CA: Sage.

McLeod, Jack M., and Byron Reeves. (1980). "On the Nature of Mass Media Effects." In *Television and Social Behavior,* edited by Stephen B. Withey and Ronald P. Abeles, 17–54. Hillsdale, NJ: Lawrence Erlbaum.

McLeod, Jack M., Jessica Zubric, Heejo Keum, Sameer Deshpande, Jaeho Cho, Susan E. Stein, and Mark Heather. (2001). "Reflecting and Connecting: Testing a Communication Mediation Model of Civic Participation." Paper presented to the Association for Education in Journalism and Mass Communication annual meeting, Washington, DC.

McLuhan, Marshall. (1964). *Understanding Media.* New York: American Library.

———. (1969). *The Gutenberg Galaxy: The Making of Typographic Man.* New York: Mentor.

McNally, Peter F., ed. (1987). *The Advent of Printing: Historians of Science Respond to Elizabeth Eisenstein's the Printing Press as an Agent of Change.* Montreal: McGill University.

McPhee, Robert D., and Pamela Zaug. (2000). "The Communicative Constitution of Organizations: A Framework for Explanation." *Electronic Journal of Communication/La Revue Electronique de Communication* 10 (1–2): 1–16.

McQuail, Denis. (1986). "Is Media Theory Adequate to the Challenge of the New Communications Technologies?" In *New Communications Technologies and the Public Interest: Comparative Perspectives on Policy and Research,* edited by Marjorie Ferguson, 1–17. Newbury Park, CA: Sage.

McQuail, Denis, Jay G. Blumler, and J. R. Brown. (1972). "The Television Audience: A Revised Perspective." In *Sociology of Mass Communication,* edited by Denis McQuail, 135–165. Baltimore, MD: Penguin.

McQuail, Denis, and Sven Windahl. (1993). *Communication Models for the Study of Mass Communications Second Edition.* New York: Longman.

Menczer, Filippo, Santo Fortunato, Alessandro Flammini, and Alessandro Vespignani. (2006). "Googlearchy or Googlocracy?" *IEEE Spectrum Inside Technology,* February.

Merton, Robert K. (1936). "The Unanticipated Consequences of Purposive Social Action." *American Sociological Review* 1 (6): 894–904.

———. (1968a). "The Matthew Effect in Science." *Science* 159 (3810): 56–63.

———. (1968b). *Social Theory and Social Structure.* New York: Free Press.

Metcalfe, Robert M. (2000). *Internet Collapses and Other Infoworld Punditry.* Foster City, CA: IDG Books Worldwide.

Meyer, Katherine. (2005). "The Best of the Worst." *Wall Street Journal,* May 3.

Michels, Robert. (1962). *Political Parties: A Sociological Study of Oligarchical Tendencies of Modern Democracy.* New York: Collier Books.

Miller, Arthur, Edie Goldenberg, and Lutz Erbring. (1979). "Type-Set Politics: Impact of Newspapers on Public Confidence." *American Political Science Review* 73:67–84.

Miller, George A. (1956). "The Magical Number Seven, Plus or Minus Two: Some Limits on Our Capacity for Processing Information." *Psychology Review* 63:81–97.

Miller, Gerald R., and Mark Steinberg. (1975). *Between People: A New Analysis of Interpersonal Communication.* Chicago: Science Research Associates.

Miller, James G. (1960). "Information Input Overload and Psychopathology." *American Journal of Psychiatry* 116:695–704.

Mishler, Elliot G. (1986). *Research Interviewing: Context and Narrative.* Cambridge, MA: Harvard University Press.

Moffitt, Sean. (2011). "Wikibrands' Top 80 Social Media Monitoring Companies." *Wikibrands.* http://wiki-brands.com/.

Mokyr, Joel. (1990). *The Lever of Riches: Technological Creativity and Economic Progress.* New York: Oxford University Press.

Molotch, Harvey, and Marilyn Lester. (1974). "News as Purposive Behavior: On the Strategic Use of Routine Events, Accidents, and Scandals." *American Sociological Review* 39:101–112.

Monge, Peter R., and Joseph N. Cappella, eds. (1980). *Multivariate Techniques in Human Communication Research.* New York: Academic Press.

Monroe, Kirsten Renwick, James Hankin, and Renee Bukovchik Van Vechten. (2000). "The Psychological Foundations of Identity Politics." *Annual Review of Political Science* 3:419–447.

Morgan, Michael, and James Shanahan. (1996). "Two Decades of Cultivation Research: An Appraisal and Meta-Analysis." *Communication Yearbook* 20:1–45.

Morley, David. (1980). *The Nationwide Audience: Structure and Decoding.* London: British Film Institute.

———. (1986). *Family Television: Cultural Power and Domestic Leisure.* London: Comedia.

———. (1992). *Television, Audiences and Cultural Studies.* New York: Routledge.

Morley, David, and Charlotte Brunsdon. (1978). *Everyday Television: Nationwide.* London: British Film Institute.

———. (1999). *The Nationwide Television Studies.* New York: Routledge.

Morley, Donald Dean. (1988a). "Colloquy on Generalization Meta-Analytic Techniques When Generalizing to Message Populations Is Not Possible." *Human Communication Research* 15 (1): 112–126.

———. (1988b). "Reply to Jackson, O'Keefe, and Jacobs." *Human Communication Research* 15 (1): 143–147.

Morrison, David E. (1978). "The Beginnings of Modern Mass Communication Research." *European Journal of Sociology* 19:347–359.

Mosco, Vincent. (2005). *The Digital Sublime: Myth, Power, and Cyberspace.* Cambridge, MA: MIT Press.

Moy, Patricia, David Domke, and Keith Stamm. (2001). "The Spiral of Silence and Public Opinion on Affirmative Action." *Journalism & Mass Communication Quarterly* 78 (1): 7–25.

Mullen, Megan Gwynne. (2003). *The Rise of Cable Programming in the United States: Revolution or Evolution.* Austin: University of Texas Press.

Murdock, Graham, and Peter Golding. (1989). "Information Poverty and Political Inequality: Citizenship in the Age of Privatized Communications." *Journal of Communication* 39 (3): 180–195.

Mutz, Diana. (1998). *Impersonal Influence: How Perceptions of Mass Collectives Affect Political Attitudes.* New York: Oxford University Press.

———. (2006). "How the Mass Media Divide Us." In *Red and Blue Nation: Characteristics and Causes of America's Polarized Politics,* edited by Pietro S. Nivola and David W. Brady, 223–248. Washington, DC: Brookings Institution.

Napoli, Philip M. (2001). "The Marketplace of Ideas." In *Foundations of Communications Policy: Principles and Process in the Regulation of Electronic Media,* edited by Philip M. Napoli, 97–124. Cresskill, NJ: Hampton Press.

———. (2003). *Audience Economics: Media Institutions and the Audience Marketplace.* New York: Columbia University Press.

———. (2008). "Hyperlinking and the Forces of 'Massification.'" In *The Hyperlinked Society: Questioning Connections in the Digital Age,* edited by Joseph Turow and Lokman Tsui, 56–69. Ann Arbor: University of Michigan Press.

———. (2010). *Audience Evolution: New Technologies and the Transformation of Media Audiences.* New York: Columbia University Press.

Nassmacher, Karl-Heinz. (1993). "Comparing Party and Campaign Finance in Western Democracies." In *Campaign and Party Finance in North America and Western Europe,* edited by Arthur B. Gunlicks, 233–267. Boulder, CO: Westview.

Negroponte, Nicholas (1995). *Being Digital*. New York: Knopf.

Nelson, Bryan (2010). "McDonald's Spent 23 Cents on Advertising for Every Human on Planet Earth in 2001." brianknelson.com.

Nelson, Thomas E., Rosalee A. Clawson, and Zoe M. Oxley. (1997). "Media Framing of a Civil Liberties Conflict and Its Effect on Tolerance." *American Political Science Review* 91:567–583.

Neuman, W. Russell. (1976). "Patterns of Recall among Television News Viewers." *Public Opinion Quarterly* 40:115–123.

———. (1982). "Television and American Culture." *Public Opinion Quarterly* 46:471–487.

———. (1986). *The Paradox of Mass Politics*. Cambridge, MA: Harvard University Press.

———. (1989). "Parallel Content Analysis: Old Paradigms and New Proposals." In *Public Communication and Behavior, Volume 2*, edited by George Comstock, 205–289. Orlando, FL: Academic Press.

———. (1990). "The Threshold of Public Attention." *Public Opinion Quarterly* 54 (2): 159–176.

———. (1991). *The Future of the Mass Audience*. New York: Cambridge University Press.

Neuman, W. Russell, Bruce Bimber, and Matthew Hindman. (2011). "The Internet and Four Dimensions of Citizenship." In *The Oxford Handbook of American Public Opinion and the Media*, edited by Robert Y. Shapiro and Lawrence R. Jacobs, 22–42. New York: Oxford University Press.

Neuman, W. Russell, and Krysha Gregorowicz. (2010). "A Critical Transition in Mass Communication: From Push Media to Pull Media." Presented at the American Political Science Association Annual Conference, Washington, DC.

Neuman, W. Russell, and Lauren Guggenheim. (2011). "The Evolution of Media Effects Theory: Fifty Years of Cumulative Research." *Communication Theory* 21 (2): 169–196.

Neuman, W. Russell, Lauren Guggenheim, Seung Mo Jang, and Soo Young Bae. (2014). "The Dynamics of Public Attention: Agenda-Setting Theory Meets Big Data." *Journal of Communication* 64 (2): 193–214.

Neuman, W. Russell, Marion R. Just, and Ann N. Crigler. (1992). *Common Knowledge: News and the Construction of Political Meaning*. Chicago: University of Chicago Press.

Neuman, W. Russell, George E. Marcus, Ann N. Crigler, and Michael MacKuen, eds. (2007). *The Affect Effect: Dynamics of Emotion in Political Thinking and Behavior* Chicago: University of Chicago Press.

Neuman, W. Russell, George E. Marcus, and Michael B. MacKuen. (2012). "The Affective Resonance of Tea Party Politics." Presented at the Midwest Political Science Association Annual Meeting, Chicago.

Neuman, W. Russell, Lee McKnight, and Richard Jay Solomon. (1998). *The Gordian Knot: Political Gridlock on the Information Highway*. Cambridge, MA: MIT Press.

Neuman, W. Russell, Yong Jin Park, and Elliot Panek. (2012). "Tracking the Flow of Information into the Home: An Empirical Assessment of the Digital Revolution in the U.S. from 1960–2005." *International Journal of Communication* 6:1022–1041.

Neuman, W. Russell, and Ithiel de Sola Pool. (1986). "The Flow of Communications into the Home." In *Media, Audience and Social Structure,* edited by Sandra J. Ball-Rokeach and Muriel Cantor, 71–86. Beverly Hills: Sage.

Newhagen, John E., and Sheizaf Rafaeli. (1996). "Why Communication Researchers Should Study the Internet." *Journal of Communication* 46 (1): 4–13.

Newman, Mark E. J. (2005). "Power Laws, Pareto Distributions and Zipf's Law." *Contemporary Physics* 46:323–351.

Nie, Norman, Sidney Verba, and John R. Petrocik. (1976). *The Changing American Voter.* Cambridge, MA: Harvard University Press.

Nielsen. (1986). *1986 Nielsen Report on Television.* New York: A. C. Nielsen.

———. (2010). *Television, Internet and Mobile Usage in the U.S.* New York: A. C. Nielsen.

———. (2014). *The Total Audience Report.* New York: A. C. Nielsen.

Nielsen, Jakob. (2000). *Designing Web Usability.* Indianapolis, IN: New Riders.

Nielsen, Jakob, and Kara Pernice. (2010). *Eyetracking Web Usability.* Berkeley, CA: New Riders.

Nisbet, Matthew C., and Teresa Myers. (2007). "Twenty Years of Public Opinion about Global Warming." *Public Opinion Quarterly* 71 (3): 444–470.

Nivola, Pietro S., and David W. Brady. (2006). *Red and Blue Nation?: Characteristics and Causes of America's Polarized Politics.* Washington, DC: Brookings Institution.

Noam, Eli. (2004). *Market Failure in the Media Sector.* New York: Columbia University Center for Telecommunication and Information Studies.

———. (2009). *Media Ownership and Concentration in America.* New York: Oxford University Press.

Noelle-Neumann, Elisabeth. (1973). "Return to the Concept of Powerful Mass Media." *Studies of Broadcasting,* 102–123.

———. (1974). "The Spiral of Silence: A Theory of Public Opinion." *Journal of Communication* 24:43–51.

———. (1984). *The Spiral of Silence.* Chicago: University of Chicago Press.

———. (1993). *The Spiral of Silence,* 2nd ed. Chicago: University of Chicago Press.

Noll, Roger G., and Bruce M. Owen, eds. (1983). *The Political Economy of Deregulation: Interest Groups in the Regulatory Process.* Washington, DC: American Enterprise Institute.

Nora, Simon, and Alain Minc. (1980). *The Computerization of Society.* Cambridge, MA: MIT Press.

Norris, Pippa. (2000). *A Virtuous Circle: Political Communications in Post-Industrial Societies.* New York: Cambridge University Press.

Nunnally, Jum C., and Ira Bernstein. (1994). *Psychometric Theory.* New York: McGraw-Hill.

Oberschall, Anthony. (1973). *Social Conflict and Social Movements.* Englewood Cliffs, NJ: Prentice Hall.

Office of National Drug Control Policy. (2000). *The National Drug Control Strategy: 2000 Annual Report.* Washington, DC: Government Printing Office.

Ogden, Charles K., and Ivor A. Richards. (1923). *The Meaning of Meaning.* Orlando, FL: Harcourt Brace.

O'Keefe, Daniel J., Sally Jackson, and Scott Jacobs. (1988). "Reply to Morley." *Human Communication Research* 15 (1): 148–151.

O'Keefe, Daniel J., and Jakob D. Jensen. (2006). "The Advantages of Compliance or the Disadvantages of Noncompliance? A Meta-Analytic Review of the Relative Persuasive Effectiveness of Gain-Framed and Loss-Framed Messages." *Communication Yearbook* 30:1–43.

Orwell, George. (1949). *1984.* New York: Signet Books.

Osgood, Charles E., George J. Suci, and Percy Tannenbaum. (1957). *The Measurement of Meaning.* Urbana: University of Illinois Press.

Ostgaard, Einar. (1965). "Factors Influencing the Flow of News." *Journal of Peace Research* 2:39–63.

Owen, Bruce M., and Ronald Braeutigam. (1978). *The Regulation Game: Strategic Use of the Administrative Process.* Cambridge: Ballinger.

Owens, Lindsay A. (2012). "Trends: Confidence in Banks, Financial Institutions, and Wall Street, 1971–2011." *Public Opinion Quarterly* 76 (1): 142–162.

Packard, Vance. (1957). *The Hidden Persuaders.* New York: McKay.

Page, Benjamin I., and Robert Y. Shapiro. (1992). *The Rational Public: Fifty Years of Trends in Americans' Policy Preferences.* Chicago: University of Chicago Press.

Page, Scott E. (2007). *The Difference: How the Power of Diversity Creates Better Groups, Firms, Schools, and Societies.* Princeton, NJ: Princeton University Press.

Paisley, William. (1984). "Communication in the Communication Sciences." In *Progress in the Communication Sciences.* Vol. 5, edited by Brenda Dervin and Melvin J. Voigt, 1–43. Norwood, NJ: Ablex.

Pajares, Frank. (2004). "Albert Bandura: Biographical Sketch." http://stanford.edu /dept/psychology/bandura/bandura-bio-pajares/Albert%20_Bandura%20 _Biographical_Sketch.html.

Pan, Zhongdang, and Jack M. Mcleod. (1991). "Multilevel Analysis in Mass Communication Research." *Communication Research* 18 (2): 140–173.

Papper, Robert A., Michael E. Holmes, Mark N. Popovich, and Michael Bloxham. (2005). *Middletown Media Studies.* Muncie, IN: Ball State University. https:// insightandresearch.wordpress.com/category/other-projects-research-in-the -news/middletown-media-studies/.

Pareto, Vilfredo. (1935). *The Mind and Society.* New York: Harcourt Brace.

Parfeni, Lucian. (2009). "Jk Wedding Dance Serves YouTube's Profitability Case." *Softpedia.com,* July 31.

Pariser, Eli. (2011). *The Filter Bubble: What the Internet Is Hiding from You.* New York: Penguin Press.

Park, David W., and Jefferson Pooley, eds. (2008). *The History of Media and Communication Research: Contested Memories.* New York: Peter Lang.

Parsons, Talcott. (1942). "Propaganda and Social Control." *Psychiatry* 5 (4): 551–572.

Pasquale, Frank. (2010). "Beyond Innovation and Competition: The Need for Qualified Transparency in Internet Intermediaries." *Northwestern University Law Review* 104 (1): 105–174.

Patterson, Thomas. (1993). *Out of Order.* New York: Knopf.

Pencil, Murdock. (1976). "Salt Passage Research: The State of the Art." *Journal of Communication* 26:31–36.

Perse, Elizabeth M. (2001). *Media Effects and Society.* Mahwah, NJ: Lawrence Erlbaum.

Peters, John Durham. (1999). *Speaking into the Air: A History of the Idea of Communication.* Chicago: University of Chicago Press.

Peters, John Durham, and Peter Simonson, eds. (2004). *Mass Communication and American Social Thought: Key Texts 1919–1968.* Lanham, MD: Rowman & Littlefield.

Pettigrew, Thomas F., and Linda R. Tropp. (2006). "A Meta-Analytic Test of Intergroup Contact Theory." *Journal of Personality and Social Psychology* 90 (5): 751–783.

Pew Center for the People and the Press. (2006). *Online News, Pew Research Center for the People and the Press.* www.people-press.org/2006/07/30/online-papers-modestly-boost-newspaper-readership/.

Pew Project for Excellence in Journalism. (2010). *New Media, Old Media: How Blogs and Social Media Agendas Relate and Differ from the Traditional Press.*

Pew Research Center. (2015). *State of the News Media 2015.* Annual Report. www.journalism.org/2015/04/29/state-of-the-news-media-2015/.

Philo, Greg. (2008). "Active Audiences and the Construction of Public Knowledge." *Journalism Studies* 9 (4): 535–544.

Pinker, Steven. (2004). "Why Nature and Nurture Won't Go Away." *Daedalus* 133:1–13.

Pool, Ithiel de Sola, ed. (1959). *Trends in Content Analysis.* Urbana: University of Illinois Press.

———. (1973). "Communication in Totalitarian Societies." In *Handbook of Communication,* edited by Ithiel de Sola Pool and Wilbur Schramm, 462–511. Chicago: Rand McNally.

———. (1983). *Technologies of Freedom.* Cambridge, MA: Harvard University Press.

Pool, Ithiel de Sola, Hiroshi Inose, Nozomu Takasaki, and Roger Hurwitz. (1984). *Communications Flows: A Census in the United States and Japan.* Amsterdam: Elsevier North Holland.

Pooley, Jefferson. (2008). "The New History of Mass Communication Research." In *The History of Media and Communication Research: Contested Memories,* edited by David W. Park and Jefferson Pooley, 43–70. New York: Peter Lang.

Poor, Nathaniel. (2005). "Mechanisms of an Online Public Sphere: The Website Slashdot." *Journal of Computer-Mediated Communication* 10 (2). onlinelibrary .wiley.com/journal/10.1111/%28ISSN%291083-6101.

Porat, Marc U. (1977). *The Information Economy*. Washington, DC: U.S. Government Printing Office.

Potter, W. James. (1994). "Cultivation Theory and Research: A Methodological Critique." *Journalism Monographs* 147:1–34.

Pratkanis, Anthony R. (1992). "The Cargo-Cult Science of Subliminal Persuasion." *Skeptical Inquirer* 16 (3): 260–272.

Preacher, Kristopher J., and Andrew F. Hayes. (2008). "Contemporary Approaches to Assessing Mediation in Communication Research." In *The Sage Sourcebook of Advanced Data Analysis Methods for Communication Research*, edited by Andrew F. Hayes, Michael D. Slater, and Leslie B. Snyder, 13–54. Thousand Oaks, CA: Sage.

Preiss, Raymond W., Barbara Mae Gayle, Nancy Burrell, Mike Allen, and Jennings Bryant, eds. (2007). *Mass Media Effects Research: Advances through Meta-Analysis*. Mahwah, NJ: Lawrence Erlbaum.

Price, Derek de Solla. (1963). *Little Science, Big Science*. New York: Columbia University Press.

Price, Vincent, and Scott Allen. (1990). "Opinion Spirals, Silent and Otherwise: Applying Small-Group Research to Public Opinion Phenomena." *Communication Research* 17 (3): 369–392.

Price, Vincent, David Tewksbury, and Elizabeth Powers. (1997). "Switching Trains of Thought: The Impact of News Frames on Reader's Cognitive Responses." *Communication Research* 24 (5): 481–506.

Priest, Susanna H. (2009). *Doing Media Research: An Introduction*. Thousand Oaks, CA: Sage.

Prior, Markus. (2007). *Post-Broadcast Democracy: How Media Choice Increases Inequality in Political Involvement and Polarizes Elections*. New York: Cambridge University Press.

Procter, James. (2004). *Stuart Hall*. New York: Routledge.

Puddington, Arch. (2011). *Freedom in the World*. New York: Freedom House.

Putnam, Linda L., and Anne M. Nicotera, eds. (2009). *Building Theories of Organization: The Constitutive Role of Communication*. New York: Routledge.

Putnam, Robert D. (2000). *Bowling Alone: The Collapse and Revival of American Community*. New York: Simon & Schuster.

Radway, Janice A. (1984). *Reading the Romance: Women, Patriarchy, and Popular Literature*. Chapel Hill: University of North Carolina Press.

———. (1988). "Reception Study: Ethnography and the Problems of Dispersed Audiences and Nomadic Subjects." *Cultural Studies* 2:359–376.

———. (1991). "Writing Reading the Romance." In *Reading the Romance: Women, Patriarchy, and Popular Literature*, edited by Janice A. Radway, 1–18. Chapel Hill: University of North Carolina Press.

Rahim, M. Afzalur. (2010). *Managing Conflict in Organizations*. New Brunswick, NJ: Transaction Publishers.

Raymond, Eric S., and Bob Young. (2001). *The Cathedral and the Bazaar: Musings on Linux and Open Source by an Accidental Revolutionary*. Sebastopol, CA: O'Reilly.

Reynolds, Vernon, Vincent Falger, and Ian Vine, eds. (1987). *The Sociobiology of Ethnocentrism: Evolutionary Dimensions of Xenophobia, Discrimination, Racism, and Nationalism*. Athens: University of Georgia Press.

Richerson, Peter J., and Robert Boyd. (2005). *Not by Genes Alone: How Culture Transformed Human Evolution*. Chicago: University of Chicago Press.

Richtel, Matt. (2010). "Hooked on Gadgets, and Paying a Mental Price." *New York Times*, June 6.

Riesman, David. (1950). *The Lonely Crowd: A Study of the Changing American Character*. New Haven, CT: Yale University Press.

Ripberger, Joseph T. (2011). "Capturing Curiosity: Using Internet Search Trends to Measure Public Attentiveness." *Policy Studies Journal* 39 (2): 239–259.

Robinson, Glen O. (1989). "The Federal Communications Act: An Essay on Origins and Regulatory Purpose." In *A Legislative History of the Communications Act of 1934*, edited by Max Paglin, 3–24. New York: Oxford University Press.

Robinson, John P. (1971). "The Audience for National TV New Programs." *Public Opinion Quarterly* 35 (3): 403–405.

Robinson, John P., and Geoffrey Godbey. (1997). *Time for Life: The Surprising Ways Americans Use Their Time*. University Park: Pennsylvania State University Press.

Robinson, John P., and Mark Levy. (1986). *The Main Source*. Beverly Hills, CA: Sage.

Rodden, John. (1989). *The Politics of Literary Reputation: The Making and Claiming of "St. George" Orwell*. New York: Oxford University Press.

Rogers, Everett. (1973). "Mass Media and Interpersonal Communications." In *Handbook of Communication*, edited by Ithiel de Sola Pool and Wilbur Schramm, 290–310. Chicago: Rand McNally.

———. (1985). "The Empirical and Critical Schools of Communication Research." In *The Media Revolution in America and in Western Europe*, edited by Everett M. Rogers and Francis Balle, 219–235. Norwood, NJ: Ablex.

———. (1986). *Communication Technology: The New Media in Society*. New York: Free Press.

———. (1992). "On Early Mass Communication Study." *Journal of Broadcasting and Electronic Media* 36 (4): 467–471.

———. (1994). *A History of Communication Study: A Biographical Approach*. New York: Free Press.

———. (2003). *Diffusion of Innovations*, 5th ed. New York: Free Press.

Rogin, Michael Paul. (1967). *The Intellectuals and McCarthy: The Radical Specter*. Cambridge, MA: MIT Press.

Rojek, Chris. (2003). *Stuart Hall.* Cambridge: Polity.

Romney, Kimball, and Susan C. Weller. (1984). "Predicting Informant Accuracy from Patterns of Recall among Individuals." *Social Networks* 6 (1): 59–77.

Rosen, Stanley. (1981). "The Economics of Superstars." *American Economic Review* 71:845–858.

Rosenberg, Milton J., Sidney Verba, and Philip E. Converse. (1970). *Vietnam and the Silent Majority: The Dove's Guide.* New York: Harper & Row.

Rosenberg, Nathan, and L. E. Birdzell Jr. (1986). *How the West Grew Rich: The Economic Transformation of the Industrial Revolution.* New York: Basic Books.

Rosenthal, Morris. (2011). "Understanding Sales Rankings for Books." www .fonerbooks.com.

Rothley, K. D., Oswald J. Schmitz, and Jared L. Cohon. (1997). "Foraging to Balance Conflicting Demands: Novel Insights from Grasshoppers under Predation Risk." *Behavioral Ecology* 8 (5): 551–559.

Rowland, Willard D., Jr. (1983). *The Politics of TV Violence: Policy Uses of Communication Research.* Beverly Hills, CA: Sage.

Rubin, Alan. (1986). "Uses, Gratifications, and Media Effects Research." In *Perspectives on Media Effects,* edited by Jennings Bryant and Dolf Zillmann, 281–301. Hillsdale, NJ: Lawrence Erlbaum.

Rubin, Alan M., Elizabeth M. Perse, and Donald S. Taylor. (1988). "A Methodological Examination of Cultivation." *Communication Research* 15 (2): 107–134.

Rubin, Rebecca B., Alan M. Rubin, and Paul M. Haridakis. (2009). *Communication Research: Strategies and Sources 7th Edition.* Belmont, CA: Thomson.

Rudolph, Frederick. (1993). *Curriculum: A History of the American Undergraduate Course of Study since 1636.* San Francisco: Josey-Bass.

Ryan, Timothy J. (2012). "What Makes Us Click? Demonstrating Incentives for Angry Discourse with Digital-Age Field Experiments." *Journal of Politics* 74 (4): 1138–1152.

Samuelson, Pamela. (2006). "The Generativity of Sony v. Universal: The Intellectual Property Legacy of Justice Stevens." *Fordham Law Review* 74:1831–1876.

———. (2010). "Google Book Search and the Future of Books in Cyberspace." *Minnesota Law Review* 94:1308–1374.

Santo, Alysia. (2011). "Occupy Wall Street's Media Team." *Columbia Journalism Review,* October 7. www.cjr.org/.

Scannell, Paddy. (1990). "Public Service Broadcasting: The History of a Concept." In *Understanding Television,* edited by Andrew Goodwin and Garry Whannel, 11–29. New York: Routledge.

———. (2007). *Media and Communication.* Thousand Oaks, CA: Sage.

Schauer, Frederick (2005). "The Exceptional First Amendment." Working Paper Series from Harvard University, John F. Kennedy School of Government (RWP05-021).

Schein, Edgar H. (1971). *Coercive Persuasion: A Socio-Psychological Analysis of the "Brainwashing" of American Civilian Prisoners by the Chinese Communists.* New York: Norton.

Scheufele, Dietram A. (1999). "Framing as a Theory of Media Effects." *Journal of Communication* 49 (1): 103–122.

Scheufele, Dietram A., and Patricia Moy. (2000). "Twenty-Five Years of the Spiral of Silence: A Conceptual Review and Empirical Outlook." *International Journal of Public Opinion Research* 12 (1): 3–28.

Schiller, Dan. (2000). *Digital Capitalism: Networking the Global Market System.* Cambridge, MA: MIT Press.

Schlesinger, Philip. (1991). *Media, State and Nation: Political Violence and Collective Identities.* Newbury Park, CA: Sage.

Schneider, David J. (2004). *The Psychology of Stereotyping.* New York: Guilford.

Schrad, Mark Lawrence. (2010). *The Political Power of Bad Ideas: Networks, Institutions, and the Global Prohibition Wave.* Oxford: Oxford University Press.

Schramm, Wilbur, ed. (1948). *Communication in Modern Society.* Urbana: University of Illinois Press.

———. (1980). "The Beginnings of Communication Study in the United States." In *Communication Yearbook, Volume 4,* edited by Dan Nimmo, 73–82. New Brunswick, NJ: Transaction.

———. (1997). *The Beginnings of Communication Study in America.* Thousand Oaks, CA: Sage.

Schrecker, Ellen. (1998). *Many Are the Crimes: McCarthyism in America.* New York: Little Brown.

Schroder, Kim Christian. (1987). "Convergence of Antagonistic Traditions? The Case of Audience Research." *European Journal of Communication* 2 (1): 7–31.

Schudson, Michael. (1982). "The Politics of Narrative Form: The Emergence of News Conventions in Print and Television." *Daedalus* 111:97–112.

———. (1984). *Advertising, the Uneasy Persuasion.* New York: Basic Books.

———. (1997). "Why Conversation Is Not the Soul of Democracy." *Critical Studies in Mass Communications* 14:297–309.

———. (1998). *The Good Citizen: A History of American Civic Life.* New York: Free Press.

Schulz, Winfried Friedrich. (1982). "News Structure and People's Awareness of Political Events." *International Communication Gazette* 30 (3): 139–153.

Schuman, Howard, and Stanley Presser. (1981). *Questions and Answers in Attitude Surveys: Experiments on Question Form, Wording, and Context.* New York: Academic Press.

Schuman, Howard, Charlotte Steeh, Lawrence D. Bobo, and Maria Krysan. (1997). *Racial Attitudes in America: Trends and Interpretations.* Cambridge, MA: Harvard University Press.

Schuman, Howard S., Stanley Presser, and Jacob Ludwig. (1981). "Contextual Effects on Survey Responses to Questions about Abortion." *Public Opinion Quarterly* 41:216–223.

Schwartz, Barry. (2004). *The Paradox of Choice: Why More Is Less.* New York: HarperCollins.

Segev, Elad. (2010). *Google and the Digital Divide: The Basis of Online Knowledge.* Cambridge: Chandos.

Shah, Dhavan V., Joseph Cappella, and W. Russell Neuman, eds. (2015). *Toward Computational Social Science: Exploiting Big Data in the Digital Age.* Philadelphia: ANNALS of the American Academy of Political and Social Science.

Shanahan, James, Carroll Glynn, and Andrew Hayes (2007). "The Spiral of Silence: A Meta-Analysis of Its Impact." In *Mass Media Effects Research: Advances through Meta-Analysis,* edited by Raymond W. Preiss, Barbara Mae Gayle, Nancy Burrell, Mike Allen, and Jennings Bryant, 415–428. Mahwah, NJ: Lawrence Erlbaum.

Shannon, Claude E., and Warren Weaver. (1949). *The Mathematical Theory of Communication.* Urbana: University of Illinois.

Shapiro, Carl, and Hal R. Varian. (1999). *Information Rules: A Strategic Guide to the Network Economy.* Boston: Harvard Business School Press.

Shenk, David. (1998). *Data Smog: Surviving the Information Glut.* New York: Harper.

Sherif, Muzafer. (1936). *The Psychology of Social Norms.* New York: HarperCollins.

Sherif, Muzafer, O. J. Harvey, B. Jack White, William R. Hood, and Carolyn W. Sherif. (1961). *Intergroup Conflict and Cooperation: The Robbers Cave Experiment.* Norman: University of Oklahoma Book Exchange.

Shils, Edward. (1957). "Daydreams and Nightmares: Reflections on the Criticism of Mass Culture." *Sewanee Review* 65 (4): 586–608.

Shirky, Clay. (2008). *Here Comes Everybody: The Power of Organizing without Organizations.* New York: Penguin.

Shoemaker, Pamela J., and Tim P. Vos. (2009). *Gatekeeping Theory.* Newbury Park, CA: Sage.

Shrum, L. J. (2007). "Cultivation and Social Cognition." In *Communication and Social Cognition: Theories and Methods,* edited by David. R. Roskos-Ewoldsen and Jennifer M. Monahan, 245–272. Mahwah, NJ: Lawrence Erlbaum.

Simmel, Georg. (1950). *The Sociology of Georg Simmel.* New York: The Free Press.

Simon, Herbert A. (1956). "Rational Choice and the Structure of the Environment." *Psychological Review* 63 (2): 129–138.

———. (1971). "Designing Organizations for an Information-Rich World." In *Computers, Communication, and the Public Interest,* edited by Martin Greenberger, 37–72. Baltimore, MD: Johns Hopkins University Press.

Simonson, Peter. (1999). "Mediated Sources of Public Confidence: Lazarsfeld and Merton Revisited." *Journal of Communication* 49 (2): 109–122.

———. (2010). *Refiguring Mass Communication: A History*. Urbana: University of Illinois Press.

Simonson, Peter, and Gabriel Weimann. (2003). "Critical Research at Columbia: Lazarsfeld and Merton's 'Mass Communication. Popular Taste and Organized Social Action.'" In *Canonic Texts in Media Research: Are There Any? Should There Be? How about These?*, edited by Elihu Katz, John Durham Peters, Tamar Liebes, and Avril Orloff, 12–38. Cambridge: Polity.

Singletary, Michael W., ed. (1994). *Mass Communication Research: Contemporary Methods and Applications*. New York: Longman.

Slater, Michael D. (1991). "Use of Message Stimuli in Mass Communication Experiments: A Methodological Assessment and Discussion." *Journalism Quarterly* 68 (3): 412–421.

———. (2007). "Reinforcing Spirals: The Mutual Influence of Media Selectivity and Media Effects and Their Impact on Individual Behavior and Social Identity." *Communication Theory* 17 (3): 281–303.

Small, Gary, Teena Moody, Prabha Siddarth, and Susan Bookheimer. (2009). "Your Brain on Google: Patterns of Cerebral Activation during Internet Searching." *American Journal of Geriatric Psychiatry* 17:116–126.

Small, Gary W., and Gigi Vorgan. (2008). *iBrain: Surviving the Technological Alteration of the Modern Mind*. New York: Collins Living.

Smith, Alfred G., ed. (1966). *Communication and Culture: Readings in the Codes of Human Interaction*. New York: Holt, Rinehart, and Winston.

Smith, Eliot R., Daniel A. Miller, Angela T. Maitner, Sara A. Crump, Teresa Garcia-Marques, and Diane M. Mackie. (2006). "Familiarity Can Increase Stereotyping." *Journal of Experimental Social Psychology* 42 (4): 471–478.

Sniderman, Paul, Richard Brody, and Philip Tetlock. (1991). *Reasoning and Choice: Explorations in Political Psychology*. New York: Cambridge University Press.

Soroka, Stuart, Blake Andrew, Toril Aalberg, Shanto Iyengar, James Curran, Sharon Coen, Kaori Hayashi, et al. (2012). "Auntie Knows Best? Public Broadcasters and Current Affairs Knowledge." *British Journal of Political Science*, December.

Spangler, Todd. (2009). "YouTube May Lose $470 Million in 2009: Analysts Credit Suisse Report Estimates Video Site Will Generate $240 Million in Revenue." *Multichannel News*, April 4.

Sparks, Glenn G. (2010). *Media Effects Research: A Basic Overview Third Edition*. New York: Wadsworth.

Sparrow, Betsy, Jenny Liu, and Daniel M. Wegner. (2011). "Google Effects on Memory: Cognitive Consequences of Having Information at Our Fingertips." *Science*, 776–778.

Staab, Joachim Friedrich. (1990). "The Role of News Factors in News Selection: A Theoretical Reconsideration." *European Journal of Communication* 5 (4): 423–443.

Starr, Paul. (2008). "Democratic Theory and the History of Communications." In *Explorations in Communication and History*, edited by Barbie Zelizer, 35–45. New York: Routledge.

Steiner, George, and Robert Boyers. (2009). *George Steiner at the New Yorker*. New York: New Directions.

Stempel, Guido H., and Bruce H. Westley, eds. (1989). *Research Methods in Mass Communication*, 2nd ed. Englewood Cliffs, NJ: Prentice Hall.

Stimson, James A. (1991). *Public Opinion in America: Moods, Cycles and Swings*. Boulder, CO: Westwood.

———. (2004). *Tides of Consent: How Public Opinion Shapes American Politics*. New York: Cambridge University Press.

Strasburger, Victor C., and Barbara J. Wilson. (2002). *Children, Adolescents, and the Media*. Thousand Oaks, CA: Sage

Streeter, Thomas. (1996). *Selling the Air: A Critique of the Policy of Commercial Broadcasting in the United States*. Chicago: University of Chicago Press.

Stroud, Natalie Jomini. (2011). *Niche News: The Politics of News Choice*. New York: Oxford University Press.

Summerfield, Derek. (1997). "The Social, Cultural and Political Dimensions of Contemporary War." *Medicine Conflict and Survival* 13:3–25.

Sunstein, Cass. (2001). *Republic.Com*. Princeton, NJ: Princeton University Press.

———. (2006). *Infotopia: How Many Minds Produce Knowledge*. Oxford: Oxford University Press.

Tajfel, Henri. (1982). "Social Identity and Intergroup Relations." *Annual Review of Psychology* 33:1–39.

Tajfel, Henri, and Michael Billig. (1974). "Familiarity and Categorization in Intergroup Behavior." *Journal of Experimental Social Psychology* 10 (2): 159–170.

Tajfel, Henri, Michael Billig, Robert P. Bundy, and Claude Flament. (1971). "Social Categorization and Intergroup Behaviour." *European Journal of Social Psychology* 2:149–178.

Tajfel, Henri, and John Turner. (1986). "The Social Identity Theory of Intergroup Behavior." In *Psychology of Intergroup Relations*, edited by Stephen Worchel and William G. Austin, 7–24. Chicago: Nelson Hall.

Tapscott, Don, and Anthony D. Williams. (2006). *Wikinomics: How Mass Collaboration Changes Everything*. New York: Portfolio.

Tetlock, Philip. (2005). *Expert Political Judgment: How Good Is It? How Can We Know?* Princeton, NJ: Princeton University Press.

Theberge, Leonard J. (1981). *Crooks, Con Men and Clowns*. Washington, DC: Media Institute.

Theroux, Louis. (2012). "How the Internet Killed Porn." *Guardian*, June 5.

Tilly, Charles. (1970). "Clio and Minerva." In *Theoretical Sociology; Perspectives and Developments*, edited by John C. McKinney and Edward A. Tiryakian, 433–466. New York: Appleton-Century-Crofts.

———. (1998). *Durable Inequality*. Berkeley: University of California Press.

———. (2002). *Stories, Identities, and Political Change*. Lanham, MD: Rowman & Littlefield.

————. (2004). *Contention and Democracy in Europe, 1650–2000.* New York: Cambridge University Press.

Toffler, Alvin. (1980). *The Third Wave.* New York: Morrow.

Tomasello, Michael. (1999). *The Cultural Origins of Human Cognition.* Cambridge, MA: Harvard University Press.

Traber, Michael. (1986). *The Myth of the Information Revolution.* Newbury Park, CA: Sage.

Trepte, Sabine. (2006). "Social Identity Theory." In *Psychology of Entertainment,* edited by Jennings Bryant and Peter Vorderer, 255–272. Mahwah, NJ: Lawrence Erlbaum.

Tuch, Hans N. (1990). *Communicating with the World: U.S. Public Diplomacy Overseas.* New York: St. Martin's Press.

Tufekci, Zeynep, and Christopher Wilson. (2012). "Social Media and the Decision to Participate in Political Protest: Observations from Tahrir Square." *Journal of Communication* 62 (2): 363–379.

Turner, Charles E., and Elizabeth Martin, eds. (1984). *Surveying Subjective Phenomena.* New York: Sage.

Turow, Joseph. (1997). *Breaking up America: Advertisers and the New Media World.* Chicago: University of Chicago Press.

United States Court of Appeals for the Ninth Circuit. (2004). "Yahoo! versus La Ligue Contre Le Racisme Et L'antisemitisme and L'union Des Opinion Etudiants Juifs De France." San Francisco. http://openjurist.org/379/f3d/1120/yahoo-inc-v-la-ligue-contre-le-racisme-et-lantisemitisme.

U.S. Securities and Exchange Commission. (2012). "Google Inc. Form 10-K." Washington, DC. www.sec.gov/.

Valkenburg, Patti M., Holli A. Semetko, and Claes H. De Vreese. (1999). "The Effects of News Frames on Readers' Thoughts and Recall." *Communication Research* 26 (5): 550–569.

Vallone, Robert P., Lee Ross, and Mark R. Lepper. (1985). "The Hostile Media Phenomenon: Biased Perception and Perceptions of Media Bias in Coverage of the Beirut Massacre." *Journal of Personality and Social Psychology* 49 (3): 577–585.

van Dijk, Jan A. G. M. (1999). *The Network Society: Social Aspects of New Media.* Thousand Oaks, CA: Sage.

Van Liere, Kent D., and Riley E. Dunlap. (1980). "The Social Bases of Environmental Concern: A Review of Hypotheses, Explanations and Empirical Evidence." *Public Opinion Quarterly* 44 (2): 181–197.

Veronis, Suhler, & Associates. (2011). *Communications Industry Report.* New York: Veronis, Suhler, & Associates.

Veysey, Laurence R. (1965). *The Emergence of the American University.* Chicago: University of Chicago Press.

Virzi, Anna Maria. (2001). "Hate or History?" *Forbes,* July 3.

Wahl-Jorgensen, Karin. (2004). "How Not to Found a Field: New Evidence on the Origins of Mass Communication Research." *Journal of Communication* 54 (3): 547–564.

Walker, Phillip L. (2001). "A Bioarchaeological Perspective on the History of Violence." *Annual Review of Anthropology* 30:573–596.

Walker, Samuel. (1994). *Hate Speech: The History of an American Controversy.* Lincoln: University of Nebraska Press.

Walther, Joseph B. (1996). "Computer-Mediated Communication: Impersonal, Interpersonal, and Hyperpersonal Interaction." *Communication Research* 23:3–43.

Wanta, Wayne, and Salma Ghanem. (2007). "Effects of Agenda Setting." In *Mass Media Effects Research: Advances through Meta-Analysis,* edited by Raymond W. Preiss, Barbara Mae Gayle, Nancy Burrell, Mike Allen, and Jennings Bryant, 37–52. Mahwah, NJ: Lawrence Erlbaum.

Wartella, Ellen, and Byron Reeves. (1985). "Historical Trends in Research on Children and the Media: 1900–1960." *Journal of Communication* 35 (2): 118–133.

Waterman, David. (2005). *Hollywood's Road to Riches.* Cambridge, MA: Harvard University Press.

Watts, Duncan J. (2004). "The 'New' Science of Networks." *Annual Review of Sociology* 30:243–270.

Weaver, David H., Dons A. Graber, Maxwell McCombs, and Chaim H. Eyal. (1981). *Media Agenda-Setting in a Presidential Election.* New York: Praeger.

Webb, Eugene J., Donald T. Campbell, Richard D. Schwartz, and Lee Sechrest. (2000). *Unobtrusive Measures.* Thousand Oaks, CA: Sage.

Weber, Max. (1910). "Towards a Sociology of the Press" (1910). *Journal of Communication* 26 (3): 96–101.

Webster, Frank. (2008). *Theories of the Information Society.* New York: Routledge.

Weerakkody, Niranjala. (2008). *Research Methods for Media and Communication.* New York: Oxford University Press.

Weigel, David. (2011). "Occupy Wall Street Starts Off with Favorable Ratings." *Slate,* October 5.

Weprin, Alex. (2011). "Al Jazeera English Gets Social." *TVNewwer,* April 21.

West, Darrell M. (1997). *Air Wars: Television Advertising in Election Campaigns, 1952–1996.* Washington, DC: Congressional Quarterly.

Westoff, Charles F., Emily C. Moore, and Norman B. Ryder. (1969). "The Structure of Attitudes toward Abortion." *Milbank Memorial Fund Quarterly* 47 (1): 11–37.

White, Aidan, and Ernest Sagaga. (2011). *Gunning for Media: Journalists and Media Staff Killed in 2010.* Brussels: International Federation of Journalists.

Wilensky, Harold L. (1964). "Mass Society and Mass Culture." *American Sociological Review* 29 (2): 173–197.

Williams, Raymond. (1958). *Culture and Society.* London: Chatto and Windus.

———. (1961). *The Long Revolution.* New York: Columbia University Press.

———. (1974). *Television: Technology and Cultural Form.* New York: Schocken Books.

Williamson, Oliver E. (1975). *Markets and Hierarchies: Analysis and Antitrust Implications.* New York: Free Press.

———. (1985). *The Economic Institutions of Capitalism: Firms, Markets, Relational Contracting.* New York: Free Press.

Willis, Paul. (1980). "Notes on Method." In *Culture, Media, Language: Working Papers in Cultural Studies, 1972–79*, edited by Stuart Hall, Dorothy Hobson, Andy Lowe, and Paul Willis, 68–74. London: Routledge.

Wilson, Edmund (1962). *Patriotic Gore; Studies in the Literature of the American Civil War.* New York: Oxford University Press.

Wilson, Edward O. (1975). *Sociobiology: The New Synthesis.* Cambridge, MA: Harvard University Press.

———. (1998). *Consilience: The Unity of Knowledge.* New York: Knopf.

Winn, Philip. (2008). "State of the Blogosphere." Technorati.Com.

Wirth, Louis. (1948). "Consensus and Mass Communication." *American Sociological Review* 13:1–13.

Wlezien, Christopher. (2005). "On the Salience of Political Issues: The Problem with 'Most Important Problem.'" *Electoral Studies* 24 (4): 555–579.

Wober, J. Mallory, and Barrie Gunter. (1982). "Television and Personal Threat: Fact or Artifact? A British Survey." *British Journal of Social Psychology* 21:239–247.

Wohn, D. Yvette, and Eun-Kyung Na. (2011). "Tweeting about TV: Sharing Television Viewing Experiences via Social Media Message Streams." *First Monday* 16 (3). firstmonday.org.

Wolf, Mauro. (1988). "Communication Research and Textual Analysis: Prospects and Problems of Theoretical Convergence." *European Journal of Communication* 3:135–149.

Wright, Charles R. (1960). "Functional Analysis and Mass Communication." *Public Opinion Quarterly* 24:605–620.

Wu, Shaomei, Jake M. Hofman, Winter A. Mason, and Duncan J. Watts. (2011). "Who Says What to Whom on Twitter." In *Proceedings of the International World Wide Web Conference.* www.w3.org.

Wu, Tim. (2003). "Network Neutrality, Broadband Discrimination." *Journal of Telecommunications and High Technology Law* 2:141–176.

———. (2010). *The Master Switch: The Rise and Fall of Information Empires.* New York: Knopf.

Yaros, Ronald A. (2006). "Is It the Medium or the Message? Structuring Complex News to Enhance Engagement and Situational Understanding by Non-experts." *Communication Research* 33 (4): 285–309.

Yonelinas, Andrew P. (2002). "The Nature of Recollection and Familiarity: A Review of 30 Years of Research." *Journal of Memory and Language* 46 (3): 441–517.

Yoo, Christopher S. (2006). "Network Neutrality and the Economics of Congestion." *Georgetown Law Journal* 94:1849–1907.

Yu, Jason J., and Deb Aikat, (2005). *News on the Web: Agenda Setting of Online News in Web Sites of Major Newspapers, Television and Online News Services*. New York: International Communication Association.

Yule, G. Udny, (1925). "A Mathematical Theory of Evolution, Based on the Conclusions of Dr. J. C. Willis, F.R.S." *Philosophical Transactions of the Royal Society of London* 213:21–87.

Yzer, Marco C., et al. (2003). "The Effectiveness of Gateway Communications in Anti-Marijuana Campaigns." *Journal of Health Communication* 8 (2): 129–143.

Zajonc, Robert B. (1968). "Attitudinal Effects of Mere Exposure." *Journal of Personality and Social Psychology* 9 (2): 1–27.

Zaller, John. (1991). *The Nature and Origins of Mass Opinion*. New York: Cambridge University Press.

———. (1999). *A Theory of Media Politics: How the Interests of Politicians, Journalists and Citizens Shape the News*. Chicago: University of Chicago Press.

———. (2003). "A New Standard of News Quality: Burglar Alarms for the Monitorial Citizen." *Political Communication* 20 (2): 109–130.

Zaller, John, and Stanley Feldman. (1992). "A Simple Theory of the Survey Response: Answering Questions Means Revealing Preferences." *American Journal of Political Science* 36:579–618.

Zhou, Shuhua, and William David Sloan. (2001). *Research Methods in Communication*, 2nd ed. Fair Lawn, NJ: Vision Press.

Zhu, Jian-Hua. (1992). "Issue Competition and Attention Distraction: A Zero-Sum Theory of Agenda-Setting." *Journalism and Mass Communication Quarterly* 69 (4): 825–836.

Zillmann, Dolf. (2000). "Humor and Comedy." In *Media Entertainment: The Psychology of Its Appeal*, edited by Dolf Zillmann and Peter Voderer, 37–58. Mahwah, NJ: Lawrence Erlbaum.

Zillmann, Dolf, and Jennings Bryant. (1985a). "Affect, Mood, and Emotion as Determinants of Selective Exposure." In *Selective Exposure to Communication*, edited by Dolf Zillmann and Jennings Bryant, 157–190. Hillsdale, NJ: Lawrence Erlbaum.

———, eds. (1985b). *Selective Exposure to Communication*. Hillsdale, NJ: Lawrence Erlbaum.

Zillmann, Dolf, and Peter Vorderer, eds. (2000). *Media Entertainment: The Psychology of Its Appeal*. Mahwah, NJ: Lawrence Erlbaum.

Zimmer, Carl. (2011). "It's Science, but Not Necessarily Right." *New York Times*, June 25.

Zimmermann, Manfred. (1989). "The Nervous System in the Context of Information Theory." In *Human Physiology*, edited by Richard F. Schmidt and Gerhard Thews, 166–173. Berlin: Springer.

Zipf, George Kingsley. (1949). *Human Behavior and the Principle of Least Effort: An Introduction to Human Ecology*. New York: Hafner.

Zittrain, Jonathan L. (2006). "A History of Online Gatekeeping." *Harvard Journal of Law and Technology* 19 (2): 253–298.

———. (2008). *The Future of the Internet and How to Stop It*. New Haven, CT: Yale University Press.

Zucker, Harold G. (1978). "The Variable Nature of News Media Influence." In *Communication Yearbook II*, edited by Brent D. Ruben, 225–240. New Brunswick, NJ: Transaction.

Zuckerman, Ethan. (2013). *Rewire: Digital Cosmopolitans in the Age of Connection*. New York: Norton.

Acknowledgments

This project has been under way for a number of years, and more than a few people kindly took the time to talk, help, explain, ask, break bread, encourage, give permission, discourage, suggest, point out, answer, probe, correct, and otherwise respond to my half-developed ideas as the work proceeded. The errors that remain are mine, of course. I have no doubt overlooked some friends and colleagues who have also tried to help, so here is a partial list in no particular order: Manuel Castells, Shanto Iyengar, Dale Kunkel, Paul Sniderman, Klaus Bruhn Jensen, Jay Blumler, Elihu Katz, Eli Noam, Denis McQuail, Paddy Scannell, Susan Neuman, Matt Hindman, Lewis Friedland, Eszter Hargittai, Erik Bucy, Jack McLeod, Mike Traugott, Bruce Bimber, Michael Slater, Dan O'Keefe, John Zaller, Michael X. Delli Carpini, Sally Jackson, Ron Rice, Joe Cappella, George Marcus, Pam Shoemaker, Paul DiMaggio, Bob Craig, Randy Collins, Bob Shapiro, Pablo Boczkowski, Susan Douglas, Wolf Donsbach, Phil Napoli, Rowell Huesmann, Silvia Knobloch-Westerwick, Christian Sandvig, Roei Davidson, Klaus Krippendorff, Byron Reeves, Nojin Kwak, Dhavan Shah, Lauren Guggenheim, Lance Bennett, Albert Bandura, Stuart Hall, danah boyd, Andy Lippman, Lance Holbert, Lloyd Morrisett, Len Lodish, Michael MacKuen, Max McCombs, Diana Mutz, Markus Prior, Ben Page, Michael Schudson, Wayne Wanta, Sung-Hee Joo, Yochai Benkler, Yong Jin Park, Andy Chadwick, Ann Crigler, Daniel Dayan, Eliot Panek, Elizabeth Eisenstein, Josh Pasek, Marko Skoric, Matt Richtel, Richard Jay Solomon, Mo Jang, Sandra Braman, Soo Young Bae, and Doris Graber.

Index

speech on, 264–265; Google Revolution, 271–272; shift to digital transmission, 272–273; network neutrality, 273–277, 292–293; origins and design of, 275–276; digital property rights, 277–282; and Pareto Principle and long tail, 283–285; audience activity, 301

Internet Corporation for Assigned Names and Numbers (ICANN), 293

Internet service providers (ISPs), 248–249, 292–293

Interpersonal communication, mass communication and, 1–2

Interpretation, selective, 66–67, 71–73

Iron Law of Oligarchy, 16, 211, 214–216, 305

Issue-attention cycle, 211, 218–219, 233–237

Issue endurance in public sphere, 234–236

Issue framing, 173–174

Issue publics, 233

Iyengar, Shanto, 3–4, 227, 236–237

Jackson, Sally, 83–85

Jensen, Klaus Bruhn, 172–173

Jonscher, Charles, 116–117

Journalists, murder or imprisonment of, 263

Just, Marion, 174

Kahneman, Daniel, 79

Katz, Daniel, 194–195

Katz, Elihu, 20, 171–172, 291–292

Keeley, Lawrence H., 184

Keller, Bill, 71–72

Key, V. O. Jr., 52

Klapper, Joseph, 49, 50, 65

Klingberg, Torkel, 106

Krippendorff, Klaus, 148–149

Kuhn, Thomas, 97, 259

Kuhnian model, 7, 8–17, 52

Lang, Annie, 20, 43

Language, 117, 118, 153, 180

Lasswell, Harold, 30, 40–41, 50, 55, 199

Law of Small Numbers, 217–218, 231–233

Lazarsfeld, Paul: Marshall and, 27, 30; leaves communication field, 40–41; and modern-day critical theory and analysis, 50; and research paradigm for media effects, 54; and disconnect between cultural studies and media effects traditions, 151–152; Adorno and, 181–182; on media and criticism, 292

Leeds-Hurwitz, Wendy, 14, 141

Lessig, Lawrence, 243, 302–303

Liberal, survey respondents' understanding of term, 168–169

Liebes, Tamar, 171–172

Limbaugh, Rush, 206–207

Lindlof, Thomas, 179

Lippmann, Walter, 72–73, 193

Lipset, Seymour Martin, 214–215

Literacy, 120–124

Literary studies, 156, 157–158, 163, 166

Lobbying business, philosophical purists in, 244–245

Lodish, Leonard, 147

"Long tail," 230–231, 282–286

"Looking for Trouble" (Katz), 291–292

Luther, Martin, 202

Machlup, Fritz, 110

Madison River Communications, 276

Management Information Systems (MIS), 108–109

Manchurian Candidate, The (Condon), 25, 31

Margin, variance at, 58–59

Marketplace of ideas, 211–213, 305–308; and digital revolution, 47–48; and Iron Law of Oligarchy, 216; openness of digital public sphere to, 242; as grounding principle for public policy, 243; government intervention in, 244–245; regulation of, 245–251; special properties and importance of, 286; as issue of global policy, 293